HETEROCYCLES IN LIFE AND SOCIETY

HETEROCYCLES IN LIFE AND SOCIETY

An Introduction to Heterocyclic Chemistry and
Biochemistry and the Role of Heterocycles in Science,
Technology, Medicine and Agriculture

By

Alexander F. Pozharskii
Soros Professor of Chemistry, Rostov-on-Don University, Russia

Anatoly T. Soldatenkov
*Professor of Chemistry, Russian People's Friendship University,
Moscow, Russia*

Alan R. Katritzky
*Kenan Professor of Chemistry, University of Florida, Gainesville, FL
USA*

JOHN WILEY & SONS
Chichester · New York · Weinheim · Brisbane · Singapore · Toronto

Copyright © 1997 by John Wiley & Sons Ltd,
Baffins Lane, Chichester,
West Sussex PO19 1UD, England

National   01243 779777
International   (+44) 1243 779777
e-mail (for orders and customer service enquiries): cs-books@wiley.co.uk
Visit our Home Page on http://www.wiley.co.uk
or http://www.wiley.com

Other Wiley Editorial Offices

John Wiley & Sons, Inc., 605 Third Avenue,
New York, NY 10158-0012, USA

VCH Verlagsgesellschaft mbH, Pappelallee 3,
D-69469 Weinheim, Germany

Jacaranda Wiley Ltd, 33 Park Road, Milton,
Queensland 4064, Australia

John Wiley & Sons (Canada) Ltd, 22 Worcester Road,
Rexdale, Ontario M9W 1L1, Canada

John Wiley & Sons (Asia) Pte Ltd, Clementi Loop #02-01,
Jin Xing Distripark, Singapore 0512

Library of Congress Cataloging-in-Publication Data

Pozharskiĭ, A. F. (Aleksandr Fedorovich)
   [Molekuly-perstni. English]
   Heterocycles in life and society : an introduction to heterocyclic chemistry and biochemistry and the role of heterocycles in science, technology, medicine, and agriculture / by Alexander F. Pozharskii, Anatoly T. Soldatenkov, Alan R. Katritzky.
      p.   cm.
   Pozharskii's name appears first in the Russian original.
   Includes bibliographical references (p.   –   ) and index.
   ISBN 0-471-96033-0 (cloth : alk. paper). — ISBN 0-471-96034-9 (pbk. : alk. paper)
      1. Heterocyclic chemistry.   I. Soldatenkov, A. T. (Anatoliĭ Timofeevich)   II. Katritzky, Alan R.   III. Title
   QD400.P6713   1996
   547'.59—dc20   96–25878
                                                        CIP

British Library Cataloguing in Publication Data

A catalogue record for this book is available from the British Library

ISBN 0 471 96033 0 (cloth)
ISBN 0 471 96034 9 (paper)

Typeset in 10/12pt Times by Mackreth Media Services, Hemel Hempstead, Herts
Printed and bound in Great Britain by Biddles Ltd, Guildford, Surrey

This book is printed on acid-free paper responsibly manufactured from sustainable forestation, for which at least two trees are planted for each one used for paper production.

CONTENTS

PREFACE

Heterocyclic compounds as a group dominate modern organic chemistry, with at least 55% of organic chemistry publications dedicated to this field. Heterocyclic chemistry is indeed taught worldwide at most universities and its scope is reflected in many fine textbooks and scientific monographs. Regrettably, many general chemistry, and even organic chemistry, texts fail to include heterocycles and the significance of their chemistry, or at most discuss them only in a nonsystematized manner. Furthermore, time constraints often prevent teachers of chemistry from elaborating on the applications of heterocycles. The objectives of the present book were determined by the above-mentioned circumstances. Our book is intended to bridge this gap and to help students navigate the intricacies of contemporary chemistry. Emphasis is placed not so much on the reactions of different classes of heterocycles as on their practical importance in life and society, especially their scientific applications in various branches of technology, medicine and agriculture. We hope that this approach will inspire the student to become involved in an immensely important and exciting field of modern chemical science and technology.

While this book is intended for university level chemistry and biochemistry students and their instructors, it should be of interest to researchers over the whole of the chemical, medical and agricultural sciences as well as in adjacent branches of science and technology. These assertions are well founded because the majority of known pharmaceutical preparations (antibiotic, neurotropic, cardiovascular, anticarcinogenic) are heterocyclic in nature; because the agricultural use of new plant development regulators and pesticides based on heterocyclic structures becomes more widespread each year; and because great attention is being paid to the synthesis and production of new kinds of thermostable polymers, highly durable fibers, fast pigments and colorants, organic metals and superconductors containing heterocyclic fragments.

This book consists of 11 chapters. The first two chapters present the elements of the structure and properties of heterocycles and are a useful introduction to the fundamentals of their chemistry. The next four chapters deal in a general way with the key role of heterocyclic molecules in such life processes as the transfer of hereditary information (3), the manner in which enzymes function (4), the storage and transfer of bioenergy (5) and photosynthesis (6). Chapters 7–9 are dedicated to the applications of heterocycles in medicine, agriculture and industry, respectively. Finally, Chapters 10 and 11 consider the future and the past: modern trends in the development of heterocyclic chemistry, the latest discoveries and the prospects of finding new spheres of use for heterocycles are presented in 10, while the emergence of heterocyclic molecules on primordial Earth is discussed in 11.

Throughout this text the student will learn to apply the knowledge gained by working on problems related to the topics covered in each chapter. Many of the 100 problems have been chosen from scientific journals and represent areas of recent significant interest. The scientists who solved these mysteries were yesterday's students. Thus, the approach to the problems will give today's students further insight into nature and a preview of what is scientifically possible. Each chapter also contains suggested further reading.

The authors have tried to organize this book in as simplified a form as possible in as far as the scientific language is concerned. Each chapter is preceded by a poem written by a Russian poet (translated into English by E. N. Sokolyuk). The selected verses may suggest subtle links with the concepts and contents of each Chapter, and have been introduced with the hope of fruitful cross-pollination between the natural sciences and humanities, so much needed in our modern world.

A. F. POZHARSKII
A. T. SOLDATENKOV
A. R. KATRITZKY

PREFACE TO ENGLISH EDITION

The book presents an updated translation of the Russian original 'Молекулы-перстни' by A. F. Pozharskii and A. T. Soldatenkov, published in 1993 by Khimiya. It has been a great pleasure to accept the invitation of my long-standing friend Sasha Pozharskii to join him and Professor Soldatenkov in producing the present English version, which follows closely the concepts and objectives of the original. We hope that this book may kindle for its readers some of the passion for heterocyclic chemistry which we the authors possess and help to repair the neglect of heterocyclic chemistry on the US academic scene. This neglect contrasts with the high importance awarded heterocyclic chemistry and biochemistry by American industry, as well as by academic and industrial chemists alike in Europe, Japan and all over the world.

This volume could not have been produced without the help of many people. Dr Daniel Brown (Cambridge) read the whole text and made very helpful suggestions. Among many other colleagues who read parts of the work, I would like to acknowledge particularly Dr Phil Cote, Dr Alastair Monro, Dr Emil Pop, Dr Nigel Richards, Dr Eric Scriven and Dr John Zoltewicz. It is a pleasure to thank also Ms Jacqui Wells, Dr Olga Denisko and Ms Cynthia Lee for all the help they gave me in producing and finalizing the manuscript.

Alan R. Katritzky
Gainesville, Florida
April 1996

1 MOLECULAR RINGS STUDDED WITH JEWELS

> Fortune Goddess, in your glory,
> In your honor, stern Kama,
> Bangles, finger-rings and bracelets
> I will lay before your Temple.
>
> V. Bryusov

Readers of this book, whether or not they are students of organic chemistry, will all be aware of the vital role of proteins, fats and carbohydrates in life processes. Experience has shown that considerably less is usually known about another class of compounds which have a similar importance in the chemistry of life, namely the heterocyclic compounds or, in short, heterocycles. What are heterocycles?

1.1 From Homocycle to Heterocycle

It is rumored that the Russian scientist Beketov once compared heterocyclic molecules to jewelry rings studded with precious stones. Several carbon atoms thus make up the setting of the molecular ring, while the role of the jewel is played by an atom of another element, a heteroatom. In general, it is the heteroatom which imparts to a heterocycle its distinctive and sometimes striking properties. For example, if we change one carbon atom in cyclohexane for one nitrogen atom, we obtain a heterocyclic ring, piperidine, from a homocyclic molecule. In the same way we can derive pyridine from benzene, or 1,2,3,4-tetrahydropyridine from cyclohexene (Figure 1.1).

A great many heterocyclic compounds are known. They differ in the size and number of their rings, in the type and number of heteroatoms, in the positions of the heteroatoms and so on. The rules of their classification help to orient us in this area.

Cyclic hydrocarbons are divided into cycloalkanes (cyclopentane, cyclohexane and so on), cycloalkenes (for example, cyclohexene) and aromatic hydrocarbons (with benzene as the main representative). The most basic general classification of heterocycles is similarly into heterocycloalkanes (e.g. piperidine), heterocycloalkenes (e.g. 1,2,3,4-

1

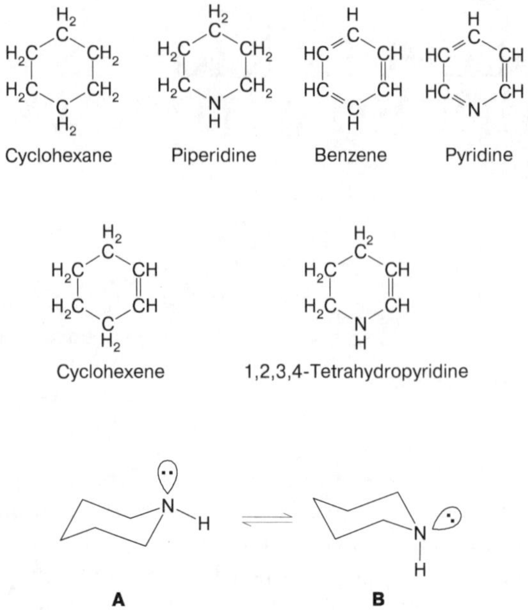

Cyclohexane   Piperidine   Benzene   Pyridine

Cyclohexene   1,2,3,4-Tetrahydropyridine

A   B

Figure 1.1 The relationship between cyclic hydrocarbons and heterocycles and the two chair conformations of piperidine.

tetrahydropyridine) and heteroaromatic systems (e.g. pyridine, etc.). Subsequent classification is based on the type of heteroatom. On the whole, the heterocycloalkanes and the heterocycloalkenes show comparatively small differences when compared with the related noncyclic compounds. Thus, piperidine possesses chemical properties very similar to those of aliphatic secondary amines, such as diethylamine, and 1,2,3,6-tetrahydropyridine resembles both a secondary amine and an alkene.

An interesting feature of heterocycloalkanes and heterocycloalkenes is the possibility of their existence in several geometrically distinct nonplanar forms which can quite easily (without bond cleavage) equilibrate with each other. Such forms are called conformations. For instance, piperidine exists mainly in a pair of chair conformations in which the internal angle between any pair of bonds is close to tetrahedral (109° 28′) to minimize steric strain. In these two chair conformations (Figure 1.1) the N–H proton is in either the equatorial (**A**) or axial (**B**) position, the first being slightly preferred.

By contrast, the heteroaromatic compounds, as the most important group of heterocycles, possess highly specific features. Historically, the name 'aromatic' for derivatives of benzene, naphthalene and their numerous analogs came from their characteristic physical and chemical properties. Aromatic compounds differ from other groups by an increased stability toward temperature and light. They tend to be oxidized and reduced with difficulty. On treatment with electrophilic, nucleophilic and radical agents they mainly undergo substitution rather than the addition reactions to the multiple bonds which are typical for ethylene and other alkenes. Such behavior results from the peculiar electronic configuration of the aromatic ring. We consider in the next section the structure of the benzene molecule.

1.2 The Detailed Structures of Benzene and Pyridine

Each carbon atom in the benzene molecule contributes one unpaired electron to four atomic orbitals which participate in four molecular orbitals. Three hybrid atomic sp^2-orbitals are oriented all in the same plane with their axes directed toward each other at an angle of 120°. These atomic orbitals overlap the similar orbitals of adjacent carbon atoms or the s-orbitals of hydrogen atoms, thereby forming the ring framework of six carbon–carbon bonds and six carbon–hydrogen bonds (Figure 1.2a). The molecular orbitals and bonds thus formed are called σ-orbitals and σ-bonds, respectively. The fourth electron of the carbon atom is located in an atomic p-orbital, which is dumbbell shaped and has an axis perpendicular to the ring plane. If the p-orbitals merely overlapped in pairs, the benzene molecule would possess the cyclohexatriene structure with three single and three conjugated double bonds, as reflected in the classic representation of benzene—the Kekulé

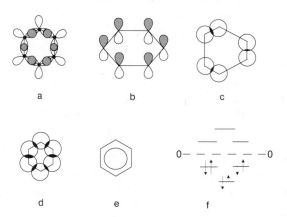

Figure 1.2 The electronic structure of the benzene molecule: (a) framework of σ-bonds; (b) p-orbital orientation; (c) overlap of p-orbitals forming localized π-bonds (view from above); (d) overlap of p-orbitals forming delocalized π-bonds; (e) representation of the benzene ring reflecting the equivalence of all carbon–carbon bonds and the equal distribution of π-electrons; and (f) energy levels of molecular π-orbitals showing electron occupation of the three orbitals of lower energy.

structure. However, in reality, the benzene ring is a regular hexagon, which indicates equal overlap of each p-orbital with its two neighboring p-orbitals, resulting in the formation of a completely delocalized π-electron cloud (see Figure 1.2d).

Thus, in the benzene molecule as well as in the molecules of other aromatic compounds, we observe a new type of carbon–carbon bond called 'aromatic', which is intermediate in length between a single and double bond. Standard aromatic C–C bond lengths are close to 1.40 Å, whereas the C–C distance is 1.54 Å in ethane and 1.34 Å in ethylene.

The high stability of the benzene molecule is explained by the energetic picture available from quantum mechanics. Benzene has six molecular π-orbitals. Three of these π-orbitals lie below the standard energy level (bonding orbitals) and are occupied by six electrons with a large energy stabilization. The remaining three are above the standard energy (antibonding orbitals). Occupation of the bonding orbitals leads to the formation of strong bonds and stabilizes the molecule as a whole. Incomplete occupation of bonding orbitals, and especially the occupation of antibonding orbitals, results in considerable destabilization. Figure 1.2f shows that all three bonding orbitals in benzene are completely occupied. Hence, it is often said that benzene has a stable aromatic π-electron sextet, a concept that can be compared in its importance to the inert octet cloud of neon or the F⁻ anion.

In addition to the π-electron sextet, stable aromatic arrangements can also be formed by two, 10, 14, 18 and $(4n + 2)$ π-electrons. Such molecules contain cyclic sets of delocalized π-electrons. For example, the aromatic molecule naphthalene possesses 10 π-electrons. The number of electrons required for a stable aromatic configuration can be calculated by the $4n + 2$ 'Hückel rule', where $n = 0, 1, 2, 3$ and so on, which was suggested by the German scientist Hückel in the early 1930s.

The electronic configuration of the pyridine molecule is very similar to that of benzene (Figure 1.3). Both compounds contain an aromatic π-electron sextet. However, the presence of the nitrogen heteroatom in the case of pyridine results in significant changes in the cyclic molecular structure. Firstly, the nitrogen atom has five valence electrons in the outer shell, in contrast with the carbon atom which has only four. Two take part in the formation of the skeletal carbon–nitrogen σ-bonds, and a third electron is utilized in the aromatic π-cloud. The two remaining electrons are unshared, their sp^2-orbitals lying in the plane of the ring. Owing to the availability of this unshared pair of electrons, the pyridine molecule undergoes many additional reactions over and above those which are characteristic of benzene or other aromatic hydrocarbons (see Chapter 2). Secondly, nitrogen is a more electronegative element than carbon and therefore attracts electron density. The distribution of the π-electron cloud in the pyridine ring is thus distorted.

Figure 1.3 The orientation of π-electron orbitals and unshared electron pairs in (a) pyridine and (b) pyrrole (C-H bonds are omitted).

Heterocyclic compounds include examples containing many other heteroatoms such as phosphorus, oxygen, sulfur, etc. By substitution of a ring carbon atom we may formally transform benzene into phosphabenzene (or phosphorine) or pyrylium and thiapyrylium cations (Figure 1.4). Note that a six-membered ring which includes oxygen or another group VI element can only be aromatic if the heteroatom bears a formal positive charge $(+1)$. Such cationic rings exist only in association with counteranions like ClO_4^- or BF_4^-. As for the nitrogen atom in pyridine, phosphorus, oxygen and sulfur atoms donate one π-electron to the aromatic electron cloud. Such heteroatoms are often called 'pyridine-like'.

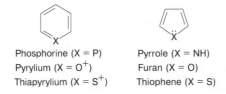

Phosphorine (X = P)   Pyrrole (X = NH)
Pyrylium (X = O⁺)   Furan (X = O)
Thiapyrylium (X = S⁺)   Thiophene (X = S)

Figure 1.4 Examples of heterocycles with pyridine-like and pyrrole-like heteroatoms.

Formally, pentagonal aromatic heterocycles can also be derived from benzene by a heteroatom taking the place of one complete CH=CH group. Two electrons of the heteroatom p-orbital must now be involved in the π-system in order to obtain an aromatic sextet. This type of heteroatom is called 'pyrrole-like' in contrast to 'pyridine-like' nitrogen which donates only one electron to the sextet. The corresponding five-membered heterocycles containing nitrogen, oxygen or sulfur atoms are named pyrrole, furan and thiophene, respectively (Figure 1.4). The difference between the pyridine-like heteroatom and pyrrole-like heteroatom is obvious: the first participates in one double bond in the Kekulé structure, while the second is involved in single bonds only.

A heterocycle can contain several heteroatoms. Pyridazine, pyrimidine, pyrazine and 1,3,5-triazine are heterocyclic compounds with a single ring but two or three identical heteroatoms (see Figure 1.5a). Together with pyridine and many other analogs they form the family of azines.

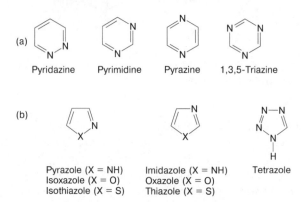

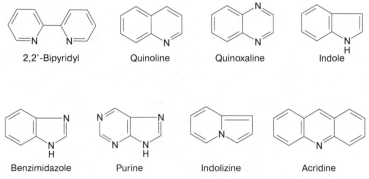

Figure 1.5 Heterocycles of (a) the azine class and (b) the azole class.

Five-membered heterocyclic compounds containing both pyridine-like and pyrrole-like nitrogen or other heteroatoms are called azoles. Pyrazole, imidazole and their oxygen and sulfur analogs belong to the azole series (Figure 1.5b).

Two or more rings are encountered in many heterocyclic compounds. The cycles may be connected to each other by a single bond (as in the case of 2,2'-bipyridyl) or may be fused as shown in Figure 1.6 to form condensed systems. For example, two fused cycles exist in quinoline, quinoxaline, indole and benzimidazole. In some cases a heteroatom may belong simultaneously to two (e.g. indolizine) or even three rings. Such a heteroatom is denoted a 'bridge-head' atom.

Figure 1.6 Examples of bi- and polycyclic heterocycles.

The comparison of heterocycles with jewel-studded rings is most appropriate for five- and six-membered systems which are frequently natural products and which have become commonplace in many research laboratories. However, polymembered cycles or macrocycles have recently drawn much attention. They resemble not so much finger-rings but rather molecular bracelets or bangles. For example, aza[18]annulene is an 18-membered analog of pyridine, and aza[17]annulene is a 17-membered analog of pyrrole (Figure 1.7a). We focus our attention on macrocycles in Chapter 10.

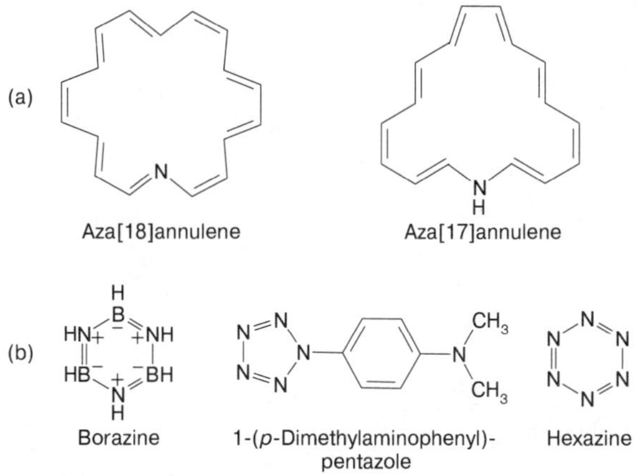

Aza[18]annulene   Aza[17]annulene

Borazine   1-(p-Dimethylaminophenyl)-   Hexazine
pentazole

Figure 1.7 Examples of (a) macroheterocycles and (b) rings without cyclic carbon atoms.

How many heteroatoms may be included in one ring? As many as one can imagine. A ring may, in principle, be completely constructed from non-carbon atoms. Borazine, a well-known example of such a compound, was baptized 'inorganic benzene' because of its high stability. 1-(p-Dimethylaminophenyl)pentazole, which has been prepared recently, contains a five-membered heterocycle composed only of nitrogen atoms. The curiosity of many chemists has long been excited by a theoretical substance named 'hexazabenzene' or 'hexazine' (Figure 1.7b). Numerous attempts to prepare this compound have so far ended in failure, supposedly because of its great instability and tendency to decompose to give nitrogen: $N_6 \rightarrow 3N_2$.

Of course, the examples given above by far do not cover all of the heterocyclic systems possible. In the following chapters we will become acquainted with many new ones.

1.3 Problems

1. How many chair conformations are possible for unsubstituted piperidine? How many for a 1,4-disubstituted piperidine? Draw their structures.

2. The boat conformation for saturated six-membered rings is energetically unfavorable. Account for this fact. Design the structure of a substituted piperidine in which the boat conformation is fixed.

3. Phosphacyclohexane (phosphorinane) exists almost completely in a chair conformation with the P–H bond axial. Discuss possible reasons for the stabilization of this conformation compared with the analogous piperidine conformation.

4. Indicate which of the heterocycles listed below can be formally regarded as aromatic. Explain your choices.

5. Provide an explanation for the instability of hexazine.

6. Draw all of the possible isomeric imidazopyridines, i.e. the heterocycles which consist of fused pyridine and imidazole nuclei.

7. What is the orientation of the nitrogen lone pair of electrons in aza[18]annulene (Figure 1.7)? Is any alternative orientation possible? Discuss the orientation of the N–H bond in aza[17]annulene.

1.4 Suggested Reading

1. Joule, J. A. and Smith, G. F., *Heterocyclic Chemistry,* Chapman & Hall, London, 1995.

2. Gilchrist, T. L., *Heterocyclic Chemistry,* Pitman Press, London, 1985; 2nd Edn, J. Wiley, New York, 1992.

3. Davies, D. T., *Aromatic Heterocyclic Chemistry,* Oxford University Press, Oxford, 1992.

4. Pozharskii, A. F., *Theoretical Basis of Heterocyclic Chemistry* (in Russian), Khimia, Moscow, 1985.

5. Katritzky, A. R., *Handbook of Heterocyclic Chemistry,* Pergamon Press, Oxford, 1985.
6. Katritzky, A. R. and Rees, C. W. (eds), *Comprehensive Heterocyclic Chemistry,* Vols 1–8, Pergamon Press, Oxford, 1984.
7. Acheson, R. M., *An Introduction to the Chemistry of Heterocyclic Compounds,* 3rd Edn, Wiley-Interscience, New York, 1976.
8. Elguero, J., Marzin, C., Katritzky, A. R. and Linda, P., in *Advances in Heterocyclic Chemistry,* Suppl. 1, Academic Press, New York, 1976.
9. Blackburne, I. D., Katritzky, A. R. and Takeuchi, Y., *Acc. Chem. Res.,* 1975, **8**, 300.

2 WHY NATURE PREFERS HETEROCYCLES

Ties subtle, full of power exist
Between the shape and flavor of a flower.
So is a brilliant unseen, until comes hour
To facet it from diamond mist.

V. Bryusov

All biological processes are chemical in nature. Such fundamental manifestations of life as the provision of energy, transmission of nerve impulses, sight, metabolism and the transfer of hereditary information are all based on chemical reactions involving the participation of many heterocyclic compounds. Why does nature utilize heterocycles? To answer this question we first describe the basic physical and physicochemical properties of the fundamental heterocyclic types.

2.1 Reactions for All Tastes

Heterocycles are involved in an extraordinarily wide range of reaction types. Depending on the pH of the medium, they may form anions or cations. Some interact readily with electrophilic reagents, others with nucleophiles, yet others with both. Some are easily oxidized, but resist reduction, while others can be readily hydrogenated but are stable toward the action of oxidizing agents. Certain amphoteric heterocyclic systems simultaneously demonstrate all of the above-mentioned properties. The ability of many heterocycles to produce stable complexes with metal ions has great biochemical significance. Such versatile reactivity is linked to the electronic distributions in heterocyclic molecules. Let us consider pyridine.

We have already seen that the nitrogen atom in pyridine induces π-electron withdrawal from the carbon atoms. As a result of this electronic shift the carbon atoms in the *ortho* and *para* positions (in relation to the nitrogen atom) acquire a partial positive charge (Figure 2.1). Thus, a π-electron deficit on the carbon skeleton is characteristic of all heterocycles containing pyridine-like heteroatoms. Such heterocycles are called π-deficient.

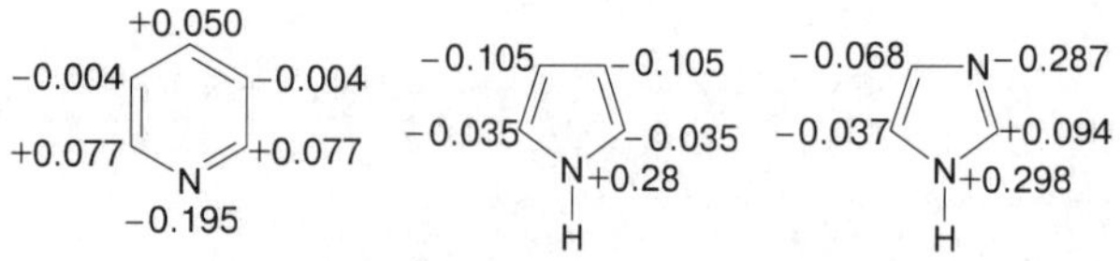

Figure 2.1   The π-electron charges in pyridine, pyrrole and imidazole.

A unique feature of π-deficient heterocycles is their facile interaction with negatively charged nucleophilic reagents. As a typical example, the reaction of pyridine with sodamide gives 2-aminopyridine in good yield:

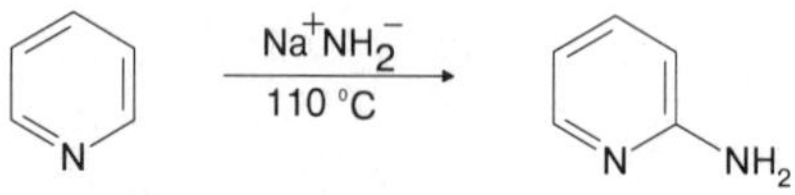

Substitution of the hydrogen atom under the action of positively charged (electrophilic) agents proceeds with difficulty or does not occur at all in π-deficient heterocycles. However, electrophiles add readily to the pyridine nitrogen owing to its unshared pair of electrons. Pyridine thus forms pyridinium and *N*-alkylpyridinium salts with acids and alkyl halides, respectively, and a zwitterionic addition compound or Lewis salt with BF$_3$:

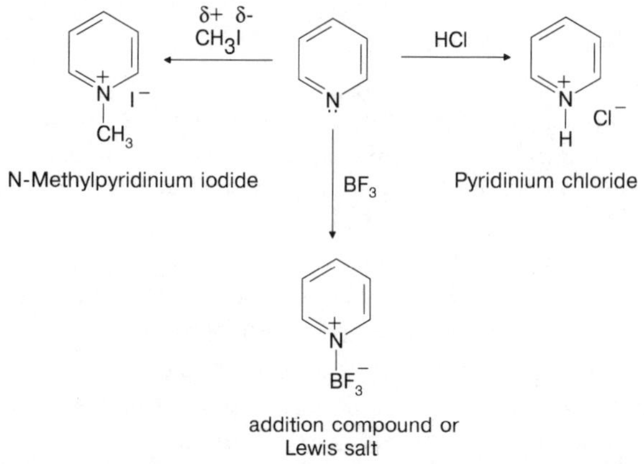

Pyridine and other heterocycles containing a pyridine-like nitrogen atom behave as bases in these and similar reactions (see Section 2.2).

The introduction of electron-accepting groups into an organic compound lowers the energy of all molecular orbitals. Hence, such compounds donate electrons with difficulty and are thus poorly oxidized. By contrast, their ability to accept additional electrons enables such compounds to be readily reduced. Pyridine-like heteroatoms are electron acceptors, and hence π-deficient heterocycles are reduced with ease. This is found to be the case, especially in relation to compounds which have a positively charged heteroatom, like salts of pyrylium, pyridinium, etc. For example, 1-benzyl-3-carbamoylpyridinium chloride is reduced by sodium dithionite to the corresponding 1,4-dihydropyridine derivative:

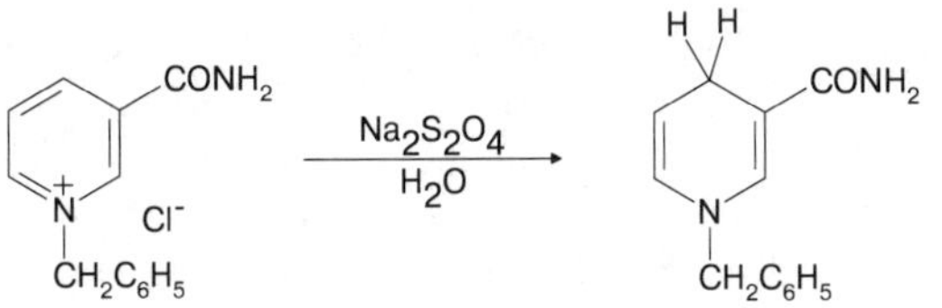

We shall see elsewhere (Sections 4.2.1 and 5.2) that nature has used this apparently simple reaction to drive a great many biologically important processes.

Quite a different situation is encountered in the case of pyrrole, furan and thiophene. Since the heteroatoms of these compounds each contribute two electrons to the π-aromatic ensemble, the cyclic system of five atoms formally has six π-electrons. As a result, in spite of the higher intrinsic electronegativity of the heteroatom, all of the carbon atoms possess excess negative charge (Figure 2.1). Such compounds are named π-excessive heterocycles. Reactions with nucleophiles agents are not common but they readily interact with electrophiles. Thus, pyrrole is almost instantly halogenated even under very mild conditions to give the tetrahalogenopyrrole, and these reactions cannot be stopped at the monosubstitution stage:

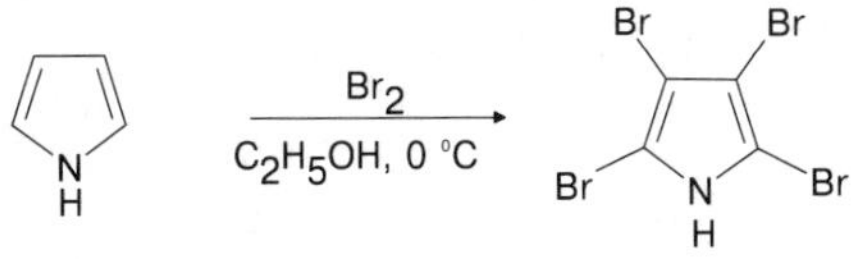

Two-electron donation to the aromatic system by the pyrrole-like heteroatom imparts a partial positive charge to the heteroatom (Figure 2.1). In the case of pyrrole and related NH-heterocycles, the N–H bond reactivity increases. N-Anions, which are readily alkylated, acylated and arylated, are thus formed under the action of bases. Such reactions are commonly used for the synthesis of various N-derivatives (note that a nonionized NH group does not, as a rule, undergo these conversions):

The molecular orbitals in π-excessive heterocycles are of high energy, and consequently these compounds are reduced with difficulty but are readily oxidized. Compounds with both pyridine-like and pyrrole-like heteroatoms, as expected, can show both π-deficient and π-excessive properties with one or the other dominant. Thus, imidazole contains two carbon atoms with a negative and a third with a positive π-charge (Figure 2.1); its high reactivity towards halogenation is attributed to a dominant π-excessive character of the imidazole anion.

2.2   Heterocycles as Acids and Bases

In the preceding section we noted the capability of nitrogen heterocycles to behave as acids or bases, the acidic properties being inherent to heterocyclic compounds containing a pyrrole-like NH group, whereas the basic properties are characteristic for those with pyridine-like nitrogen. We describe this in more detail because acid–base properties play a vital role not only in general reactivity but in many biochemical processes as well.

The acid dissociation constant (K_a) is universally used as the quantitative measure of acidity. Dissociation constants are obtained by application of the law of mass action to the acid–base equilibrium

$$H\text{-}A \rightleftharpoons A^- + H^+$$

The dissociation constant K_a is equal to the anion concentration multiplied by the proton concentration, divided by the concentration of the nondissociated acid

$$K_a = [A^-][H^+]/[HA] \tag{2.1}$$

In practice, following the analogous use of pH, it is more convenient to use the negative logarithm of K_a, the so-called acidity index pK_a

$$pK_a = -\log K_a = -\log[A^-] - \log[H^+] + \log[HA]$$

As the value of $-\log[H^+] = pH$, then

$$pK_a = \log\{[HA]/[A^-]\} + pH \tag{2.2}$$

It is clear from equation (2.2) that the value of the pK_a is equal to the value of the pH when the nondissociated acid (HA) content and the anion (A^-)

content are equal, i.e. when the degree of dissociation is 50%.

We see that the stronger the acid, the greater the numerator and, consequently, the larger the K_a value; a larger K_a value corresponds to a smaller pK_a. Vice versa, in a series of compounds, the pK_a increases as the acidity decreases. It should be emphasized that pK_a values, which are essentially acid ionization constants, are also employed for the measurement of basicity. As a consequence of the reversibility of the dissociation process, any acid which donates its proton is thus converted to the conjugate base; similarly, a base which accepts a proton becomes the conjugate acid. Stronger acids obviously correspond to weaker conjugate bases and vice versa. Thus, for bases, the order of the pK_a changes in the opposite sense: the larger the pK_a of the conjugate acid, the stronger the base, and the weaker bases have correspondingly lower pK_a values.

The acid dissociations of pyrrole and imidazole (Figure 2.2a) are used as an example. The corresponding pK_a values are 17.5 and 14.2, respectively.[†] As pK_a is a logarithmic scale, pyrrole is a weaker acid than imidazole by a factor of $10^{3.3}$, i.e. by a factor of 2000. This also indicates that the pyrrole anion is a stronger base than the imidazole anion by the same factor.

Whereas both pyrrole and imidazole are very weak acids, some heterocycles have pK_a values close to those of conventional acids. Tetrazole (Figure 1.5) has a pK_a of 4.89, almost equal to that of acetic acid (pK_a 4.76).

Under ordinary conditions a neutral pyrrole-like nitrogen is unlikely to add a proton because of the tendency to preserve the aromaticity of the

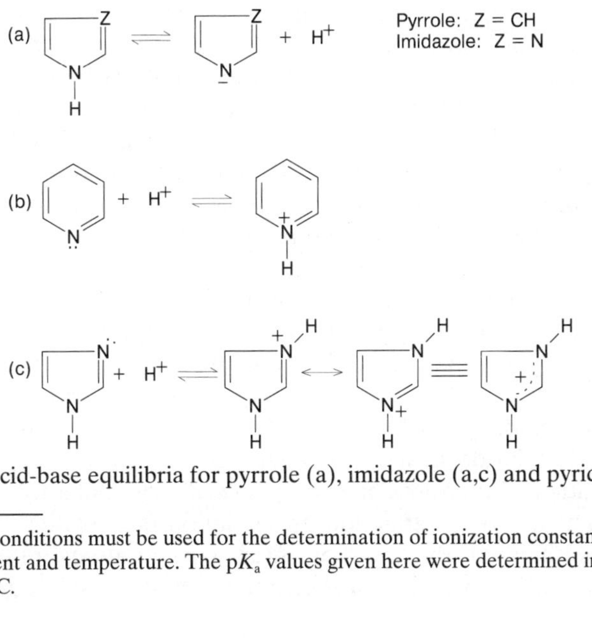

Figure 2.2 Acid-base equilibria for pyrrole (a), imidazole (a,c) and pyridine (b).

[†]Standardized conditions must be used for the determination of ionization constants as the latter depend on solvent and temperature. The pK_a values given here were determined in aqueous solutions at 20 °C.

heterocycle. In contrast, the lone electron pair of a pyridine nitrogen does not participate in the formation of the aromatic sextet and readily adds a proton to form a heteroaromatic cation. Thus, pyridine has a pK_a of 5.23. This value formally reflects the acidity of the pyridinium ion (Figure 2.2b), but is more often used to assess the basicity of pyridine. It can be seen that the proton of the pyridinium cation is 12 orders of magnitude more acidic than the NH of pyrrole. This is readily explained by the facile loss of a proton from the positively charged nitrogen atom in the pyridinium cation.

Obviously, heterocycles such as imidazole have amphoteric properties: imidazole is both an NH acid and a strong neutral base with a pK_a of 6.95. The imidazole ring system is frequently encountered in proteins (see Section 4.1) and is one of the strongest of all bases found in biological systems. The imidazole unit, therefore, plays an active role in proton transfer processes and the various catalytic events accompanying them. The enhanced basicity of imidazole is due to electron donation from the pyrrole nitrogen, thus favoring proton addition. The stabilized imidazolium ion can be represented by two equivalent resonance structures in which the positive charge is isolated on one nitrogen atom in one and on the other in the second representation, or by an average structure with delocalized charge (Figure 2.2c).

2.3   Heterocycles and Metals

It is well known that minute quantities of different metals are necessary for the normal development of all living organisms. In addition to the widespread sodium, potassium, magnesium, calcium, iron and zinc, the group of 'essential metals' also includes more exotic members such as molybdenum, cobalt, chromium and others. Metals exist in organisms in the form of cations linked with various basic ligands by coordination bonds. The basic functionality may involve the amino, hydroxy or thiol groups of amino acids as well as nitrogen heterocycles (azines and azoles). The ability to form stable metallic complexes seems to be 'preprogrammed' into the structure of the heterocycles.

The fixed and outwardly directed unshared pair of electrons of a pyridine nitrogen atom is available for coordination with practically all metal ions. Thus, pyridine gives complexes of various types: a linear arrangement with the silver ion, a tetrahedral structure with aluminum chloride, a square planar coordination compound with copper(II) chloride and a dianionic complex with cobalt(II) chloride, as shown in Figure 2.3.

The formation of coordination compounds is very similar to the production of pyridinium salts by protonation or alkylation (see Section 2.1), although some peculiarities in the electron shell configurations of some metals diversify the spatial structures of the complexes. It should be noted that the

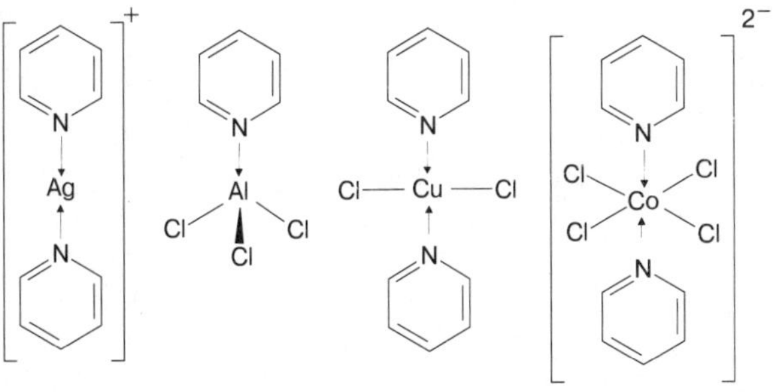

Figure 2.3 Pyridine as a ligand in complexes.

oxidation–reduction potentials and other properties of metal ions can be markedly changed by coordination. Such changes can, in turn, significantly affect the functioning of biological systems.

The number of ligands coordinated to a metal ion depends on the type and number of unfilled orbitals in the outer electron shell. For example, in order to complete its outer shell of eight electrons, aluminium(III) requires one additional pair of electrons. This electron pair can be donated by many bases, such as pyridine. The four valence bonds formed by aluminium in the $C_5H_5N{:}AlCl_3$ ensemble are formally built utilizing one $3s$-orbital and three $3p$-orbitals. Quantum mechanics ascertains that such bonding is achieved from the more energetically favorable mixed (hybrid) sp^3-orbitals. The best arrangement of four sp^3-orbitals in space is achieved when their axes are directed toward the corners of a regular tetrahedron. This provides for minimal interelectronic repulsion and results in a tetrahedral configuration of the complex.

A pyridine molecule can donate only one electron pair for coordination with a metal ion. Such a ligand is described as monodentate. Imidazole seems to be the most important heterocyclic monodentate ligand in biochemical processes (see Section 4.2). Polydentate ligands are able to provide several electron pairs and are highly effective. 2,2'-Bipyridyl, an example of a bidentate heterocyclic ligand (Figure 1.6), forms stable complexes with many metals, in particular with iron(II) (see Section 9.6).

The porphyrin system is a very important natural tetradentate macrocyclic ligand composed of four pyrrole rings which are linked to each other via carbon bridges (Figure 2.4). Two of the pyrrole rings in porphyrin are in the oxidized state: their nitrogen atoms are of the pyridine type, with the unshared electron pairs oriented toward the inside of the macrocycle. If both the N–H bonds in the porphyrin molecule are ionized, a highly symmetrical dianion is formed in which all four nitrogens become equal because of delocalization of the negative charges. In this dianion all four unshared pairs of electrons are directed toward the inside of the macrocyclic cavity. The

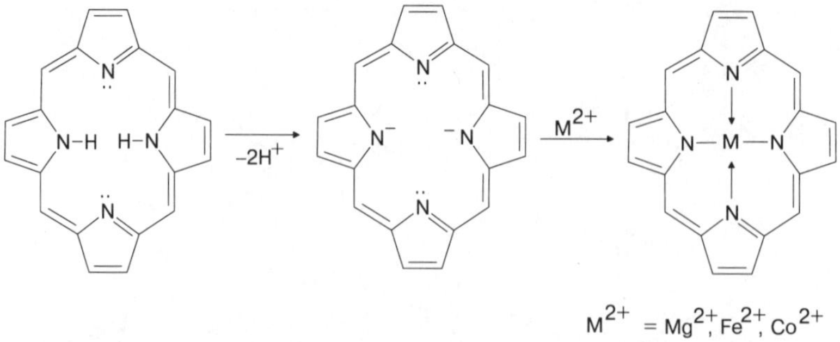

$M^{2+} = Mg^{2+}, Fe^{2+}, Co^{2+}$

Figure 2.4   The porphyrin molecule, its dianion and complexes with metals.

ionic radii of many metals allow them to fit within this cavity and the metal ions can be fixed in space by coordination bonds with the four porphyrin nitrogen atoms. Such complexes have considerable stability and are deeply colored. A porphyrin system which includes magnesium is part of the green plant pigment chlorophyll (see Section 6.1). A porphyrin system containing an entrapped iron(II) ion is of primary importance in respiratory and metabolic processes as it is a constituent of the red pigment of blood, hemoglobin (Section 4.2.2). An analogous complex containing the cobalt(II) ion is a structural fragment of vitamin B_{12}.

2.4   'There are Subtle Ties of Power...'

The reactions described above are accompanied by the cleavage and formation of covalent donor–acceptor bonds which are polar to a large extent. These bonds are known to be particularly strong, their energies being between 60 and 100 kcal mol^{-1}. By the sharing of electrons and the formation of covalent bonds, stable molecules from which the living organism constructs numerous structures, for example the cell membrane, are produced. Moreover, in living systems the covalent bond serves as a reservoir of energy to be released when needed (see Chapter 5). From a biological point of view, stable covalent bonding has both advantages and disadvantages. Since such bonds are difficult to break, they cannot always provide the necessary flexibility and mobility required by living systems. For example, many biochemical reactions are reversible, the same molecule being able to react thousands of times as a result of regeneration. Thus, enzymes act as biological catalysts and hemoglobin as an oxygen carrier. These, and certain other compounds such as metal ion delivery systems are able to change their spatial structures in a rapid and reversible manner. This is only possible when the bonds involved are much weaker than covalent linkages—such are the van der Waals bonds or non-bonding attractions.

The energies of van der Waals bonds at $0.5–10\,\text{kcal}\,\text{mol}^{-1}$ are one to two orders of magnitude lower than those of covalent bonds. It should be noted that nonbonding attractions are ubiquitous and do not occur only in biologically significant molecules. They emerged in the universe at the same time as the appearance of atoms and molecules. As an example, while two atoms of helium cannot share their electrons to form a covalently bonded molecule, the electrons of one helium atom are subtly attracted by the nucleus of another and vice versa. The strengths of such attractions are very small, from 0.2 to $0.5\,\text{kcal}\,\text{mol}^{-1}$, but when there are many such interactions they can substantially affect the properties of a compound. Helium atoms are attracted to each other so very weakly that gaseous helium becomes liquefiable only at very low temperatures ($-269\,°\text{C}$), whereas methane, being molecular, liquefies at $-161\,°\text{C}$. n-Hexane is a liquid under ordinary conditions (boiling point $69\,°\text{C}$), while n-octadecane, a homolog of the former, is a solid with the relatively low melting point of $28\,°\text{C}$.

The weakness of nonbonding interactions hinders their observation, and consequently their measurable effects were recorded historically later than those of covalent bonds. In particular, the attraction mentioned above between the electrons of one atom and the nucleus of another was first described by London in 1928, almost 10 years after the formulation of covalent bonds by Lewis. It is clear that London forces, also called dispersion forces, are inherent to all atoms and molecules. However, there exist certain types of nonbonding interactions which are found only in compounds with certain structural characteristics. It is these specific interactions which make an essential contribution to the chemistry of living systems.

2.4.1   HYDROGEN BONDING

Hydrogen atoms bound to electronegative elements such as nitrogen, oxygen or fluorine acquire a positive charge because of the high polarity of the corresponding bond. As a result of this charge, and because the hydrogen atom is small, the hydrogen atom can approach other atoms and interact electrostatically with their unshared electron pairs. If the attractive force is high enough, a hydrogen atom nucleus (i.e. as a proton) can reversibly transfer to form a covalent bond with another atom. This exchange is the foundation of acid–base interactions

$$\text{AH} + :\text{B} \longrightarrow \text{A}^- + \text{HB}^+$$

If, however, the attraction between the proton-donating AH group and the proton-accepting base :B is not strong enough to enable proton transfer, then a hydrogen bridge AH•••B can still arise.

A classic example of hydrogen bonding is found in water, the molecules of which are aggregated in linear and three-dimensional structures. Although the strength of one hydrogen bond is not great (2–8 kcal mol^{-1}), the overall consequence of a multitude of such bonds is significant. The numerous unique properties of water (low volatility, moderate viscosity and density, specific qualities of ice, etc.) which have allowed life on Earth to become possible are the result of hydrogen bonding.

Hydrogen bonding is certainly the most important type of nonvalent interaction between biomolecules. The polypeptide helical chains of proteins and the double-helical structure of DNA are stabilized by such interactions (see Chapter 3). The ability to form hydrogen bonds is inherent in practically all nitrogen heterocycles. Some (pyridine and other azines) are proton acceptors, others (pyrrole, indole) are proton donors, and a third group of compounds includes both proton-donating and proton-accepting functionalities (imidazole, pyrazole). Imidazole, for example, forms rather stable linear associations, whereas pyrazole is inclined to give dimers because of the specific orientation of its NH group and pyridine-like nitrogen, as can be seen in Figures 2.5a and 2.5b.

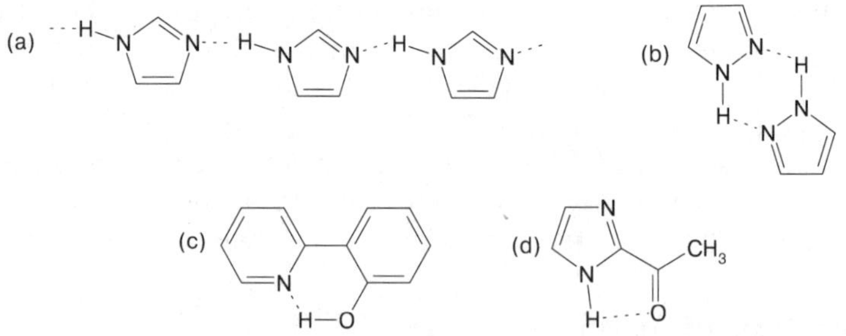

Figure 2.5 Intermolecular and intramolecular hydrogen bonding: (a) imidazole association; (b) dimer of pyrazole; and (c, d) hydrogen bonding in 2-(o-hydroxyphenyl)pyridine and 2-acetylimidazole.

Certain heterocyclic compounds with suitably oriented functional groups can form intramolecular bonds. Such is the case for 2-(o-hydroxyphenyl)pyridine and 2-acetylimidazole shown in Figures 2.5c and 2.5d. Intramolecular hydrogen bonding is much stronger when it involves the construction of a six-membered ring.

2.4.2 ELECTROSTATIC INTERACTIONS

Electrostatic attractions or repulsions of charged particles are a type of nonvalent interaction as widespread as hydrogen bonding. At one time, it was thought that electrostatic interactions were characteristic only of ions. In reality, many neutral molecules engage in similar interactions especially when their electron clouds are polarized. Such molecules behave as if composed of two oppositely charged poles. These dipolar molecules can attract each other or ions (Figure 2.6).

Figure 2.6   Nonbonding interactions: (a, b) ion–dipole and (c, d) dipole–dipole.

Almost all heterocyclic molecules are dipoles in addition to being capable of ion formation. Electrostatic interactions exert a marked influence upon heterocyclic behavior. For example, pyridine, pyrrole and 1-methylimidazole have molecular weights close to that of benzene. However, benzene demonstrates much greater volatility: it boils at 80 °C, while the heterocycles mentioned have boiling points of 115, 130 and 196 °C, respectively. The heterocyclic molecules are, to a significant degree, polar[†] and are subject to dipole–dipole associations (Figures 2.7a–2.7c). In order to transform these associates into the vapor state, considerable energy is obviously needed, from which a decrease in the volatility of the compound results. Electrostatic interactions play an essential role in biology where they participate, in particular, in optimizing the spatial arrangements of complex biomolecules. Such three-dimensional configurations can endow highly specialized biological activity. For example, the imidazole ring, found in many enzymes, exists at physiological pH to the extent of about 50% as the positively charged imidazolium ion. It is clear that the negatively charged ionized carboxylate group at the end of the protein chain will be attracted to the imidazolium cation and thus induce the relevant portion of the macromolecule to form a coil, as shown in Figure 2.7d. By contrast, if an ammonium ion is present in the chain, the repulsive force between this ion and the imidazolium ion prevents the protein chain from coiling (Figure 2.7e).

[†]The polarity of a molecule may be estimated from the electronic charges of the different atoms (Figure 2.1). More definitive proof of polarity can be obtained from dipole moment values calculated as the magnitude of the distance between the centers of positive and negative charges, multiplied by the average charge. Numerical values of dipole moment are expressed in debye units (D). The higher the polarity of a molecule, the higher its dipole moment. The dipole moments of pyridine, pyrrole and 1-methylimidazole are 2.2, 1.8 and 3.8 D, respectively.

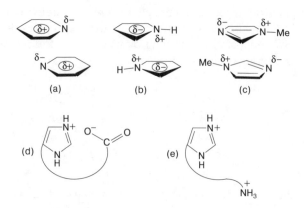

Figure 2.7   Electrostatic interactions of heterocyclic molecules: (a–c) dipole–dipole associations; (d) attraction; and (e) repulsion between charged groups of a protein chain.

2.4.3   MOLECULAR COMPLEXES

In many chemical reactions, the cleavage of an existing bond and the formation of a new one are preceded by electron transfer between the molecules of the reacting compounds. As a rule, an electron is transferred from the highest occupied molecular orbital (HOMO) of the donor to the lowest unoccupied molecular orbital (LUMO) of the acceptor. As a result of the transfer the donor becomes a cation-radical, and the acceptor is converted into an anion-radical. Both particles are ions as they acquire charge by contributing or accepting an electron. At the same time they can be considered as radicals, for they have an odd number of electrons. Like all oppositely charged ions these cation-radicals and anion-radicals are attracted to each other and form ion-radical pairs

$$D: \ + \ A \ \longrightarrow \ D^{+\cdot}A^{-\cdot}$$

In principle, any molecule can act as a donor or an acceptor because it has both a HOMO and LUMO, which are also called frontier orbitals. In practice, however, most compounds display a tendency toward either donating or accepting electrons. Polycyclic aromatic hydrocarbons (anthracene, benzpyrene, etc.), aromatic amines, phenols, thiophenols and other compounds with accessible unshared pairs of electrons (alcohols, esters, ketones, tertiary amines, sulfides and so on) are typical donors. Almost all π-excessive heterocycles, particularly polynuclear compounds such as indole, carbazole, phenothiazine and others (Figure 2.8), possess good donor properties.

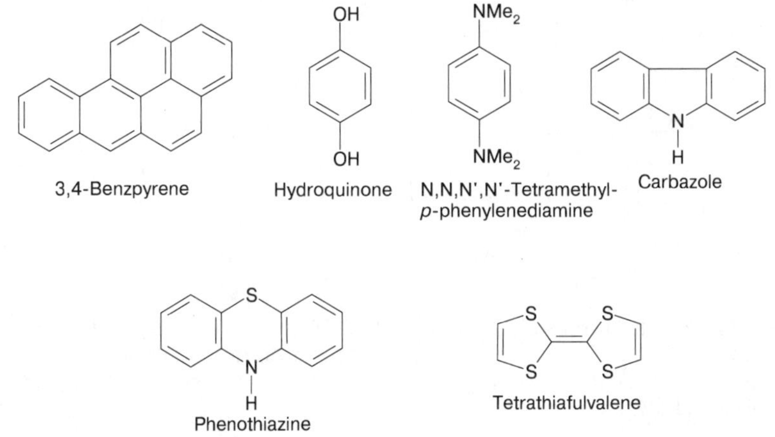
OH
OH
NMe2
NMe2
N
H
3,4-Benzpyrene
Hydroquinone
N,N,N',N'-Tetramethyl-
p-phenylenediamine
Carbazole
S
N
H
Phenothiazine
S
S
S
S
Tetrathiafulvalene

Figure 2.8 Examples of electron donors.

Aromatic polynitro compounds, quinones, conjugated polycyanides, inorganic substances such as molecular iodine, bromine, interhalogen compounds (ICl, IBr), heavy metal ions and so on are a few of the many types of electron acceptors commonly encountered. The acceptor class also includes all π-deficient heteroaromatic compounds, and heteroaromatic cations in particular. Some examples of acceptors are shown in Figure 2.9.

Both a strong donor and a strong acceptor are a necessary condition for the formation of ion-radical salts. This ensures a small energy gap between the donor's HOMO and the acceptor's LUMO which allows an electron to

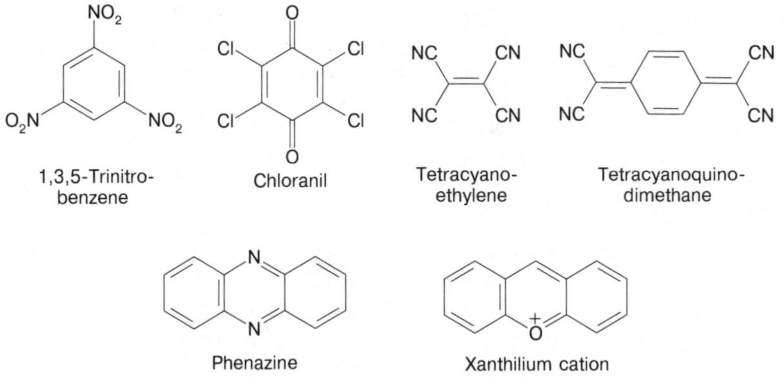
NO2
O2N
NO2
1,3,5-Trinitro-
benzene
O
Cl
Cl
Cl
Cl
O
Chloranil
NC
CN
NC
CN
Tetracyano-
ethylene
NC
CN
NC
CN
Tetracyanoquino-
dimethane
N
N
Phenazine
O
Xanthilium cation

Figure 2.9 Examples of electron acceptors.

leap from one orbital to another, as in the case of ion-radical salt formation from tetrathiafulvalene and tetracyanoquinodimethane (Figures 2.10 and 2.11a; see also Section 10.5).

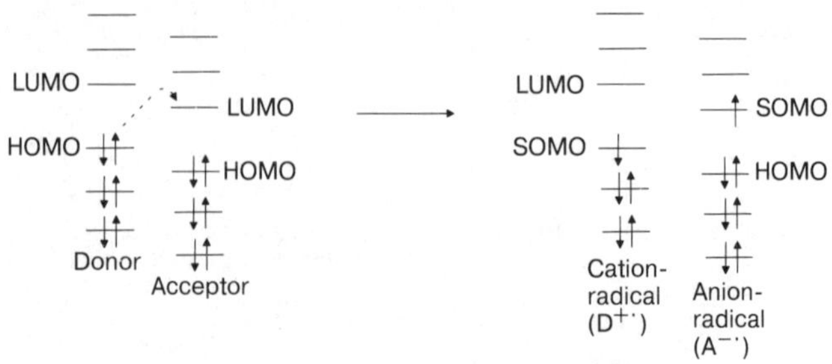

Figure 2.10   Electron transfer in the formation of an ion-radical salt: the energy gap between the HOMO of the donor and the LUMO of the acceptor is smaller than that between the HOMO and LUMO of either the donor or the acceptor itself (SOMO = singly occupied MO).

If the donor and acceptor are not sufficiently powerful, the energy difference between their frontier orbitals increases and the transfer of an electron becomes difficult. The possibility of these orbitals overlapping still remains, although such overlapping is unlikely.

In such cases with weak donors and acceptors, partial transfer of electron density or charge transfer may occur. The charges emerging as a result of this transfer (positive on the donor and negative on the acceptor) weakly bind the two molecules to form an association named a molecular complex or a charge–transfer complex (CTC). Complexation of indole with chloranil generates a typical molecular complex (Figure 2.11b).

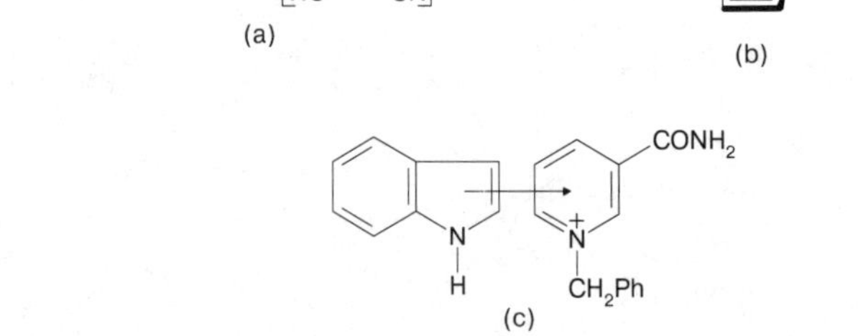

Figure 2.11   Molecular complexes: (a) an ion-radical salt from tetrathiafulvalene and tetracyanoquinodimethane; (b) a charge transfer complex between indole and chloranil; and (c) a CTC between the 1-benzyl-3-carbamoylpyridinium cation and indole.

The CTC composition is not always in a simple 1:1 stoichiometric ratio. Two molecules of a donor may be linked with one molecule of an acceptor or vice versa. Binding energies in molecular complexes are normally below $6\,kcal\,mol^{-1}$ and facile cleavage occurs in solution so that a rapid equilibrium exists with partial dissociation of the complexes into the donor and acceptor molecules.

The linkage between the components of a complex is symbolized by either a point or an arrow directed from the donor toward the acceptor

$$D \; + \; A \; \rightleftharpoons \; D\cdot A \;\; or \;\; D{\rightarrow}A$$

Donor and acceptor molecules tend to configure themselves into oriented layers such that maximum overlap occurs between their π-orbitals in the complex. The most indicative feature of CTC or ion-radical salt formation is the appearance of color in the reaction mixture. For example, tetrathiafulvalene, tetracyanoquinodimethane, indole and chloranil are all practically colorless compounds, but their ion-radical salts and molecular complexes shown in Figure 2.11 are black-green and red, respectively.

Although we have distinguished between ion-radical salts and molecular complexes, we should emphasize that there is no fundamental difference between them. In ion-radical salts the electron transfer is never complete and rarely exceeds 60%.[†] In other words, an ion-radical salt is a molecular complex in which electron transfer is quite pronounced.

Many biologically important heterocyclic compounds possess significant electron-donor or electron-acceptor ability. For example, metalloporphyrins, indoles and nucleic acid purine bases are all good donors. Electron-accepting properties are inherent in isoalloxazine, the main component of flavin systems (see Figure 4.6), and 1-benzyl-3-carbamoylpyridinium chloride, which is used as a model for the respiratory coenzyme NAD^+. In test-tube (*in vitro*) experiments, these compounds react with various acceptors and donors to give molecular complexes. So, if indole is mixed with 1-benzyl-3-carbamoylpyridinium chloride, a yellow 1:1 molecular complex is afforded (Figure 2.11c). Such results suggest that molecular complexes also occur *in vivo* (in living tissues). Indeed, in the last 25 years, conclusive evidence has been obtained for the participation of ion-radical salts and molecular complexes in photosynthetic and respiratory processes. Electron transport may also play an important role in the action of some drugs, especially neurotropics.

---

† Here we draw attention to an analogy with ionic and covalent bonds. It is known that purely ionic bonds do not exist in condensed phases and that even in such a typical ionic compound as NaCl the electron transfer from sodium to chlorine does not exceed about 80%. Thus, a bond is formally referred to as ionic if the electron transfer exceeds 50%.

2.4.4 HYDROPHOBIC FORCES

This type of nonbonding interaction is not generally intrinsic to heterocyclic compounds. However, hydrophobic interactions do influence the behavior of heterocycles, especially in various life processes. If water is shaken with a nonpolar liquid, for example octane, a dispersion of tiny droplets of octane in water, i.e. an emulsion, is formed. When the agitation is stopped the octane droplets coalesce rapidly and the emulsion is converted into two liquid layers. This clearly demonstrates the presence of certain repulsive forces between the apolar octane molecules and water. These forces are named hydrophobic (Greek: *hydro*, water; *phobos*, fear).

As described above, hydrogen bonding is responsible for many of the specific properties of water. In order to dissolve a substance in water we need to break a large number of hydrogen bonds which create the association of water molecules with themselves. A substance can be dissolved in water only if it supplies the necessary energy for these processes to occur. Various salts, for example sodium chloride (or cooking salt), dissolve in water because the energetic expenditure is compensated for by the energy released from the interactions between the Na^+ and Cl^- ions with the water dipole (solvation). Thus, 'supply and demand' is also evident in nature.

Nonionic compounds such as sugars, the lower alcohols, ketones, carboxylic acids and pyridine are soluble in water because of the formation of new hydrogen bonds between the molecules of these compounds and the water molecules. On the other hand, the insolubility of octane, benzene and other nonpolar organic substances in water is caused by the fact that the attractive forces between the organic molecules and water are considerably weaker than those between the water molecules.

Surfactants such as trimethyloctylammonium chloride consist of a long hydrocarbon chain ('tail') with an ionic group at one end ('head') and display some very curious properties. When these substances dissolve in water, the water molecules repel the apolar hydrocarbon tails but are simultaneously attracted to the ionic head, thus solvating it. As a result of these contradictory tendencies the molecules of amphoteric compounds hide their 'tails' from the water dipoles by exposing only their 'heads' to the water. This curious situation results in the formation of spherical particles, namely micelles (Figure 2.12b). The formation of micelles aids the dissolution of amphoteric compounds and the solutions thus formed are somewhat turbid and opalescent because micelles are much larger in size than normal molecules.

Hydrophobic interactions therefore describe the fact that water molecules are more attracted to polar than to nonpolar compounds. The repulsive interactions with nonpolar compounds force them to gather together in specific aggregations. The useful rule of thumb 'like dissolves like' embraces this phenomenon.

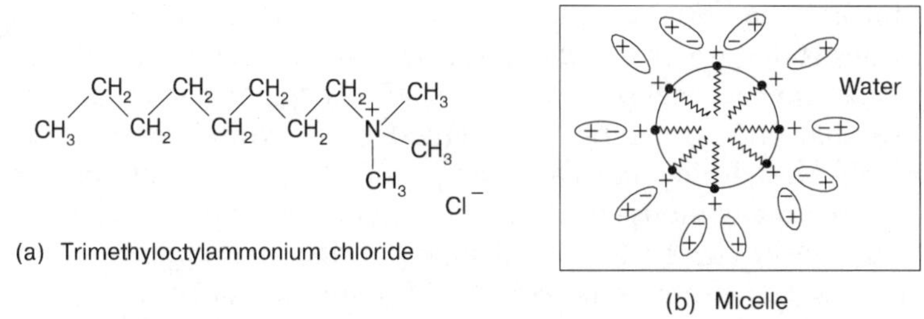

(a) Trimethyloctylammonium chloride

(b) Micelle

Figure 2.12   (a) A surfactant and (b) a representation of micelle formation.

How much energy is associated with hydrophobic forces? The association of two methyl groups has been calculated as a gain of $0.3\,kcal\,mol^{-1}$ and that of two isobutyl groups is as much as $1.5\,kcal\,mol^{-1}$ The same quantity of energy is released during the association of two phenyl groups (Figure 2.13) with a coplanar disposition of their rings ('stacking'). Stacking is a typical phenomenon for all planar rings including heterocycles. The specific role of stacking is evident in the stabilization of the DNA helical structure (see Section 3.2).

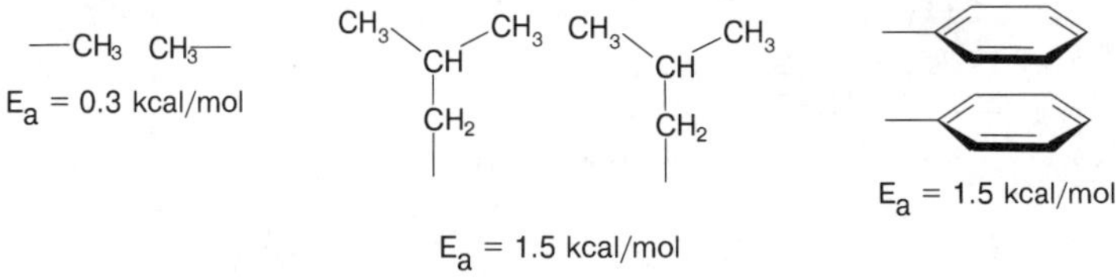

Figure 2.13   Energies of association (E_a) of some hydrophobic groups.

At first glance, hydrophobic forces appear to be very weak. But this is true only for the association of several small molecules. A many-fold increase in hydrophobic interaction energy is observed when hundreds of hydrocarbon groups of large biomolecules, such as proteins having molecular weights of up to hundreds of thousands of daltons, are involved. Associations of the hydrocarbon moieties of amino acid residues invoke the formation of hydrophobic clefts, pockets and cavities in the three-dimensional structures. Small molecules of other compounds, attracted by the same hydrophobic forces, can enter such structural clefts, and sometimes fit together like a 'lock and key.' These intriguing properties of proteins determine their specific type of biological activity whether as an enzyme, hormone or antibody.

A further important consideration connected with hydrophobic interactions is that water is our omnipresent natural solvent and the medium for all biochemical reactions. If all biologically important compounds were water soluble, life would be represented by a broth-like structure, as it presumably was during the early stages of chemical evolution. However, today, we see many thousands of highly organized forms of living matter. A vital prerequisite to such development is the hydrophobicity of many biomolecules: fats, proteins, polysaccharides, steroids and so on.

Hydrophobicity led to the initial structural organization of living matter at the cellular level. Cells became protected from the environment by a semipermeable membrane which vital nutrients could cross to enter the cell and metabolic wastes could pass to exit. The membrane structure (Figure 2.14a) and its functions are dependent, to a great extent, on hydrophobic forces. The membrane has a trilayer structure and consists of fats (about 40%, mainly phospholipids) and proteins (about 60%).

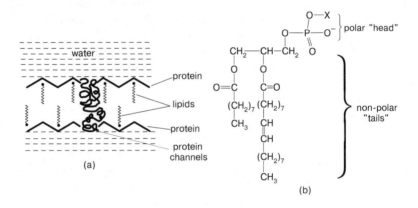

Figure 2.14 (a) Model of the cell membrane. (b) Structure of phosphoglycerides, one of the groups of lipids which constitute cell membranes, (X) = different low molecular weight polar groups).

The external and internal walls of a cell membrane are composed of proteins, whereas the middle section is formed from both proteins and polar phospholipids having an amphoteric character (Figure 2.14b). Some protein molecules span the whole width of the membrane forming highly specific channels. Through these channels polar and ionic substances can pass into a cell.[†] The phospholipids in turn are organized in two sublayers resulting from

---

†Polar moieties such as K^+ or Ca^{2+} ions are normally delivered into the cell by specialized molecular transporters which possess affinity both for the ions and also for the nonpolar lipid phase (see Section 10.1). The lipid layer of the cell membrane also serves as a matrix into which certain key compounds can be incorporated.

the orientation of the lipids with their nonpolar tails toward each other as in micelles. The wall proteins are associated with the polar heads of the lipids through hydrogen bonding and electrostatic attractions.

The presence of the extended lipid layer in the membrane (with a thickness of 60–70 Å, while the total thickness of the membrane is about 90 Å) allows various non-polar molecules necessary for the functioning of the cell to penetrate the membrane.

2.5 Tautomerism: Heterocycles and Their 'Masks'

Like other classes of organic compounds, heterocycles contain compounds with the same elemental composition and molecular weight but different spatial arrangements. Such substances, called isomers, are able to exist independently, and often have quite different physicochemical properties. Imidazole and pyrazole form a good example: structurally, these two isomers differ only in the arrangement of their two nitrogen atoms, but imidazole is a stronger base than pyrazole by a factor of 40 000.

When we refer to isomers, we most frequently mean two or more compounds which do not interconvert, or do so only with great difficulty. However, in chemistry there are a variety of reversible transformations in which a compound can exist in several isomeric forms in equilibrium. This type of isomerism is called tautomerism. The equation showing the equilibrium caused by the interconversion between cyanic and isocyanic acid is a simple example of tautomerism

$$H—O—C \equiv N \rightleftharpoons O{=}C{=}N—H$$

As can be seen, the two isomers (or tautomers) differ in the position of proton attachment and also in the multiplicity of the bonds between oxygen, carbon and nitrogen. The migration of a proton from one heteroatom to another (sometimes to a carbon atom) often results in tautomeric interconversion. The ease of such interconversion results from the rather high acidity of the protons attached to heteroatoms. It is not surprising that the presence of different heteroatoms makes tautomerism ubiquitous in the heterocyclic series. Thus, imidazole normally exists with the proton interconverting between the nitrogen atoms at great speed (Figure 2.15a). In this case the two tautomers are indistinguishable because of the symmetry of the imidazole ring. But if we distort the symmetry by the introduction, for example, of a nitro group at position 4 or 5, the tautomers become nonequivalent and their equilibrium concentrations will be different. Figure 2.15b illustrates that the equilibrium shifts towards 1H-4-nitroimidazole, as indicated by the longer arrow (the ring atoms are numbered starting from the pyrrole-like nitrogen). This shift is explained by the fact that the strong acceptor nitro group decreases the electron density on all the ring atoms, but more strongly on the nitrogen atom nearest to the nitro group. The proton, therefore, is held more tightly at the nitrogen atom remote from the nitro group.

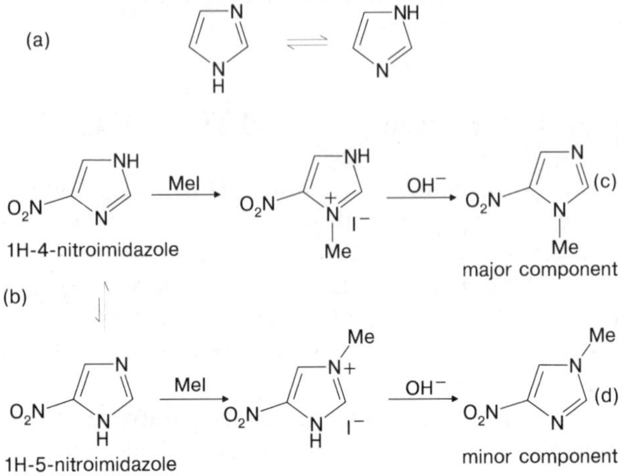

Figure 2.15 Tautomerism of (a) imidazole and (b) 4(5)-nitroimidazole. (c, d) Methylnitroimidazoles, fixed forms of tautomers.

The study of tautomerism is very important since the structures of reaction products depend on the tautomeric equilibrium. For example, the methylation of 4(5)-nitroimidazole with methyl iodide in neutral conditions affords a mixture of 1-methyl-4-nitro- and 1-methyl-5-nitroimidazole, the latter being produced in markedly greater quantity (Figures 2.15c and 2.15d). Experimental evidence suggests that the 1H-4-nitro tautomer prevails in the initial mixture and that alkylation proceeds only at the pyridine-like nitrogen in neutral media. The heterocyclic bases are liberated by the action of aqueous hydroxide on the initially produced mixture of the two quaternary salts. The alkylated products are not tautomers but as fixed forms provide models for the study of tautomerism as they have some properties very similar to those of the original individual tautomers. Individual tautomers, as a rule, cannot be isolated owing to their rapid interconversions.

To account for the behavior of tautomeric compounds, we need to realize that tautomers are, in effect, masks under which the same compounds are hidden. The name 'tautomerism' was proposed by Laar more than 100 years ago, derived from the Greek word meaning 'part of the same' (Gr.: *tauto*, same; *meros*, part). One tautomeric compound can have many similar masks, which can often change depending on the media. For example, the biologically important compound purine can be written in four reasonable tautomeric forms (Figure 2.16). The fixed *N*-methyl models of all four tautomers have been prepared. In practice, however, only two tautomeric forms occur in measurable amounts: the 7*H*-tautomer and 9*H*-tautomer. Their ratio in water is near 1:1. Purine crystallizes exclusively in the 7*H*-form, whereas in a dimethyl sulfoxide solution the 9*H*-tautomer dominates.

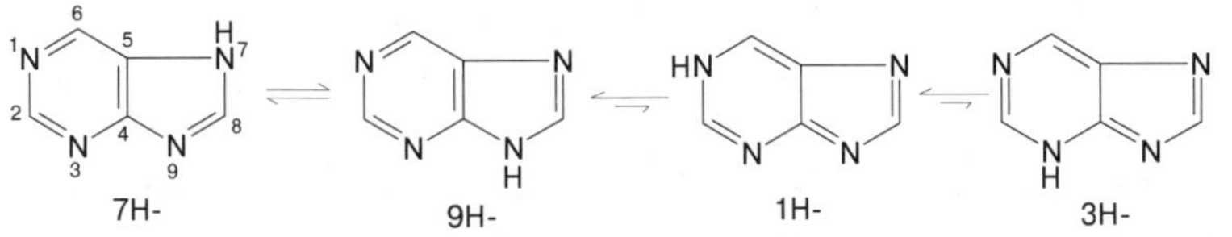

Figure 2.16 Purine tautomers.

In purine and imidazole the proton migrates between the ring nitrogen atoms. Tautomeric conversions in which ring heteroatoms and functional groups participate are no less widespread. 2-Hydroxypyridine, seen in Figure 2.17a, may exist in the hydroxy as well as in the oxo form. In the vapor phase and in highly dilute hexane solutions both exist, with the hydroxy form predominating by a small factor. From purely bond energy considerations, the amide structure of the 2-pyridone form would be expected to be more stable than the imidol structure of 2-hydroxypyridine. However, in low polarity media the higher aromaticity of the latter is decisive. Nevertheless, the second form, 2-pyridone, exists in the crystalline phase and dominates in all solvents more polar than hexane. The vapor phase equilibrium assesses the stability of tautomers in the absence of extraneous effects. The preponderance of the hydroxy tautomer in the gas phase is explained by its aromatic character, while the aromaticity of the 2-pyridone structure obviously depends on charge transfer which is less favored in media of low dielectric constant. The driving force for the shift in the equilibrium toward the oxo tautomer in the crystalline state and in aqueous solution lies in the stabilization of the highly polar pyridone form. A particular stabilization occurs by dimeric association in which two molecules are attached to each other by hydrogen bonding and by dipole–dipole interactions (Figure 2.17b). The dimers are further stabilized by the polar environment (crystalline lattice or polar solvents). These interactions outweigh the energy losses induced by the lower aromaticity of the pyridone form.

In 2-aminopyridine, an equilibrium between the amino and imino forms is also possible (Figure 2.17c). However, only the amino structure is detectable in all phases (vapor, liquid and solid). The transition to the imino form offers no advantage in terms of bond energies and the imino form is of lower aromaticity with no offsetting stability of a dimer. The last factor is a

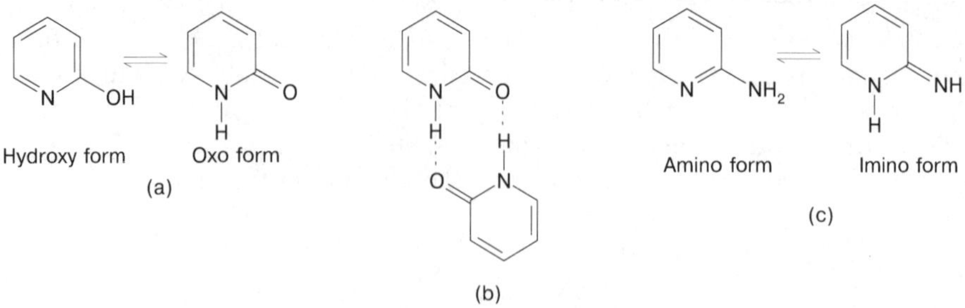

Hydroxy form   Oxo form

(a)

(b)

Amino form   Imino form

(c)

Figure 2.17 (a) Tautomerism of 2-hydroxypyridine with 2-pyridone. (b) The 2-pyridone dimer. (c) Tautomerism of 2-aminopyridine with 2-pyridonimine.

consequence of the relatively low polarity of the imine and the decreased strength of hydrogen bonding in the

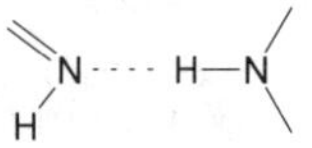

fragment as compared to that in the

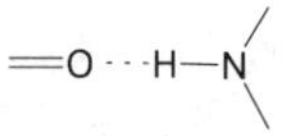

fragment, as well as geometrical considerations.

In closing this chapter, we hope that our readers will have realized the unique role of heterocycles in nature. This role is explained by the pervading influence of the heteroatoms on the reactivity, nonbonding interactions and structural modifications of heterocyclic compounds. Heterocycles have both multipurpose and specific properties which are implicit in many important chemical, biochemical and technical applications, as discussed in the following chapters.

2.6 Problems

1. Piperidine (pK_a 11.22), unlike pyrrole, is a strong base. Account for this fact. Also explain the significantly higher basicity of piperidine in comparison to pyridine.
2. Benzimidazole has two ionization constants with pK_a values of 12.9 and 5.3. To which acid–base equilibria do these correspond?
3. Purine has two ionization constants: $K_a = 4.07 \times 10^{-3}$ and $K_a = 1.17 \times 10^{-9}$.

Calculate the pK_a values and write the equations describing the corresponding acid–base equilibria.

4. Arrange the anions of imidazole, tetrazole, purine and benzimidazole in decreasing order of basicity (see Section 2.2 and Problems 2 and 3).

5. Imidazole is nitrated by a mixture of concentrated nitric and sulfuric acids at $100\,°C$ to give 4(5)-nitroimidazole in 75% yield. By contrast, it is difficult to halt the bromination of imidazole at the monosubstitution stage even at room temperature (reaction of bromine in chloroform leads directly to 2,4,5-tribromoimidazole). Account for the difference in reactivity.

6. Provide structures for compounds A–H in the following reactions

(a) [structure: 1-methylpyrrole linked to pyridine] $\xrightarrow[\text{H}_2\text{SO}_4]{\text{HNO}_3 \ (1 \ \text{equiv.})}$ A

(b) [structure: 1-methylpyrrole linked to pyridine] $\xrightarrow{\text{C}_2\text{H}_5\text{I}}$ B $\xrightarrow{\text{Na}_2\text{S}_2\text{O}_4}$ C

(c) [structure: 1-methylimidazole linked to furan] $\xrightarrow[\text{H}_2\text{SO}_4]{\text{HNO}_3 \ (1 \ \text{equiv.})}$ D

(d) [structure: 1H-1,2,3-triazole] $\xrightarrow[\text{KOH}]{\text{CH}_3\text{I} \ (1 \ \text{equiv.})}$ E + F

(e) [structure: 2-chloropyridine] $\xrightarrow{\text{CH}_3\text{ONa} \ (1 \ \text{equiv.})}$ G

(f) [structure: bipyridine/pyrimidine with Cl substituents] $\xrightarrow{(\text{CH}_3)_2\text{NH} \ (1 \ \text{equiv.})}$ H

7. Assess the polarities of the three isomeric diazines pyridazine, pyrimidine and pyrazine on the basis of the dipole moment for pyridine being equal to $2.2\,D$.

8. The dipole moment of compound I decreases as its concentration in dioxane solution is increased. At the same time the dipole moment of compound J, being equal to $18.7\,D$, remains unaffected with increasing concentration. Explain.

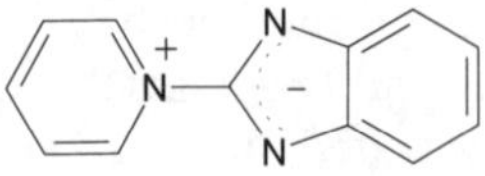

I 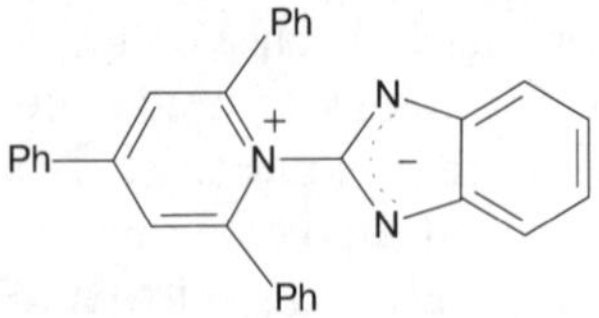   J

9. One part of benzene dissolves at 20 °C in 600 parts of water. However, pyridine is miscible with water in any proportion. Account for this fact.

10. A surprising property of nitrogen heterocycles is the decreased solubility of many derivatives when a CH is replaced by an OH group (the reverse of what is observed in the aliphatic series). Remember that such compounds exist mainly in their oxo forms. For example, one part of purine and one of each of its 2-oxo, 6-oxo, 8-oxo, 2,6-dioxo (xanthine) and 2,6,8-trioxo (uric acid) derivatives will dissolve in two, 380, 1400, 240, 2000 and 39 500 parts of water, respectively, at 20 °C. For comparison, one part of 1,3,7-trimethylxanthine (caffeine) dissolves in 70 parts of water. Explain.

11. Quinoline and naphthalene form 1:1 molecular complexes with bromine. The stability constants (K_c) of these complexes are equal to 115 and $0.231 \, mol^{-1}$ (CCl_4, 20 °C), respectively, and their long wavelength absorption maxima occur at 290 and 346 nm, respectively. Discuss the reasons for these differences (hint: the K_c values of the analogous 8-bromo-, 8-methyl- and 3-bromoquinoline complexes are 1.1, 4.8 and 12, respectively).

12. 1-Acetylimidazole has a half-life in aqueous solution at pH 7.0 and 25 °C of 41 min. By contrast, aliphatic amides (e.g. acetamide) are not hydrolyzed under these conditions. Account for this fact.

13. Indicate which of the following compounds can theoretically exist in alternative tautomeric forms containing intramolecular hydrogen bonding

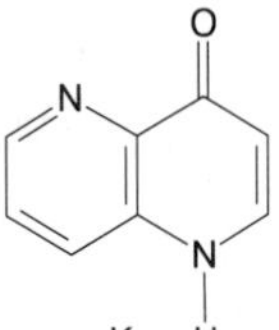

K 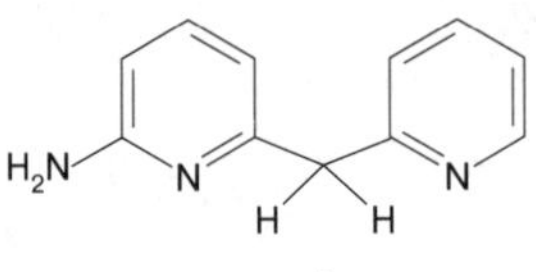   L 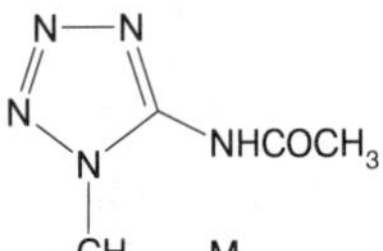   M

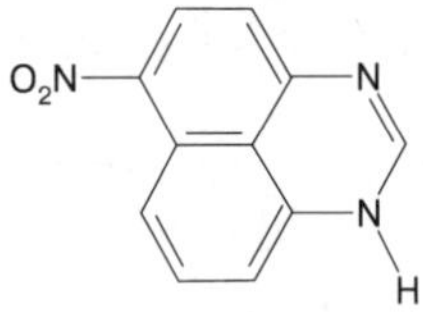

N 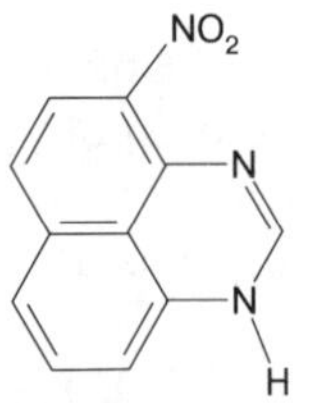   O 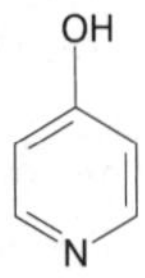   P

2.7 Suggested Reading

1. Jencks, W. P., *Catalysis in Chemistry and Enzymology*, Dover, New York, 1987.
2. Pozharskii, A. F., *Theoretical Basis of Heterocyclic Chemistry*, (in Russian), Khimia, Moscow, 1985.
3. Katritzky, A. R. and Rees, C. W., (eds), *Comprehensive Heterocyclic Chemistry*, Vols 1–8, Pergamon Press, Oxford, 1984.
4. Schofield, K., Grimmett, M. R. and Keene, B. R. T., *Heteroaromatic Nitrogen Compounds: The Azoles*, Cambridge University Press, Cambridge, 1976.
5. Matuszak, C. A. and Matuszak, A. J., *J. Chem. Educ.*, 1976, **53**, 280.
6. Elguero, J., Marzin, C., Katritzky, A. R., Linda, P., in *Advances in Heterocyclic Chemistry*, Suppl. 1, Academic Press, New York, 1976.
7. Frieden, E., *J. Chem. Educ.*, 1975, **52**, 754.
8. Slifkin, M. A., *Charge Transfer Interactions of Biomolecules*, Academic Press, London, 1971.
9. Albert, A., *Heterocyclic Chemistry: An Introduction*, 2nd Edn, Athlone Press, London, 1968.
10. Schofield, K., *Heteroaromatic Nitrogen Compounds; Pyrroles and Pyridines*, Butterworth, London, 1967.
11. Andrews, L. J. and Keefer, R. M., *Molecular Complexes in Organic Chemistry*, Holden-Day, San Francisco, 1964.
12. Pullman, B. and Pullman, A., *Quantum Biochemistry*, Wiley-Interscience, New York, 1963.

3 HETEROCYCLES AND HEREDITARY INFORMATION

> A Cossack brought with him from a campaign
> A pretty black girl and a bottle of champagne.
> Ever since, all the offspring of his kin
> Are the carriers of a somewhat darkened skin.
> Oh, that intrepid Cossack of a campaign.
>
> Ya. Kozlovsky

3.1 Nucleic Acids

For centuries the question of heredity and the related theme of destiny were surrounded with a mystic secrecy which provided imaginative scope for astrologers, prophets and fortune tellers. Ancient Greco-Roman mythology offers numerous tales of the Fates or Parcae, fortune goddesses who were portrayed as very old women spinning and severing the threads of human life at will. Many poets have paid them tribute and the Russian symbolist Merezhkovsky wrote the following quatrain 100 years ago

> I do not care, be what there will.
> The weird sisters, with timeless skill,
> Keep their wheels spinning to generate,
> The tangled threads of our fate.

The significant associations made by ancient mythologists and poets, surprisingly enough, anticipated the scientific fact that the threads of life are represented in reality by very long macromolecules. These macromolecules, not tangled but exquisitely organized, indeed store all the hereditary information which predetermines the development of all life forms, be it plants, animals or mankind.

The mechanism of the transfer of hereditary information from one generation to the next was discovered in 1953 by Watson and Crick in one of the most impressive achievements of science in the twentieth century, appropriately earning a Nobel prize. Avery in 1944 established that

37

hereditary information, in other words, the genetic code, is encrypted in the huge deoxyribonucleic acid (DNA) molecules (with molecular weights of several million) contained in each living cell. It was Watson and Crick who demonstrated that cytosine, thymine, uracil, adenine and guanine (Figure 3.1), derivatives of the well-known nitrogen heterocycles pyrimidine and purine, participate directly in the encoding of all genetic information and proposed how the information is transferred from generation to generation.

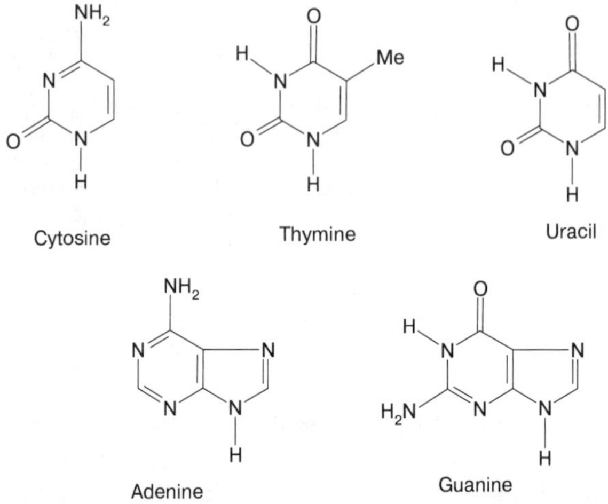

Cytosine   Thymine   Uracil

Adenine   Guanine

Figure 3.1   Nucleic acid pyrimidine and purine bases.

Cells also contain ribonucleic acid (RNA). The chief function of RNA is to control the synthesis of proteins. The DNA code is transferred to the RNA and is thus used to determine the amino acid sequence of proteins. Thus, in one sense RNA plays a secondary role relative to DNA.

We first examine the primary structures of DNA and RNA. By 'primary structure' is meant the nature of the components and their arrangement in the long molecular sequence. Complete hydrolysis of nucleic acids leads to the isolation of three types of subunits: (i) phosphoric acid, (ii) a pentose sugar and (iii) the heterocyclic bases enumerated above. At this level there are two substantial differences between DNA and RNA. Firstly, whereas both DNA and RNA contain cytosine, guanine and adenine, DNA contains thymine and no uracil, while RNA contains uracil and no thymine. Secondly, the sugar component of RNA is D-ribose, whereas the comparable subunit in DNA is D-2-deoxyribose. The names of RNA and DNA are in fact derived from the names of the corresponding sugars.

As is the case for most monosaccharides, ribose exists not in the linear form but as a mixture of cyclic hemiacetals. Intramolecular nucleophilic addition of the C-4 or the C-5 hydroxy group to the carbonyl group affords respectively the five-membered furanose or six-membered pyranose ring, (Figure 3.2). In aqueous solution D-ribose comprises 76% of the two pyranose forms and 24% of the two furanose forms. The latter are less stable because of the angle strain inherent in five-membered rings. In both furanose and pyranose forms, the hydroxy groups at C-1 are those of hemiacetals; compounds of type $R_2C(OH)OR$ are denoted hemiacetals to differentiate them from acetals of the general formula $R_2C(OR)_2$.

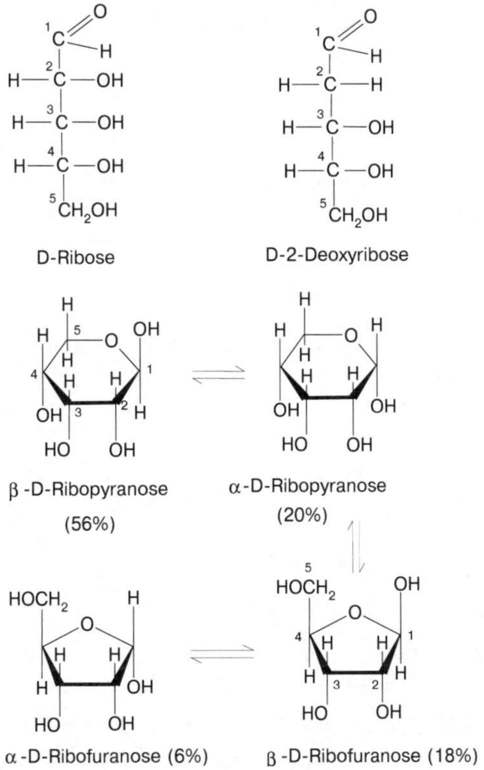

Figure 3.2 The linear forms of ribose and deoxyribose and the cyclic forms of ribose.

As implicit in Figure 3.2 and the discussion above, the hemiacetal hydroxy group of cyclic sugars occupies one or the other of two different positions, either above or below the plane of the ring. These are denoted the α- and β-pyranose and the α- and β-furanose forms. In spite of the lower strain energy of pyranose forms, the ribose moieties in nucleic acids are exclusively β-D-ribofuranoses. This configuration apparently facilitates construction of long and stable (from the point of view of electronic and geometric factors) polymer chains.

The way in which the units are joined together to form nucleic acids was solved by enzymatic hydrolysis. Four crystalline acids were isolated from the hydrolysis of DNA: deoxyadenylic, deoxyguanylic, deoxycytidylic and deoxythymidylic acids (Figure 3.3). Each acid was composed of one molecule each of a heterocyclic base, a sugar and phosphoric acid. One of the nitrogen atoms of the base was attached to the furanose ring, thus forming a nucleoside. The phosphoric ester units derived from a nucleoside and phosphoric acid are called nucleotides (or deoxyribonucleotides). The analogous enzymatic hydrolysis of RNA generates four different acids (ribonucleotides): adenylic, guanylic, cytidylic and uridylic acids (Figure 3.3).

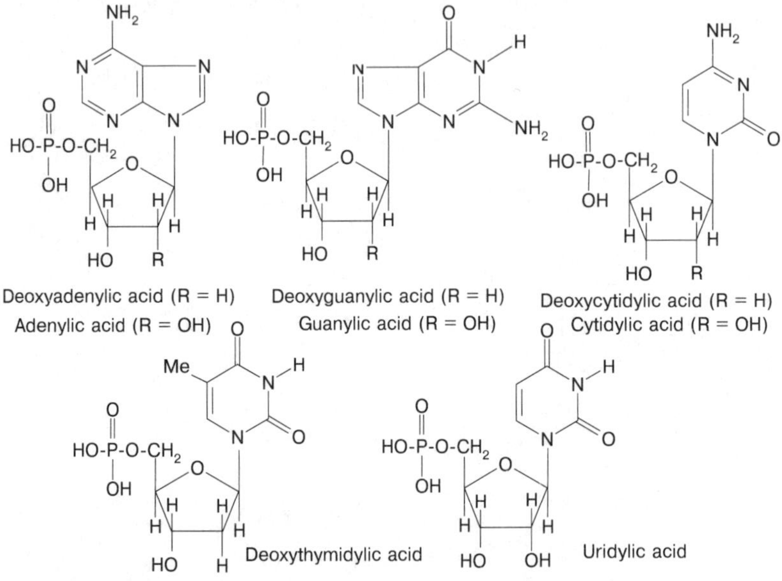

Figure 3.3 Nucleotides formed from the enzymatic hydrolysis of DNA and RNA.

The nucleotides of DNA and RNA have the following structural features. (i) The pentose residue is attached to the N-9 atom of purines and N-1 in pyrimidines (these positions are the sterically least hindered). (ii) The base is condensed with the sugar by displacement of the hemiacetal hydroxy group at the pentose C-1 atom. (iii) The pentose CH_2OH group is esterified by phosphoric acid. Since the other hydroxy groups of pentose can also be esterified, the names of the nucleotides shown in Figure 3.3 need further clarification. For example, a more precise name for adenylic acid is adenosine-5′-phosphate, 5′ designating the number of the carbon atom in the sugar unit which carries the phosphoric acid residue (see also Figure 3.2).

Further chemical investigations have established that nucleotides are linked one to another by a second phosphate ester bond. The bond is formed between the phosphoric acid residue of one nucleotide and the 3'-pentose hydroxy group of the other. Thus, polynucleotide chains contain 3'-5'-phosphodiester linkages. A section of the sequence is represented in Figure 3.4.

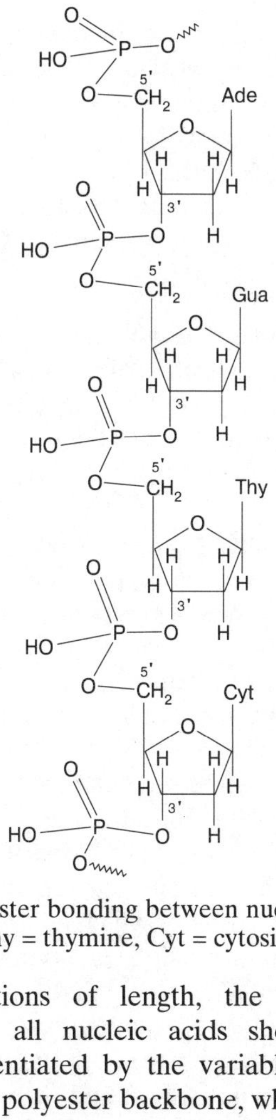

Figure 3.4 3'-5'-Phosphodiester bonding between nucleotides in a DNA chain (Ade = adenine, Gua = guanine, Thy = thymine, Cyt = cytosine).

Apart from considerations of length, the general structures of the polynucleotide chains of all nucleic acids show considerable similarity. However, they are differentiated by the variable sequences of purine and pyrimidine residues on the polyester backbone, which are of great significance. Indeed, these sequences express the genetic code. Therefore, elucidation of the primary structure of DNA and RNA above all necessitates elucidation of the precise sequence of the heterocyclic bases. Reliable and effective methods for such determination now exist. Moreover, methods for the synthesis of sequence-specific polynucleotides have been elaborated, and today one can use an automated synthesizer. Such computer-controlled instruments assemble high molecular weight nucleic acids from nucleotide units, in one sense imitating the natural assembly lines which exist in cellular organisms.

3.2 The Double Helix

The fascinating history of the discovery of the structure and mode of action of DNA was described in 1968 by Watson in his book *The Double Helix*. His account reveals that the most difficult aspect of the characterization was the elucidation of the spatial, the so-called secondary, structure of DNA. The now well-known double-helix backbone structural model of DNA was consistent with all of the experimental data then available, including X-ray analysis results. As demonstrated by Watson and Crick, the main features of the secondary DNA structure are as follows.

(i)   The DNA molecule consists of two right-handed polynucleotide chains which are intertwined to form a double helix (Figure 3.5). The helices run antiparallel to each other; that is, if one moves along them in the same direction, the sequence of bonding in one sugar-phosphate chain will be -5', 3'-5', 3'-5', 3'-, whereas in the other it will be the reverse configuration -3', 5'-3', 5'-3', 5'-.

(ii)   The purine and pyrimidine base units lie in the center of the double helix with their planes nearly perpendicular to the main axis of the helix. By contrast, the planes of the pentose rings are almost parallel to the main axis of the helix.

(iii) One coil of each helix has a length of 34 Å and includes 10 nucleotide pairs. The distance between the planes of each pair of neighboring bases in one chain is 3.4 Å.

(iv)   Significant contributions to the secondary structure of DNA are made by several types of noncovalent bonding interactions which serve to attach the two helices together. Heterocyclic bases of one chain are bound to the appropriately situated bases of the other chain by hydrogen bonding. It is only certain specific pairs of bases which are thus attached to one another, the so-called complementary pairs, e.g. adenine–thymine (two hydrogen bonds) and guanine–cytosine (three hydrogen bonds), as depicted in Figure 3.6. It follows that the number of purine residues in a DNA molecule exactly equals the number of pyrimidine moieties. In all types of DNA this ratio is experimentally found to be 1:1, although the ratio of each type of base within the purine and pyrimidine classes can change depending on the biological species. So, in double-stranded helical DNA, the two chains are complementary to each other. For example, human sperm contains 31% adenine, 19% guanine, 31% thymine and 19% cytosine, while the DNA of tuberculosis bacilli contains 36% each of guanine and cytosine with adenine and thymine at only 14% each.

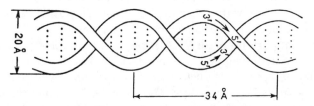

Figure 3.5   Schematic representation of the double-helix structure of DNA (hydrogen bonding of purine–pyrimidine pairs is shown by dotted lines). Reproduced with permission from Terney, *Contemporary Organic Chemistry,* Saunders, Philadelphia, 1979. © 1967 Recthed, *Concepts of Biochemistry,* McGraw-Hill Book Co.

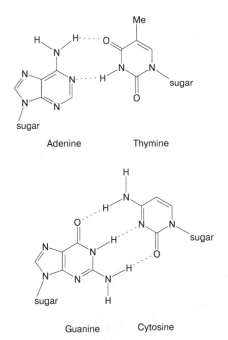

Figure 3.6   Hydrogen bonding between complementary pairs of heterocyclic bases in DNA.

The second force which stabilizes each individual DNA chain in the helical form as well as both chains together in the double-helix arrangement is the stacking interaction (see Section 2.4.4). Hydrophobic interactions exist between heterocyclic bases which are vertically stacked. The role of these forces in the formation of the double helix is believed to be even more important than that of hydrogen bonding.

We next turn from DNA to consider the secondary structure of RNA. In the overwhelming majority of cases where RNA is encountered in nature the RNA is composed of a single rather than double strand. Nevertheless, its secondary structure is determined by the same types of interactions as are encountered in DNA. Individual portions of an RNA molecule can become associated under the influences of hydrogen bonding and stacking. When a sufficiently lengthy section of the molecule has such interactions it can become partly helical, resembling the gross DNA structure. The secondary structure of an RNA, for instance that which controls the transfer of phenylalanine, is a typical example of a partially helical molecule (Figure 3.7), with a molecular form which, if depicted in a single plane, resembles a clover leaf. The bases in the short stems are complementary to each other, just as in DNA, forming hairpin loops. The three-dimensional structure is still more complex (Figure 3.8).

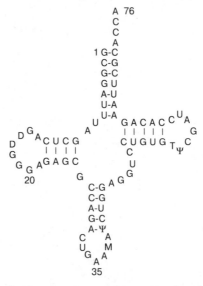

Figure 3.7 Heterocyclic base sequence in yeast phenylalanine transfer RNA (tRNA). A single line indicates a pair of bases attached by hydrogen bonding (D = 4,5-dihydrouracil, Ψ= pseudouracil, M = dimethylguanine) (adapted from Pauling, L. and Pauling, P., *Chemistry,* Freeman, San Francisco, 1975, with permission).

DNA can also have a tertiary structure. In a chromosome, DNA is wound onto proteins called histones and these large bundles can in turn be compacted into superhelices. Such superhelices are considered in mathematical topology. The function of such contortions of the DNA superstructure is the formation of more compact forms. Linear DNA molecules could not be accommodated within cellular structures: the total length of all the DNA molecules in a single cell can reach 2 m. In some microorganisms the DNA can form closed macrorings.

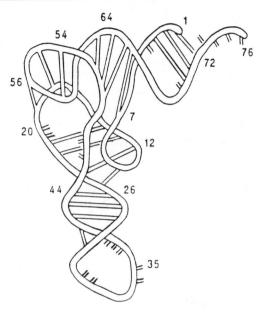

Figure 3.8 Schematic representation of the three-dimensional structure of phenylalanine tRNA. Double lines designate hydrogen bonding between heterocyclic bases (base numbers correspond to those in Figure 3.7) (reproduced with permission from Pauling, L. and Pauling P., *Chemistry*, Freeman, San Francisco, 1975).

DNA molecules exist in cells as complexes with proteins. Such complexes are called nucleoproteins. Proteins interact with the DNA in grooves formed on the external surface of the double helix. The packed DNA molecules make up the chromosomes within the cell nuclei. Under a microscope chromosomes appear as small, slightly bent rods. The number of chromosomes in the cells of different living organisms varies. For example, a single human cell contains a total of 46 chromosomes, two each of 23 types.

3.3 How DNA Functions

By the process of self-duplication, a living cell produces two new cells, each identical to the parent. The manner by which such a strikingly accurate reproduction of all of the complex biological material takes place was first suggested by Watson and Crick. They proposed that at a given time, specialized proteins direct the uncoiling of the double helix. Each of the two individual DNA chains then begins to make its complement, each effecting the biosynthesis of a new double helix identical to the original. As a result, all of the initial genetic material of the cell is doubled: the 46 initial chromosomes are converted into 92 chromosomes and the swollen parent cell divides.

It should be noted that the entire double helix of DNA cannot be completely uncoiled simultaneously, even theoretically. The helix begins to unwind at several points like a broken zipper. Because the chains in DNA are antiparallel, one strand can be made continuously, while the other strand is made in shorter bits which are joined together by the enzyme DNA ligase. DNA synthesis is catalyzed by very complex enzymes called DNA polymerases, first studied by Kornberg. This process, called replication, is shown schematically in Figure 3.9. Scientists have been able to study replication with the aid of electron microscopes and have demonstrated the accuracy of this hypothesis.

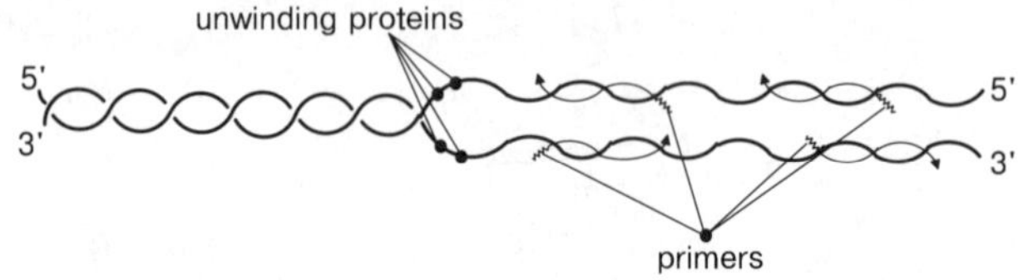

Figure 3.9   DNA replication (adapted from Barton, D. and Ollis, W. D., (eds), *Comprehensive Organic Chemistry*, Vol. 5, Pergamon Press, Oxford, 1979, with permission).

The mechanism of replication relies on the timely delivery of each necessary nucleoside-5′-triphosphate to the assembly line and its subsequent attachment to the growing DNA chain by DNA polymerase (Figure 3.10).

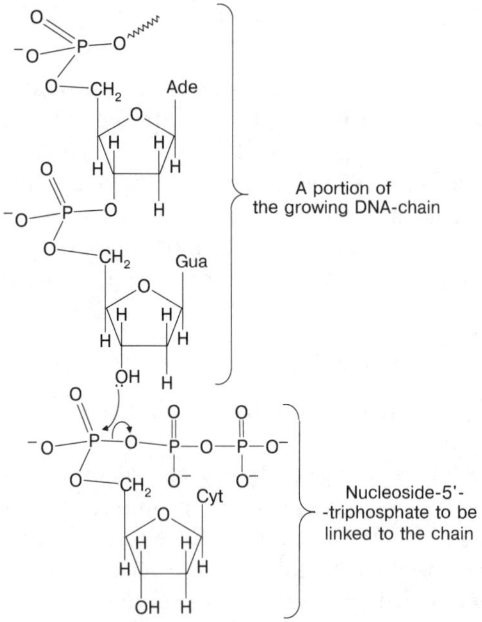

Figure 3.10   DNA chain lengthening with the assistance of the DNA polymerase delivery system.

This reaction involves nucleophilic displacement of a pyrophosphate group ($P_2O_7{}^{4-}$) resulting from attack by the terminal nucleotide 3′-hydroxy substituent at the α-phosphate group of the nucleoside-5′-triphosphate. Thus, synthesis of the chain is in the 5′-3′ direction, one chain being extended continuously, the other in short sections in the opposite direction.

The new DNA molecules are constructed with the complementary heterocyclic bases. For instance, if a 5′-deoxyadenyl residue appears on a DNA strand, it is absolutely necessary for a 5′-deoxythymidyl unit to be provided at the corresponding site of the new growing DNA molecule. Thus, one double-helical DNA structure affords two new identical double-chained DNA aggregates, each composed of one original and one newly synthesized strand of DNA. It is the high degree of fidelity which is responsible for genetic stability, together with many enzymes which repair mistakes and DNA damage.

3.4   Protein Synthesis and the Genetic Code

Overall, the proteins make up the most important class of natural compounds, because proteins perform an enormous variety of functions in the living organism. Cellular membrane walls, supporting and protective tissues, and muscle fibers are but a few examples of living structural materials constructed from proteins. However, a key role of many proteins lies in the regulation of metabolic processes (see Chapter 4). Chemically speaking, proteins are biopolymers with molecular weights ranging up to hundreds of thousands in which the polymer chains are constructed from α-amino acid residues joined by amide linkages:

When such a chain includes a relatively small number of amino acids (several dozen), the compound is named a polypeptide, or simply a peptide. A peptide with many amino acid residues is called a protein. As mentioned above, the diversity of nucleic acids stems from inclusion of only four heterocyclic bases. By contrast, proteins are derived from as many as 20 or so amino acids, giving rise to an almost infinite number of possible sequences within the polypeptide chain. These chains have different side groups (R) which include heterocyclic residues as in histidine and tryptophan (Figure 3.11). Two derivatives of pyrrolidine, namely proline and hydroxyproline, are unusual α-amino acids in that their amino groups participate in the formation of the heterocycle; they are thus secondary amines.

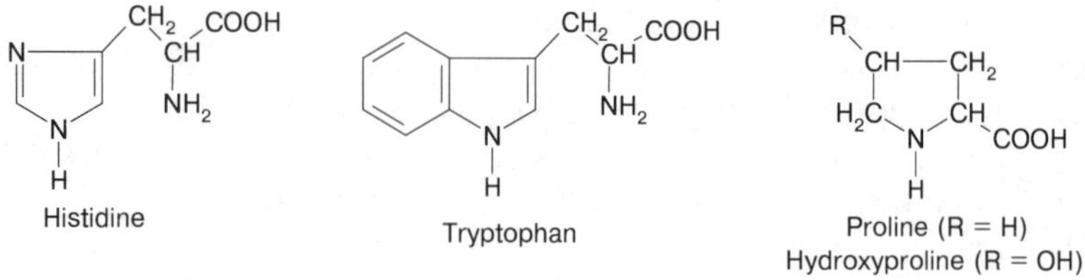

Histidine   Tryptophan   Proline (R = H)
                                    Hydroxyproline (R = OH)

Figure 3.11   Heterocyclic α-amino acids.

There are obviously infinite possibilities for changing the primary structure of proteins, i.e. the sequence of amino acid residues in the chains. In human organisms, thousands of proteins have been identified. Proteins are self-organizing molecules. Under the influence of nonbonding interactions between different portions of the polypeptide chain, the protein is arranged into a definite spatial form (tertiary structure), which determines its functions. Moreover, the ability of proteins to complex with metal ions and biomolecules such as nucleic acids, sugars, lipids, pigments and so on extends their chemical and biological functions.

Unlike DNA, proteins are not capable of self-reproduction. The genetic program for their synthesis is contained within the DNA. The cell site where a DNA molecule stores coded information for one type of protein is called a gene. One DNA strand contains many genes. Certain portions of different genes often overlap each other, particularly in microorganisms. A human organism contains at least 100 000 genes, 30–50% of which are expressed in the brain. One of the most ambitious undertakings of modern science, and one of the most expensive, is the sequencing of the complete human genome. This involves finding the whole of the nucleotide sequence and then constructing a map of the heterocyclic base sequence in each of the human genes. 'Genebank' now lists the sequences of several thousand human genes.

The synthesis of a protein is an exceedingly complex biochemical process in which three types of ribonucleic acids participate: messenger acids (mRNA), ribosomal acids (rRNA) and transfer acids (tRNA). The genetic information for protein synthesis is contained in mRNA. The synthesis of mRNA itself is directed by one of the DNA strands. The chemistry of this synthesis, called transcription, is similar to the replication of DNA, but a significant difference is the participation of a different enzyme, RNA polymerase, which introduces uracil instead of thymine into the RNA chain. The molecular weights of RNA are very varied and in some cases can be in the vicinity of 10^6.

As proteins contain approximately 20 different amino acids and DNA includes only four bases, the incorporation of one amino acid into a protein chain cannot be encoded by one or even two bases. It has been established that a particular amino acid is encoded by a sequence of three consecutive bases. Such a three-letter sequence is known as a codon. Some 64 codons are theoretically possible. The relationship between codons and α-amino acids is referred to as the genetic code (listed in Table 3.1), as it determines and guides the chemistry and biology of all organisms.

Table 3.1 Genetic Code (Adapted from: Pauling, L., Pauling, P., *Chemistry*, Freeman, San Francisco, 1975; with permission).

		Second Letter					
		U	C	A	G		
F i r s t	U	UUU ⌉ Phe UUC ⌋ UUA ⌉ Leu UUG ⌋	UCU ⌉ UCC UCA ⎬ Ser UCG ⌋	UAU ⌉ Tyr UAC ⌋ UAA * UAG *	UGU ⌉ Cys UGC ⌋ UGA * UGG Trp	U C A G	T h i r d
	C	CUU ⌉ CUC CUA ⎬ Leu CUG ⌋	CCU ⌉ CCC CCA ⎬ Pro CCG ⌋	CAU ⌉ His CAC ⌋ CAA ⌉ Gln CAG ⌋	CGU ⌉ CGC CGA ⎬ Arg CGG ⌋	U C A G	
l e t t e r	A	AUU ⌉ AUC ⎬ Ile AUA ⌋ AUG Met	ACU ⌉ ACC ACA ⎬ Thr ACG ⌋	AAU ⌉ Asn AAC ⌋ AAA ⌉ Lys AAG ⌋	AGU ⌉ Ser AGC ⌋ AGA ⌉ Arg AGG ⌋	U C A G	l e t t e r
	G	GUU ⌉ GUC GUA ⎬ Val GUG ⌋	GCU ⌉ GCC GCA ⎬ Ala GCG ⌋	GAU ⌉ Asp GAC ⌋ GAA ⌉ Glu GAG ⌋	GGU ⌉ GGC GGA ⎬ Gly GGG ⌋	U C A G	

[a]Phe = phenylalanine; Leu = leucine; Ser = serine; Tyr = tyrosine; Cys = cysteine; Trp = tryptophan; Pro = proline; His = histidine; Gln = glutamine; Arg = arginine; Ile = isoleucine; Met = methionine; Thr = threonine; Asp = aspartic acid; Lys = lysine; Val = valine; Ala = alanine; Asn = asparagine; Glu = glutamic acid; Gly = glycine. The asterisks denote codons for protein synthesis termination.

The genetic code was deciphered in the following way. Synthetic RNAs consisting of one type of codon only were prepared and introduced into a solution containing all 20 amino acids together with enzymes prepared from bacterial cells. It was found that the synthesis of a polypeptide chain consisting of units derived from a single amino acid then took place. For example, when synthetic RNA made only from uridylic acid (UUU sequence) was used, the polypeptide isolated contained exclusively phenylalanine residues. Thus, it was clear that the UUU sequence serves as a codon for phenylalanine.

Note that the genetic code is degenerate. This means that several codons can produce the same amino acid. However, in many cases the variations in the codons for any one amino acid involve only the third codon base. Moreover, three of the codons do not correspond to any amino acid. Their function is to encode for the termination of protein synthesis. When the protein molecule under construction on messenger RNA reaches a termination codon, synthesis is halted.

Amino acids locate the correct position on the messenger RNA with the help of transfer RNA. In a given organism there are as many tRNAs as there are codons. tRNAs are soluble and have relatively low molecular weights (about 20 000). Their molecules are arranged in 'clover leaf' configurations (see Figure 3.7). Specific enzymes 'charge' the adenosine residue situated at the end of the tRNA with the appropriate amino acids (Figure 3.7, base 76). The ribose fragment of the terminal adenosine residue has two free hydroxy groups at the 2-position and 3-position. One of these groups reacts with the amino acid carboxy group to form an energy-rich ester bond (Figure 3.12).

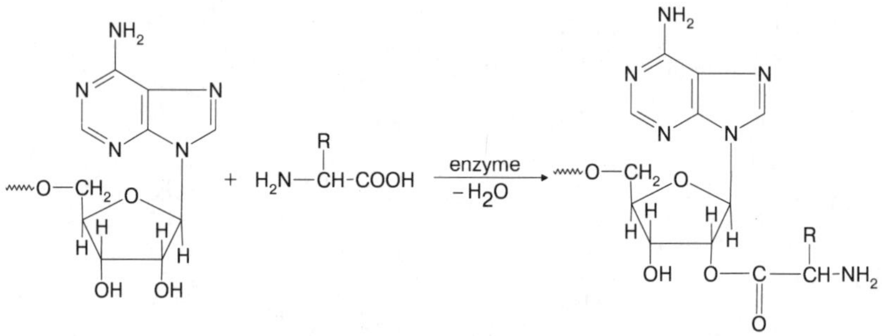

Figure 3.12   Attachment of amino acids to transfer RNA.

The amino acid laden tRNA is able to attach itself at the mRNA codon which corresponds to the amino acid through hydrogen bonding between the three heterocyclic bases of the codon and the complementary tRNA bases, named anticodons. For example, if the tryptophan codon is UGG, its tRNA anticodon will be CCA. The anticodons of the tRNA are situated in the central loop of the 'clover leaf'.

To commence protein synthesis, a ribosome, a tRNA and an mRNA are all needed. A ribosome is a complex composed of two ribosomal RNA moieties (65%) and many proteins (35%). Two subunits, one small and one large which visually resemble an inverted mushroom (Figure 3.13), combine immediately prior to the protein assembly. The small subunit contains rRNA of molecular weight around 500 000, while the large component has a molecular weight of approximately 10^6. Ribosomes play a support role in protein synthesis and provide a physical site for the process. They are therefore sometimes referred to as 'protein assembly shops'.

Protein synthesis can be represented as in Figure 3.13. The small subunit of the ribosome attaches itself to mRNA and maintains it at this site in an unfolded state. The ribosome attaches to at least two mRNA codons as both the

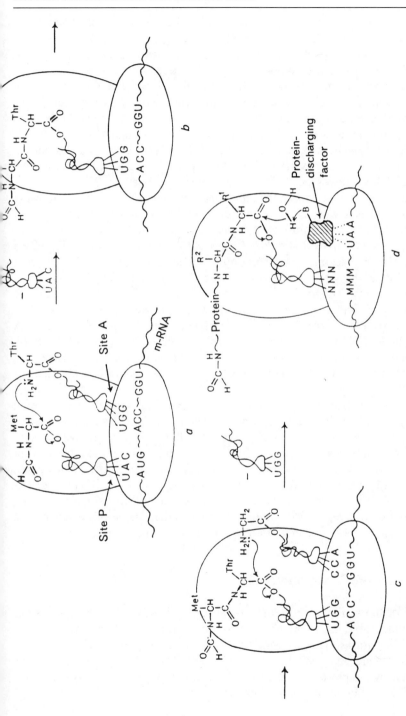

Figure 3.13 Protein synthesis on a ribosome, the addition of threonine (Thr = MeCH(OH)) and glycine residues to *N*-formylmethionine (Met = CH₂CH₂SMe): (a) initiation step, (b, c) chain lengthening and (d) termination (reproduced with permission from Barton, D. and Ollis, W. D. (eds), *Comprehensive Organic Chemistry*, Vol. 5, Pergamon Press, Oxford, 1979).

free amino acid and the growing protein chain take part in the formation of the peptide bond. Two tRNAs, one charged with a single amino acid and the other with the growing polypeptide chain, are simultaneously attached to the mRNA and to the large ribosome subunit. In the initial step, a ribosome attaches to the mRNA to open the initiating codon AUG (for the tRNA of methionine). In the following sequence, the true first phase of the synthesis, an *N*-formylmethionine tRNA is delivered to the initiation codon UAC. The protein synthesis is monitored on the surface of the big subunit at two sites designated A and P.

A free amino acid is carried to site A by its tRNA. As soon as the amino group of the acid is acylated, the whole peptide segment, including the attached tRNA, moves to site P. At this time the messenger RNA also shifts and the next site A codon thus becomes ready to attach the following complex of tRNA with a new amino acid. Figure 3.13a depicts the stage at which the formylated methionine has just moved to site P, and the threonine-loaded tRNA has arrived at site A (codon ACC and anticodon GGU). By a nucleophilic displacement, the threonine amino group attacks the adjacent methionine carbonyl group causing cleavage of the CO–O ester bond between the methionine residue and its tRNA. The latter leaves the ribosome and the newly generated dipeptide moves, together with the threonine tRNA, to site P, simultaneously inducing a shift of the mRNA (Figure 3.13b). Figure 3.13c demonstrates the colinearity of the GGU codon to glycine. Thus, the protein synthesis continues until the appearance of a termination codon at site A (Figure 3.13d shows the codon UAA). There is no tRNA for such a codon, but it is recognized by a special enzyme (protein factor) which hydrolytically cleaves the terminal ester bond. The completed protein molecule is then liberated and the ribosome dislodges the mRNA.

Each molecule of mRNA can serve as a template for the synthesis of hundreds of protein molecules. However, nature has adapted to prevent overproduction. To this end, one terminus of the mRNA strand has a site consisting of several hundreds of adenyl residues. After termination of the synthesis of each protein molecule some portion of the site (containing probably two or three nucleotides) is severed. When the site is exhausted in this way, the mRNA molecule is no longer active, and is rapidly degraded.

3.5   What are Mutations?

In spite of the stability of biological species, individual living organisms are rather vulnerable as they are subject to continual unfavorable changes induced by various factors. From a chemical point of view, nucleic acids are the most fragile components of an organism. Their long strands are easily torn even by simple agitation of a solution with a glass rod. They are also susceptible to damage by many chemical reagents.

More serious chemical disorders are caused by changes in the normal sequence of the heterocyclic bases in the DNA chains. Such changes, called mutations, lead to the synthesis of abnormal proteins and, hence, often to metabolic disorders. What are the origins of mutations? Firstly, there is a certain, purely statistical probability of an incorrect heterocyclic base being inserted into the DNA or mRNA chain undergoing synthetic assembly. We should note that this probability is extremely low—not more than 10^{-7}. Moreover, there exists a special enzyme, part of the DNA polymerase, which 'oversees' chain construction and can eliminate erroneously incorporated nucleotides and replace them with the correct bases. This is known as proofreading. Nevertheless, because of the enormous scale of nucleic acid synthesis, a certain number of incorrect bases will remain unchanged in the chain.

The appearance of some spontaneous mutations may depend on tautomerism of the heterocyclic bases. Although these bases exist very predominantly (10^4–10^5:1) in the amino or oxo form (see Figure 3.1), the less stable tautomer cannot be completely disregarded. Cytosine exists in physiological solution (water, pH 7) in equilibrium with a tiny proportion of the imino form (Figure 3.14a).

A similar, minuscule amount of the hydroxy form is present for guanine (Figure 3.14b). If cytosine and guanine exist in these minor forms at the time of DNA construction, the formation of abnormal pairs of bases becomes theoretically possible (Figures 3.14c and 3.14d). As in the normal pairs (see Figure 3.6), these abnormal base pairs can be stabilized by hydrogen bonding

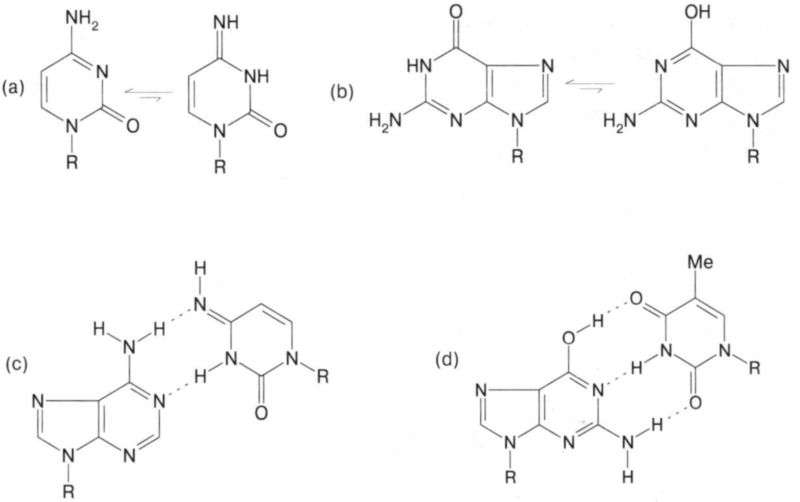

Figure 3.14 Tautomerism in (a) cytosine and (b) guanine. Formation of abnormal pairs (c) of adenine with cytosine in the imino form and (d) of thymine with guanine in the hydroxy form (R = H or sugar).

and have the same dimensions as the normal Watson–Crick base pairs. But this subject is still open to investigation.

The most important causes of mutations are environmental effects. Penetrating radiation, for instance, brings about deep destruction of nucleic acids by numerous bond cleavages. Mutations are also the result of certain chemical compounds. Insufficiently tested medicines, environmental pollutants, and food contaminants are among this class of mutagens. For example, nitrites (salts of nitrous acid) are added in very small quantities to meat products to impart a fresh appearance. Laboratory experiments have demonstrated that nitrous acid readily converts adenine into hypoxanthine and the latter forms a base-pair not with thymine but with cytosine (Figure 3.15).

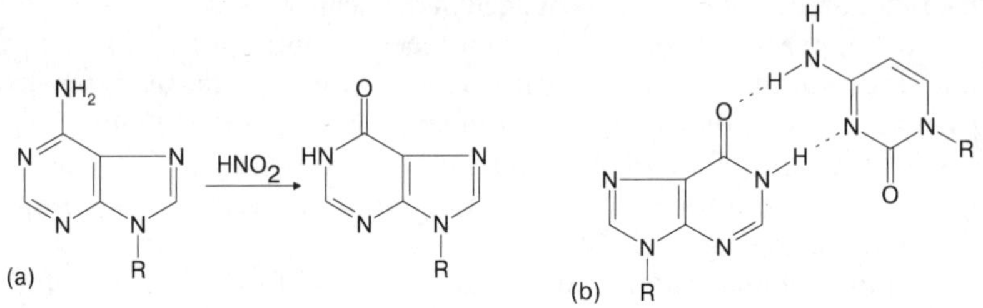

Figure 3.15   (a) Conversion of adenine into hypoxanthine. (b) Formation of the abnormal base pair hypoxanthine–cytosine (R = H or sugar).

Organic chemists know well the high toxicity of such active alkylating agents as methyl iodide, dimethyl sulfate, methyl fluorosulfonate and diazomethane. Once introduced into an organism, these compounds alkylate N–H, O–H and S–H bonds and can also form quaternary salts with the biologically important cyclic bases (Figure 3.16). In such salts the bond between the nitrogen of the

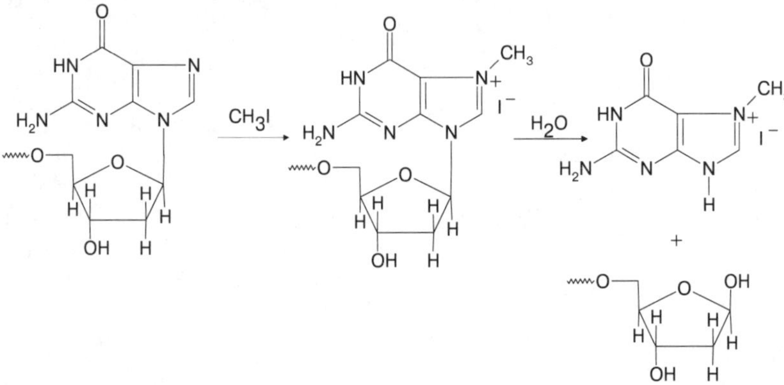

Figure 3.16   Alkylation of the imidazole ring of deoxyguanosine and the hydrolytic decomposition of the quaternary salt.

base and ribose (or deoxyribose) is readily cleaved, even by water, and the base can be lost from the RNA (or DNA) chain. Therefore, a mutagen acts like one of the mythical Parcae, cutting the thread of life.

In recent years a further class of mutagenic compounds called 'intercalators' has commanded the attention of scientists. These molecules have a planar polycyclic structure which gives them the possibility of penetrating the gap which exists between adjacent pairs of complementary bases in DNA. The guest molecules thus intercalated in the DNA tend to arrange themselves parallel to the plane of the bases. All polynuclear aromatic hydrocarbons and many heterocycles are potential intercalators. Under the effect of intercalation, the polynucleotide chain is strained like a spring and can become unwound in the vicinity of the defective site. The intercalator itself often serves as a matrix for the incorporation of a superfluous nucleotide into the DNA chain during its replication. Sometimes the effect of the intercalator can be rather specific. For example, the carcinogenic activity of 3,4-benzpyrene in living organisms is connected with its enzymatic oxidation to an epoxide which then intercalates into the DNA chain. This epoxide, being a strong alkylating agent, forms covalent bonds with amino groups in the nucleic acid (Figure 3.17).

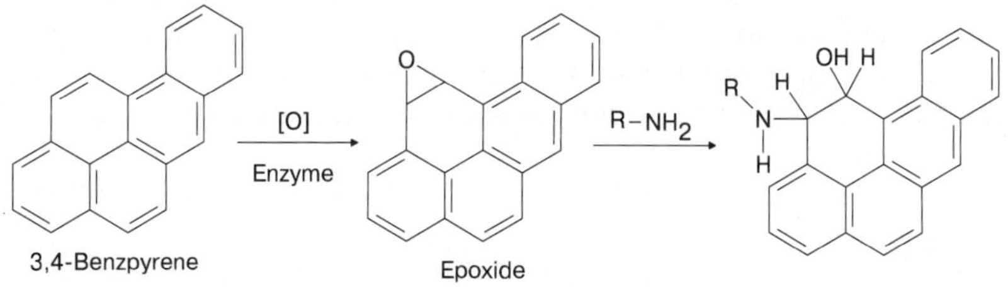

Figure 3.17 The chemistry of the mutagenic action of 3,4-benzpyrene (R = residue of a heterocyclic base of a DNA chain).

The incorporation of additional bases into the DNA chain and the loss of existing bases can both cause especially severe mutations. The net result is a frame shift, so that the wrong codon triplets are read by the protein-synthesizing machinery. The protein formed on the faulty messenger becomes unequipped to fulfill its natural biological functions or the chain is terminated early. However, under certain conditions, intercalators can be turned into life-saving drugs (see Section 7.4.3).

Milder effects are caused by point mutations when one pair of complementary bases in the DNA chain is replaced with another pair. In this case, only one defective amino acid is included in the protein. If such an inclusion takes place far from the active site of the enzymatic protein, the mutation can be innocuous (a silent mutation). However, if the incorrect amino acid plays an important role in the protein function, the mutation will induce functional disorders and, consequently, illness. However, all organisms have a battery of 'repair' enzymes which monitor the DNA continually.

Gradual accumulation of mutations is thought to be one of the major causes of aging in living organisms. Unfavorable mutations can also be transmitted from parents to their young. Such a transfer gives rise to an inherited or genetic disease. At the present time, the chemical causes of hundreds of such molecular maladies have become clear, although most of these diseases still remain incurable. It is one thing to understand the cause; it is quite another to devise a cure. However, we can hope that genetic engineering will steadily bring us closer to the desirable goal of being able to repair genetic disorders successfully.

3.6 Problems

1. (a) Draw the structure of the polypeptide which corresponds to the sequence poly(ACUGU). (b) What corresponds to poly(ACUUG)?
2. One chain of double-helical DNA which codes for the amino acid sequence in silk protein contains 1685 nucleotides (gene E2) including 588 adenine units and 484 guanine units. (a) How many codons does the DNA chain contain? How many amino acid residues does the silk protein chain incorporate? (b) What is the relative molar ratio of nucleotides in this DNA chain? In the complementary chain?
3. How long must a DNA section (called a cistron) be to code for the formation of a protein containing 200 amino acid residues?
4. A single DNA chain contains the 5'–3'-sequence
   ATCGTCGACGATGATCATCGGCTACTCGA.
   Determine: (a) the base sequence in the complementary DNA chain, (b) the base sequence in the mRNA translated from the first chain of DNA and (c) the sequence of amino acid units in the corresponding polypeptide.
5. Formaldehyde is a toxic compound which causes denaturation of DNA. Which DNA functional groups are most vulnerable to attack? Propose a mechanism for the chemical interaction.
6. What is the minimum number of nucleotides that must be changed or incorrectly translated in order to form the defective hemoglobin responsible for sickle-cell anemia? (As a clue, the β-chain in normal hemoglobin contains the following sequence of N-terminal amino acids: ValHisLeuThrProGluGluLys. In patients suffering from sickle-cell anemia the corresponding sequence is ValHisLeuThrProValGluLys.)

7. Certain proteins contain amino acid residues which have no corresponding codon in the genetic code. For example, collagen contains hydroxyproline and hydroxylysine residues. Suggest a possible mechanism for insertion of the 'noncoded' amino acids into a polypeptide chain.

8. The first totally synthetic triplex DNA (triple-stranded form), synthesized in 1957, was of no practical importance. Its potential use as 'molecular scissors' to cut DNA, which could be useful in mapping and sequencing the human genome or in blocking viral reproduction by disrupting gene transcription, is now being explored. In 1987 it was established that a short third synthetic strand (11–15 nucleotides) could be attached to the natural double DNA helix. The binding is very selective and a short strand of synthetic DNA can identify its target from millions of base pairs. Draw the triple-helical complex formed when the synthetic third strand of DNA (comprising 20 nucleotides and consisting of only two pyrimidine bases), TTTTTCTTTCTTCTTTTCTT, is attached to the following portion of the human gene only by one of its two strands: 5'...-ACGGATCCTTTTTCTTTCTTCTTTTCTTC-...3'. (Note: the binding follows normal chemical rules.)

9. Spider dragline silk has a tensile strength and elasticity much greater than those of steel. The complete sequence of 655 amino acid residues has recently been determined (A = alanine, G = glycine, L = leucine, etc.; see Xu, M. and Lewis, R. V., *Proc. Natl. Acad. Sci. USA*, 1990, **87**, 7120) as:

QGAGAAAAAAGGAGQGGYGGLGGQGAGQGGYGGLGGQGAG
QGAGAAAAAAAGGAGQGGYGGLGSQGAGRGGQGAGAAAAA
AGGAGQGGYGGLGSQGAGRGGLGGQGAGAAAAAAGGAGQ
GGYGGLGNQGAGRGGQGAAAAAAGGAGQGGYGGLGSQGAGR
GGLGGQAGAAAAAGGAGQGGYGGLGGQGAGQGGYGGLGSQ
GAGRGGLGGQGAGAAAAAAGGAGQGGLGGQGAGQGAGAS
AAAAGGAGQGGYGGLGSQGAGRGGEGAGAAAAAGGAGQG
GYGGLGGQGAGQGGYGGLGSQGAGRGGLGGQGAGAAAGGA
GQGGLGGQGAGQGAGAAAAAGGAGQGGYGGLGSQGAGRG
GLGGQGAGAVAAAAGGAGQGGYGGLGSQGAGRGGQGAGA
AAAAGGAGQRGYGGLGNQGAGRGGLGGQGAGAAAAAAG
GAGQGGYGGLGNQGAGRGGQGAAAAAGGAGQGGYGGLGSQG
AGRGGQGAGAAAAAVGAGQEGIRGQGAGQGGYGGLGSQGS
GRGGLGGQGAGAAAAAGGAGQGGLGGQGAGQGAGAAAAA
AGGVRQGGYGGLGSQGAGRGGQGAGAAAAAGGAGQGGYG
GLGGQGVGRGGLGGQGAGAAAAGGAGQGGYGGVGSGASAAS
AAAASRLSS.

(a) Find the two amino acids which occur most often in the sequence and calculate their overall percentage.

(b) Find similar or repeating units in the sequence and arrange the sequence in a scheme more suitable for analysis.
(c) Determine the first and second bases of the codons for the two most abundant residues. Calculate their overall percentage.

3.7 Suggested Reading

1. Lehninger, A. L., *Principles of Biochemistry*, 2nd Edn, Worth, New York, 1993.
2. Wen-Hsing, L. and Graur, D., *Fundamentals of Molecular Evolution*, Sinauer, Sunderland, MA, 1991.
3. Hawkins, J. D., *Gene Structure and Expression*, 2nd Edn, Cambridge University Press, Cambridge, 1991.
4. Townsend, L. B. and Tipson, R. S. (eds), *Nucleic Acid Chemistry. Improved and New Synthetic Procedures, Methods and Techniques*, Wiley, New York, 1991.
5. Townsend, L. B. (ed), *Chemistry of Nucleosides and Nucleotides*, Vol. 2, Plenum Press, New York, 1991.
6. Neurath, H. (ed), *Perspectives in Biochemistry*, Vol. 2, American Chemical Society, Washington, DC, 1991.
7. Wyman, J. and Gill, S. J., *Binding and Linkage: Functional Chemistry of Biological Macromolecules*, University Science, Mill Valley, CA, 1990.
8. Blackburn, G. M. and Gait, M. J. (eds), *Nucleic Acids in Chemistry and Biology*, 2nd Edn, Oxford University Press, Oxford, 1996.
9. Stryer, L., *Biochemistry*, 3rd Edn, Freeman, New York, 1988.
10. Watson, J. D., *The Double Helix; A Personal Account of the Discovery of the Structure of DNA*, Norton, New York, 1980.
11. Judson, H. F., *The Eighth Day of Creation. Makers of Revolution in Biology*, Simon and Schuster, New York, 1979.
12. Musil, J., Novakova, Q. and Kunz, K., *Biochemistry: A Schematic Perspective*, Avicenum, Prague, 1977.
13. Khorana, H. G., in *Nobel Lectures: Physiology and Medicine (1963–1970)*, Elsevier, New York, 1973, p. 341.
14. Nirenberg, M., in *Nobel Lectures: Physiology and Medicine (1963–1970)*, Elsevier, New York, 1973, p. 372.

4 ENZYMES, COENZYMES AND VITAMINS

Can't you see, good friend of mine,
All that's sensed with eye and ear
Is but shadow, reflection
Of th'unseen by us, my dear?

V. Solov'ev

A multitude of different chemical processes occur constantly in every living cell involving the synthesis and breakdown of millions of complex organic molecules. Scientists are fascinated by the cell as a biosynthetic machine not only because of the variety of functions carried out but also because of the amazing level of coordination between all of the different components. The rapid rates of reaction and the mild conditions under which these conversions take place in the organism (neutral aqueous media, temperatures from about 36 to 40 °C, normal pressure) are of particular interest. A further remarkable characteristic is the stereoselectivity of all of these biochemical reactions which leads specifically to the formation of molecules with a single spatial configuration.[†] To attain the synthetic sophistication of nature has long been the ultimate goal, one towards which we are moving constantly and steadily, if but slowly. The enzymes which are specific biocatalysts are of major importance in these processes. We now consider the nature of enzymes and their relationship with heterocycles.

4.1 Molecular Robots

Readers will be familiar with the manner in which robots work from real life, the cinema or TV. A robot takes an object and manipulates it according to a prearranged program. Enzymes function likewise as natural micro-manipulators and therefore can be considered as molecular robots. An

---

[†]The majority of biomolecules are chiral, i.e. their structures do not possess an axis or a plane of symmetry. Such molecules exist in the form of stereoisomers (optical isomers or mirror image isomers) which relate to each other as the left hand does to the right. Almost all chemical processes in organisms take place with the participation of single, strictly determined stereoisomers, the 'left-handed' or 'right-handed' molecules.

enzyme captures a molecule of reactant, conveys it to the reaction center, appropriately orients the molecule in space, and if necessary activates it. The enzyme then ejects the newly formed product from the reaction zone, liberating the site for another incoming molecule of starting material.

As a rule, enzymes are proteins[†] with molecular weights of up to hundreds of thousands or even millions. Enzymes can be composed of one or several polypeptide chains attached to and wound around each other in a three-dimensional manner which is determined by nonbonding interactions. This tertiary structure of enzymes means that their molecules contain clefts, pockets and/or trenches on the surface. The active site of an enzyme is one such cleft into which a specific reactant may enter 'as a key enters a lock'. Enzymes operate only with their particular 'key-molecule'; that is, the enzyme can function only with a rigidly defined type of molecule (substrate). Each chemical reaction in an organism demands a specific enzyme. It is now clear why so many types of enzymes exist; at present, several thousand have been characterized.

The histidine residue is a constituent of the active site in many enzymes (see Figure 3.11). The imidazole ring of histidine has a series of unique properties which enable it to show catalytic activity. Firstly, the rather high basicity enables histidine both to form strong hydrogen bonds and also to abstract a proton from acidic groups, such as the OH group of water and alcohols. As an RO⁻ anion is a much stronger nucleophile than a neutral ROH molecule, the imidazole ring can catalyze a nucleophilic addition to a carbonyl group (Figure 4.1). This variety of catalysis is called general base catalysis.

In the organism this type of catalysis is represented by the hydrolytic cleavage of protein amide bonds. The participating enzymes are called

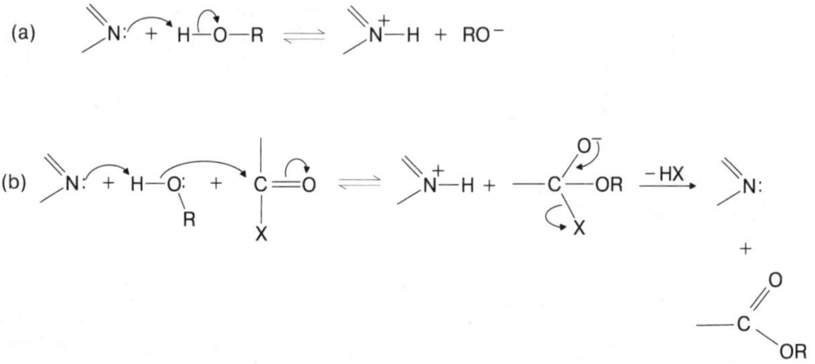

Figure 4.1   (a) Ionization of the alcoholic hydroxy group. (b) General base catalysis with participation of the imidazole ring (the pyridine-like nitrogen atom is designated).

[†]In certain microorganisms polyribonucleotides (RNA) can play the role of enzymes in the cleavage of ester bonds in other RNA molecules (for details see McCorkle, G. M. and Altman, S., *J. Chem. Educ.*, 1987, **64**, 221).

proteases. Histidine and serine residues are constituents of the protease chymotrypsin. A proposed mechanism for the action of chymotrypsin is shown in Figure 4.2. Within the enzyme, the imidazole ring of a histidine and the hydroxy group of a serine are bound by hydrogen bonding (Figure 4.2a). When a protein molecule approaches, the imidazole nitrogen abstracts a proton from the OH group, thus activating the serine oxygen atom toward attack at the carbonyl carbon atom in the polypeptide (Figure 4.2b). The unstable activated complex, shown in Figure 4.2b, is called an enzyme–substrate complex. Further conversion, very similar to that represented in Figure 4.1, brings about cleavage of the amide bond and acylation of the enzyme at its hydroxy group (Figure 4.2c). By an analogous catalytic mechanism, subsequent hydrolysis of the ester bond occurs (Figure 4.2d) with elimination of acid (RCO_2H) and regeneration of the enzyme (Figure 4.2e).

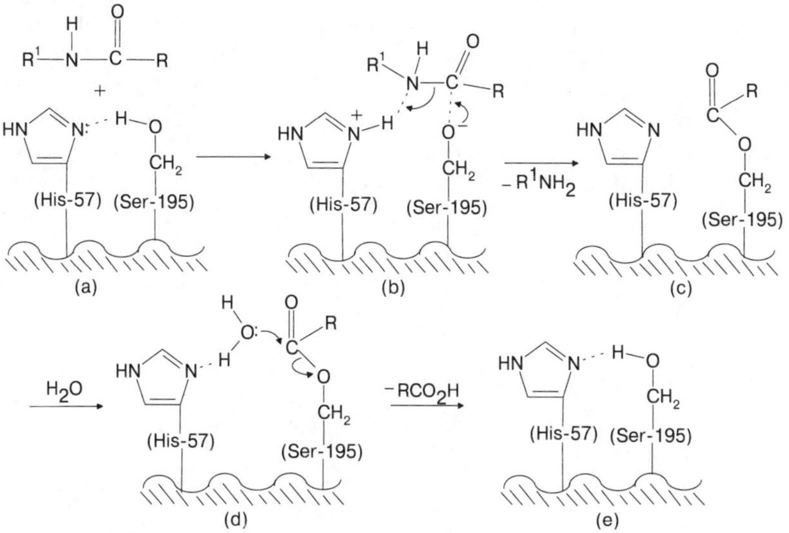

Figure 4.2 Simplified scheme of amide bond hydrolysis catalyzed by chymotrypsin. Histidine (His-57) and serine (Ser-195) take part in the enzyme active site (numbers designate the positions of the amino acids in the protein chain) (adapted from Lehninger, A. L., *Biochemistry,* Worth, New York, 1970, Chap. 9, p. 173, Figure 9-2, with permission).

The scheme in Figure 4.2 is oversimplified. In reality, a third amino acid residue is involved in the active site of chymotrypsin. This is aspartic acid (or α-aminosuccinic acid). In the normal state, its ionized carboxy group and the NH proton of a histidine imidazole ring are connected by a hydrogen bond. However, in hydrolytic reactions the NH proton is donated to aspartic acid in order to increase the negative charge on the imidazole ring. This transfer facilitates the abstraction of a proton from the serine OH group. All of the proton transfers are believed to be carried out synchronously through a transition state which might be illustrated as

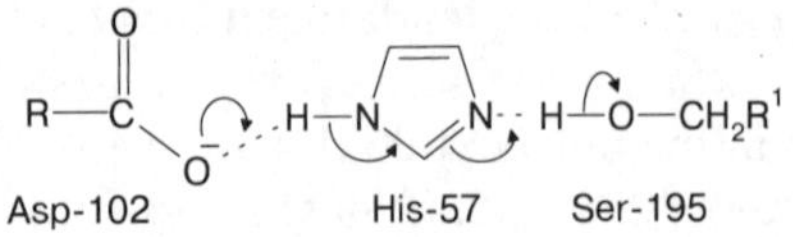

The imidazole ring in chymotrypsin functions in an amphoteric manner, i.e., both as an NH acid and as a base. Significantly, the basicity of histidine is such that it exists 50% as the neutral form and 50% as the imidazolium cation at the physiological pH of 7.4.

This peculiarity is utilized by the enzyme ribonuclease, which catalyzes P–O phosphate bond cleavage at the 3′-position of ribose (Figure 4.3). The specificity of this enzyme is such that it only severs those ribose residues which are attached to pyrimidine bases. Two histidine residues are found at the active site of ribonuclease, one in the neutral form (His-12) and the other (His-119) in the imidazolium cation form (Figure 4.3a). The neutral histidine abstracts a proton from the 2′-OH group of ribose in the enzyme–substrate complex (Figure 4.3b). The oxygen atom is thus activated and can perform

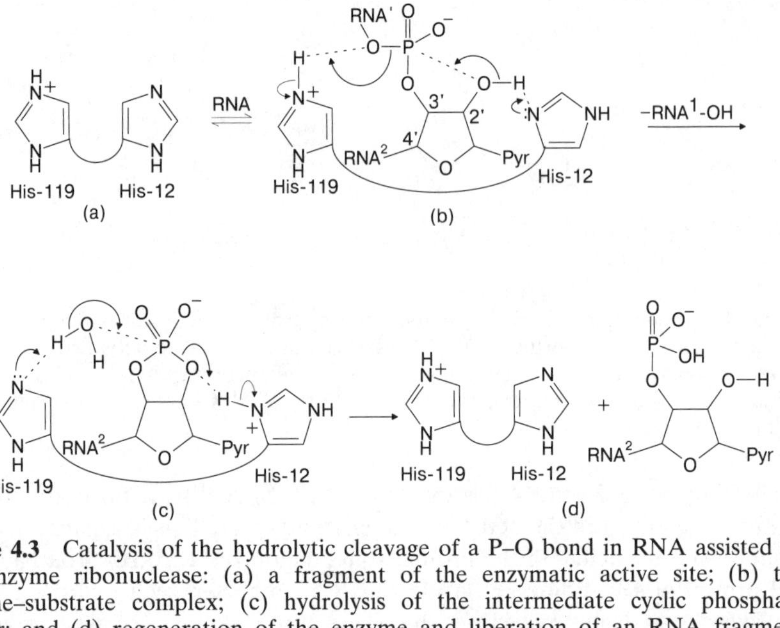

Figure 4.3 Catalysis of the hydrolytic cleavage of a P–O bond in RNA assisted by the enzyme ribonuclease: (a) a fragment of the enzymatic active site; (b) the enzyme–substrate complex; (c) hydrolysis of the intermediate cyclic phosphate diester; and (d) regeneration of the enzyme and liberation of an RNA fragment (RNA[1] and RNA[2] are different portions of the RNA chain; Pyr is a pyrimidine base).

intramolecular attack at the electrophilic phosphorus atom to form a cyclic phosphate diester. The P–O bond cleavage occurs simultaneously with elimination of an RNA^1–OH fragment. This process is catalyzed by the imidazolium ion of His-119 which donates its proton to the departing residue of RNA^1. The final step in the overall process involves hydrolysis of the cyclic phosphate diester (Figure 4.3c). By contrast with the initial reaction sequence (Figure 4.3b), this time the imidazolium ion of His-12 plays the role of proton donor in the P–O bond cleavage, and the neutral imidazole of His-119 serves as a proton acceptor which activates a water molecule for nucleophilic attack at phosphorus. At the completion of the sequence, the enzyme molecule is regenerated to its initial state and the RNA chain is one unit smaller.

4.2 Coenzymes and Enzymes as 'Joint Molecular Ventures'

In addition to enzymes which possess purely protein structures (chymotrypsin, ribonuclease and so on), there are a variety of enzymes which incorporate alongside the protein structure non-amino acid fragments, called coenzymes or cofactors. In these cases it is the coenzyme that facilitates the necessary chemical reaction by interacting with the substrate. The role of the protein moiety (called the apoenzyme) is limited to the spatial organization of the overall process. The coenzyme is positioned in a superficial cavity of the apoenzyme and is held in place by nonvalent interactions. Sometimes, in addition, a covalent sulfide (C–S–C) or disulfide (C–S–S–C) bond is formed between the coenzyme and apoenzyme.

All reactions of coenzymes can be divided into two groups: oxidation–reduction and transfer reactions. Because the majority of coenzymes are derivatives of nitrogen heterocycles, we now consider some of their reactions.

4.2.1 OXIDATIVE–REDUCTIVE COENZYMES

Many oxidative–reductive conversions occur in organic chemistry. Hydrogenation–dehydrogenation reactions form an important subclass among these (Figure 4.4a). A molecule is oxidized if it loses a hydrogen atom together with a pair of electrons which can be in the form of an H_2 molecule or a hydride ion H^-, but not a proton or hydrogen atom. If a molecule gains a hydrogen, it is reduced. In principle, hydrogen atoms can be eliminated in pairs as H_2, for example, by heating or under the action of a porous metallic catalyst like palladium, platinum or nickel. However, more frequently the hydrogen undergoes transfer to a hydrogen acceptor molecule known as an oxidant (A) or is donated by a hydrogen donor molecule known as a reductant (AH_2). The most abundant natural oxidant is molecular oxygen, which converts hydrogen into water or, less often, into hydrogen peroxide. Alternatively, incorporation of oxygen into a molecule may occur without any change in the original number of hydrogen atoms. For example, a C–H bond may be transformed into a C–OH group (Figure 4.4b). Such reactions are also considered as oxidations on the grounds that the electron pair of a C–O bond is shifted toward oxygen (in comparison with that of the C–H bond), and thus the carbon atom loses some electron density.

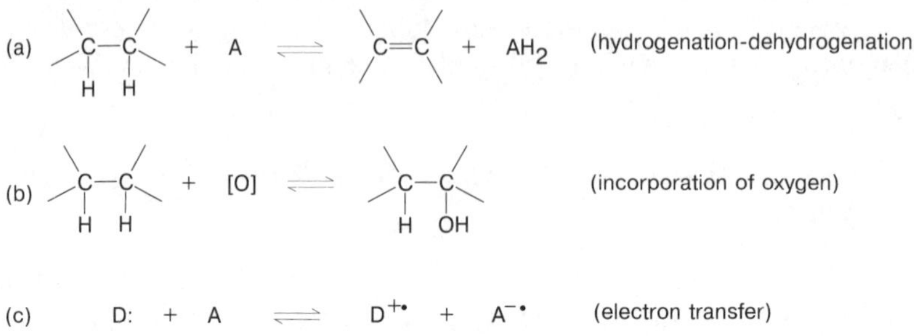

(a) $\overset{\diagup}{\underset{\diagdown}{C}}\!-\!\overset{\diagup}{\underset{\diagdown}{C}}$ + A $\rightleftharpoons$ $\overset{\diagup}{\diagdown}C\!=\!C\overset{\diagup}{\diagdown}$ + AH_2 (hydrogenation-dehydrogenation)
      H H

(b) $\overset{\diagup}{\underset{\diagdown}{C}}\!-\!\overset{\diagup}{\underset{\diagdown}{C}}$ + [O] $\rightleftharpoons$ $\overset{\diagup}{\underset{\diagdown}{C}}\!-\!\overset{\diagup}{\underset{\diagdown}{C}}$ (incorporation of oxygen)
      H H   H OH

(c)   D: + A $\rightleftharpoons$ $D^{+\bullet}$ + $A^{-\bullet}$ (electron transfer)

Figure 4.4 Types of oxidation–reduction reactions.

One further variant of oxidative–reductive conversions involves the gain or loss of electrons by a molecule without any change in the connectivity of the atoms (Figure 4.4c). This type of interaction occurs between an electron donor (D) and an electron acceptor (A). The products of such an electron transfer are a cation-radical $D^{+\bullet}$ and an anion-radical $A^{-\bullet}$ which may be bound in a molecular complex D → A. All of the above types of reactions are widely encountered in the chemistry of living organisms.

Dehydrogenases as hydrogen transporters

Two coenzymes, nicotinamide adenine dinucleotide (NAD) and the analogue phosphorylated at one of the hydroxy groups (NADP), are the most biologically important hydrogen transfer agents (Figures 4.5a and b).

The functional portion of both coenzymes, their 'chemical motor', is the nicotinamide residue which can exist as either (i) the oxidized pyridinium form (NAD^+ or $NADP^+$), as shown in Figures 4.5a and b, or (ii) the reduced dihydropyridine form (NAD-H or NADP-H), as shown in Figure 4.5c. The function of the oxidized form of the enzyme is to receive a hydride ion from a hydrogen-donor substrate (Sub-H_2). In the process, the pyridinium cation is converted into a 1,4-dihydropyridine (Figure 4.5c). The second hydrogen atom of the substrate is released as a proton into solution. As with all enzymatic reactions, this process is reversible and the equilibrium can be shifted depending on the substrate type and the conditions within the living cell.

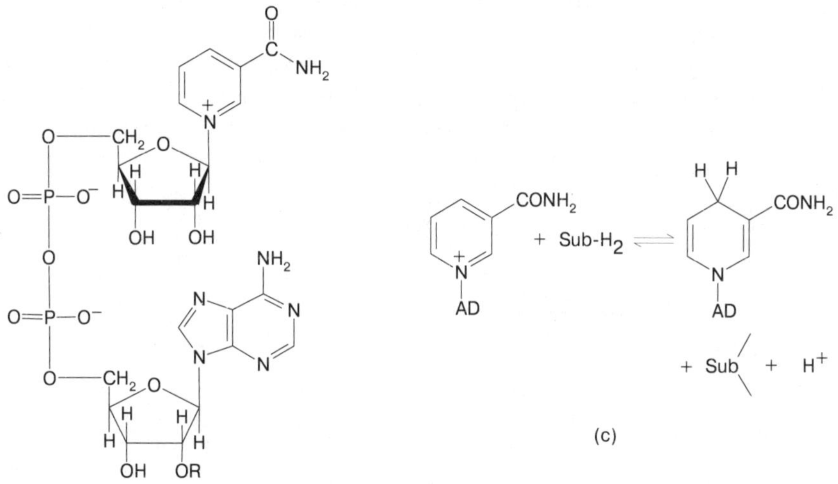

(a, R = H; b, R = PO$_3$H$_2$)

Figure 4.5 Structure of two dehydrogenating coenzymes: (a) nicotinamide adenine dinucleotide (R = H) and (b) nicotinamide adenine dinucleotide phosphate (R = PO$_3$H$_2$). (c) An oxidation–reduction process with their participation (Sub-H$_2$ = substrate).

Enzymes containing these groups (NAD or NADP) are named dehydrogenases or, more precisely, pyridine-dependent dehydrogenases.[†] More than 150 dehydrogenases are known at present, and no other coenzyme class controls so many different cellular reactions. This variety is readily explained. We have already mentioned (Section 2.1) that azines, and especially azinium cations, are readily reduced. In the case of the coenzymes NAD$^+$ and NADP$^+$, the ease of reduction is increased by the electron-accepting carbamoyl substituent CONH$_2$. On the other hand, the aromaticity of the pyridine ring is distorted in the partially reduced NAD-H. Therefore, in case of need, reverse donation by NAD-H of a hydride ion can occur readily. These diametrically opposed tendencies allow the formation of optimal redox balances whose flexibility is utilized by organisms. The activity of the pyridine-dependent dehydrogenases is significantly enhanced by several divalent metal ions (Mg^{2+}, Zn^{2+} and others). These ions complex with the coenzyme which thus becomes connected to the apoenzyme. Additional information about the NAD-H coenzyme is presented in Section 10.3.

Flavin-dependent dehydrogenases make up another important class of dehydrogenases. The biological action of flavin coenzymes results from a

---

[†]Dehydrogenating or hydrogenating coenzyme pairs such as NAD$^+$–NAD-H are called oxidoreductases.

vital heterocyclic constituent called alloxazine. Alloxazine is a tricyclic heterocycle and can be considered as being composed of two condensed fragments, quinoxaline and uracil (Figure 4.6). Several tautomeric forms are possible, and the two most important are called alloxazine and isoalloxazine, differing from one another by the position of a proton at N-1 or N-10.

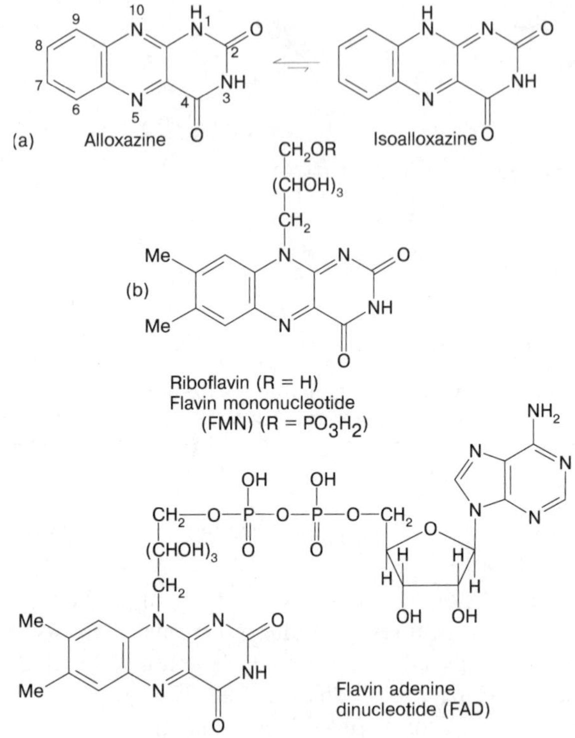

Figure 4.6 (a) Two important alloxazine tautomers. (b) The flavin coenzymes flavin mononucleotide (FMN) and flavin adenine dinucleotide (FAD).

Under ordinary conditions the alloxazine $\rightleftharpoons$ isoalloxazine equilibrium of Figure 4.6a is almost completely shifted to the left. However, Nature has chosen to use the isoalloxazine structure in flavin-dependent dehydrogenases. We have already described (see Section 2.5) the possibility of fixing unstable tautomers by substitution at the position to which the labile hydrogen atom is attached. Flavins are examples of such fixed tautomers. Thus, riboflavin is a 10-substituted isoalloxazine which has two methyl groups at the C-7 and C-8 positions and one ribityl residue (partly reduced acyclic ribose) at N-10. The monophosphate derivative, flavin mononucleotide (FMN), is one of the two biologically important flavin coenzymes. The second, flavin adenine dinucleotide (FAD), has a more complex structure (Figure 4.6b).

What advantages does the isoalloxazine form possess over the alloxazine tautomer from a biological viewpoint? It is well known that quinones, such as p-benzoquinone, are readily reduced by various reagents to hydroquinones (Figure 4.7a). The driving force for this reaction is a combination of the presence of the electron-accepting carbonyl groups and the increased aromaticity of the reduced product.

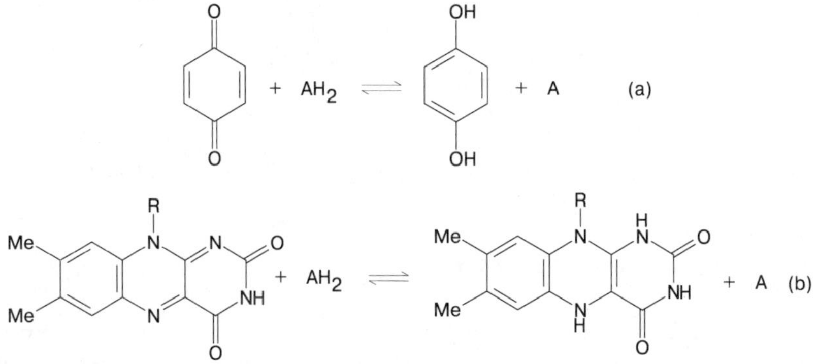

Figure 4.7 Reduction (a) of p-benzoquinone to hydroquinone and (b) of the flavin coenzyme FMN or FAD to its H-form.

The same forces are at play in the reduction of the FMN and FAD coenzymes whose structures incorporate the quinonoid bond systems. Hydrogen uptake occurs at the N-1 and N-5 positions, i.e. at the termini of the quinonoid chain. As a result, the uracil ring acquires a more stable π-electron structure. The reduced flavin coenzymes are designated as FMN-H and FAD-H. It is interesting to compare the chemical actions of pyridine-dependent and flavin-dependent dehydrogenases. While both transfer two electrons, they differ in the number of hydrogen atoms being transported. Coenzymes with the nicotinamide moiety accept or donate one hydrogen atom, whereas with flavins two hydrogen atoms are involved.

Recently, new coenzymes have been found which participate in hydrogenation–dehydrogenation processes. These new compounds were also found to be nitrogen heterocycles. One enzyme having the 5-deazaflavin structure was isolated from anaerobic methane-discharging bacteria (Figure 4.8a). As in the case of flavins, hydrogenation occurs at positions 1 and 5.

A further coenzyme (PQQ), a derivative of a pyrrolo[2,3-f]quinoline quinone (Figure 4.8b), is encountered in many important oxidoreductases such as alcohol dehydrogenase, aldehyde dehydrogenases, D-glucose dehydrogenase and others. The central o-quinone ring is reduced to the corresponding dihydric phenol. The pyrrole and pyridine rings do not directly take part in this reaction. However, the first ring is a π-donor and the second a π-acceptor. Their combined effect probably creates the required optimal oxidative–reductive potential.

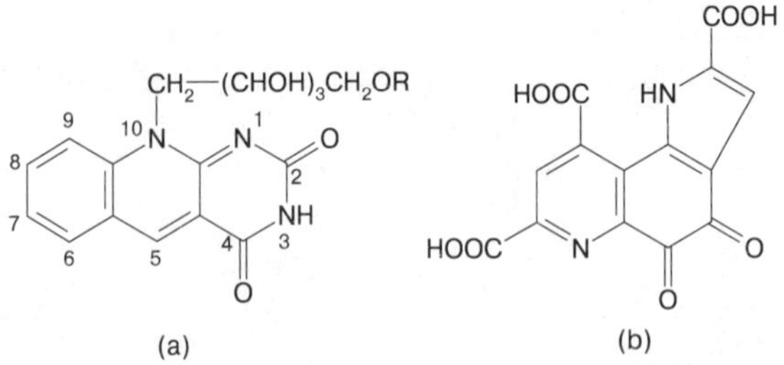

(a)   (b)

Figure 4.8 Coenzymes transporting hydrogen with (a) 5-deazaflavin and (b) pyrrolo[2,3-*f*]quinoline moieties.

Cytochromes as electronic 'postmen'

Oxidative–reductive enzymes whose only function is to transfer electrons from one molecule to another occur in all living cells. These electronic 'postmen' are named cytochromes.

Hemes, complexes of porphyrin with iron(II), serve as the coenzymes for numerous cytochromes. Hemes are deeply colored, usually red, brown or orange, and impart a corresponding color to the cellular structures containing them. The name cytochrome is a derivative of two Greek words: *kytos*, cell and *chroma*, color. The classification of cytochromes, designated by letters *a*, *b*, *c* and so on, is based on their color, or the position of their long wavelength absorption band. For example, cytochromes *a*, *b* and *c* have absorption maxima at 600, 563 and 550 nm, respectively.

We now examine in detail the porphyrins, a compound class which we will encounter many times. Porphyrins are substituted derivatives of porphin, a tetrapyrrolic macrocycle (see Figure 2.4). Depending on the type and the position of their substituents, porphyrins are classified as etioporphyrins, mesoporphyrins, coproporphyrins or protoporphyrins. The last are the most frequently encountered and contain four methyl groups, two vinyl groups and two propionic acid residues as substituents. These substituents can be attached to the four pyrrole rings in 15 different orientations, giving rise to 15 isomeric protoporphyrins. Protoporphyrin IX (Figure 4.9a) is a constituent of the cytochromes, in particular the widely studied cytochrome *c*. The coenzyme in cytochrome *c* is bound to the protein of the enzyme by two sulfide bridges which are formed by addition of the SH groups of cysteine to the vinyl groups of protoporphyrin IX (Markovnikov addition) (Figure 4.9b).

It is interesting that in other cytochromes there are no covalent bonds linking the coenzyme and apoenzyme, the stability of the complex being purely the result of nonbonding interactions. Additional nonbonding interactions evidently also exist in cytochrome c. For example, there are two coordinate bonds between the Fe^{2+} ion and two amino acid ligands situated on either side of the plane of the porphyrin ring system, and almost perpendicular to this plane. One such ligand is almost always the imidazole ring of a histidine residue. The other may also be a histidine residue or another amino acid residue such as methionine.

It is thus clear that the coordination number of iron(II) in the cytochromes is six, and therefore the electron shell is that of the nearest inert gas, krypton. Owing to complete saturation with ligands, the Fe^{2+} ion is incapable of

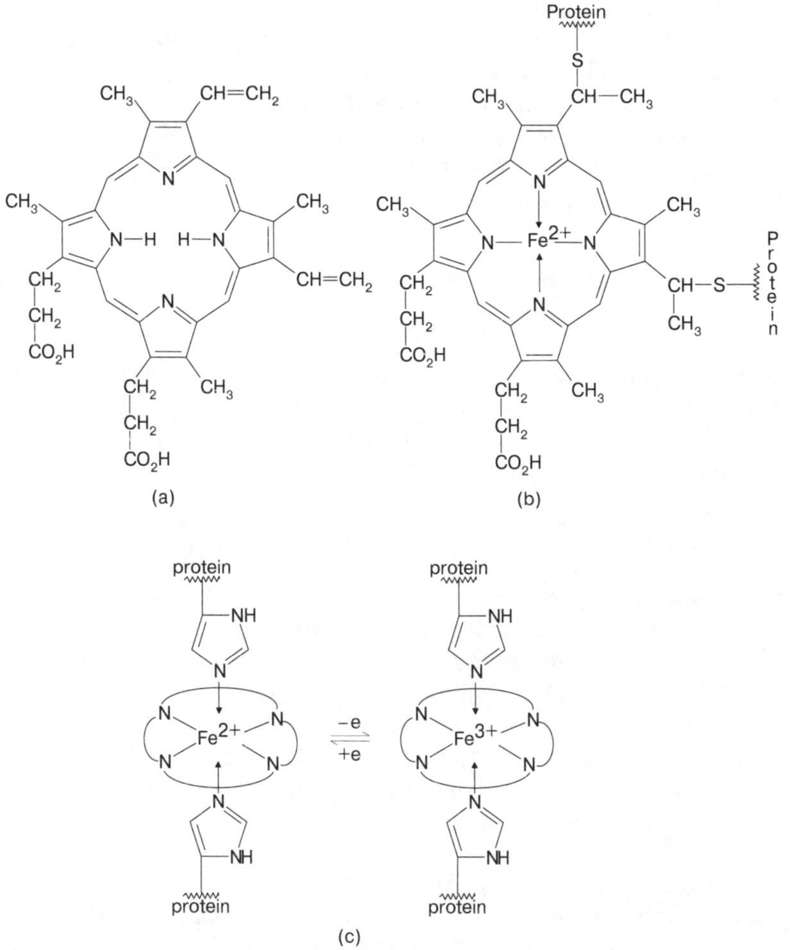

Figure 4.9 (a) Protoporphyrin IX. (b) The coenzyme in cytochrome c. (c) Oxidation–reduction equilibrium in cytochromes.

binding with other molecules. However, the metal ion is able to lose one electron thereby becoming Fe^{3+}. It is this reversible $Fe^{2+} \rightleftharpoons Fe^{3+}$ exchange which is the basis of the activity of the cytochromes (Figure 4.9c). In the next chapter we describe the interactions of cytochromes with electron donors and acceptors in a cell.

Oxygenases

With the exception of anaerobic bacteria, all living organisms require molecular oxygen. Oxygen is utilized both as an electron acceptor in the respiratory chain (Section 5.2.3) and for the biosynthesis of various substances, such as catecholamines, porphyrins, etc. Oxygen can also be supplied from water as occurs, for example, in the oxidative deamination of amino acids (Figure 4.25) or during the formation of oxo derivatives of purine. In the latter case, the water covalently adds to the electron-deficient CH=N bond with subsequent dehydrogenation of the adduct formed catalyzed by a dehydrogenase

$$-CH{=}N- \;+\; H_2O \;\rightleftharpoons\; \underset{\substack{| \\ OH}}{-CH}{-}\underset{\substack{| \\ H}}{N}- \;\xrightarrow{\;A\;}\; \underset{\substack{\| \\ O}}{-C}{-}\underset{\substack{| \\ H}}{N}- \;+\; AH_2$$

Enzymes which promote the incorporation of molecular oxygen into organic compounds are called oxygenases. There are two types: monooxygenases, which control the introduction of one oxygen atom, and dioxygenases, which monitor the incorporation of two oxygen atoms into a substrate

$$\text{substrate} \;+\; O_2 \;\longrightarrow\; \text{substrate-O} \;+\; (O) \quad \text{(monooxygenase)}$$
$$\text{substrate} \;+\; O_2 \;\longrightarrow\; \text{substrate-O}_2 \qquad\qquad \text{(dioxygenase)}$$

In the case of monooxygenases the second atom of oxygen clearly has to be converted into a water molecule. Therefore, monooxygenase activity is accompanied by the action of a further hydrogenase enzyme which donates two hydrogen atoms to bind with this oxygen atom.

An important class of reaction catalyzed by oxygenases is the cleavage of aromatic C=C bonds near a hydroxy group, between two hydroxy groups or in an indole system. Monooxygenases which hydroxylate aromatic and aliphatic compounds without C–C bond cleavage are called hydroxylases. They can also cause N-dealkylation, and the formation of epoxides, N-oxides and S-oxides. Alternative oxidations of tryptophan (Figure 4.10) provide typical examples of the action of mono- and dioxygenases. A monooxygenase converts tryptophan into 5-hydroxytryptophan which is then decarboxylated to give 5-hydroxytryptamine (serotonin), a central nervous system transmitter (see Section 7.4.1). On the other hand, tryptophan-2,3-

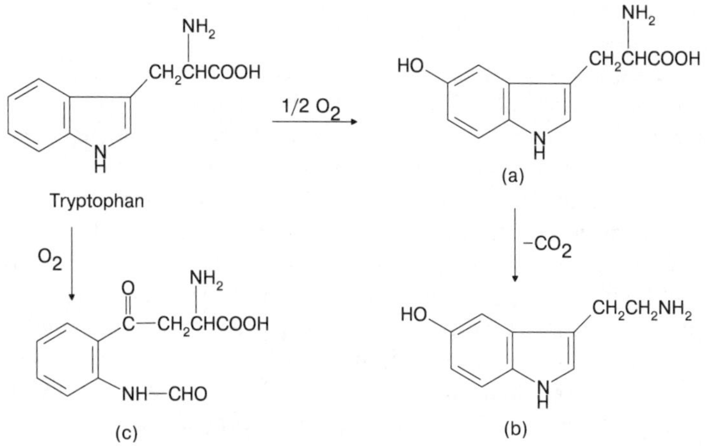

Figure 4.10 Tryptophan conversions under the action of mono- and dioxygenases into (a) 5-hydroxytryptophan, (b) serotonin and (c) N-formylkynurenine.

dioxygenase causes cleavage of the double bond between C-2 and C-3 in the pyrrole ring, and thus effects formation of N-formylkynurenine.

We now consider the mechanism of action of oxygenases. The main function is the activation of an oxygen molecule, which is achieved by the attachment of oxygen to the coenzyme in a defined manner. The function of the coenzyme is carried out in all dioxygenases by an iron cation (usually as Fe^{2+}, and less often Fe^{3+}) bound directly to the protein moiety or, occasionally, associated with the heme.

The activation can be represented by one electron transfer from the Fe^{2+} cation to the molecule of oxygen: $Fe^{2+} \cdot O_2 \rightleftharpoons Fe^{3+} \cdot O_2^-$. In reality, however, the substrate also participates in the activation. It is believed that the initial step is the formation of an enzyme–substrate complex in which the enzyme protein is reconstructed in such a manner as to make the Fe^{2+} ion more available to the oxygen molecule. The rearrangement thus facilitates the subsequent formation of the triple complex (enzyme–substrate–O_2), inside which the oxidation takes place (Figure 4.11).

Monooxygenases are of great interest as many contain heterocyclic coenzymatic fragments. Two types of heterocycle are found in monooxidases:

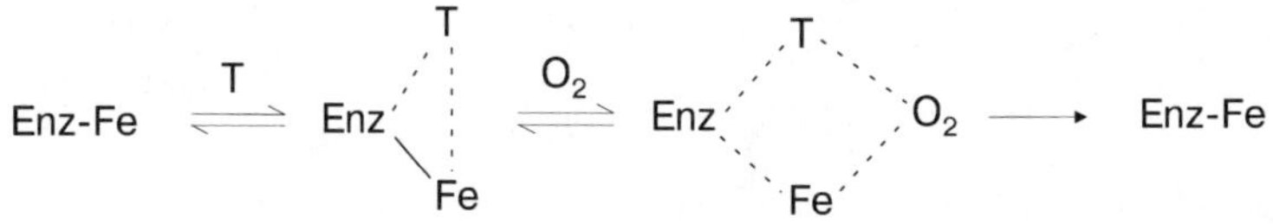

Figure 4.11 Oxidation of a substrate (T) by the action of a dioxygenase.

the dihydroflavins (Figure 4.7) and pteridines, which are widespread naturally occurring systems, each incorporating fused pyrazine and pyrimidine rings (Figure 4.12; see also Figure 9.2). Both these coenzymes are capable of adding an O_2 molecule, in each case at the bridgehead 4a-position. Thus, adducts (1) and (2) are produced in which the oxygen is in the form of an active peroxide. The conversion of phenylalanine into tyrosine is a typical example of the action of such a monooxygenase (Figure 4.12). Phenylalanine oxidase, which has a tetrahydrobiopterine fragment, takes part in this reaction. The adduct with oxygen oxidizes phenylalanine to epoxide (4), which is then aromatized by migration of hydrogen atom H_A from the *para* position of phenylalanine to the *meta* position of tyrosine. Tetrahydrobiopterine is thus transformed into the oxidized form (5) through a recyclization pathway via putative acyclic intermediate (3). In the final step the oxidized form (5) is reduced by an NAD-H-dependent hydrogenase, thus regenerating the original enzyme.

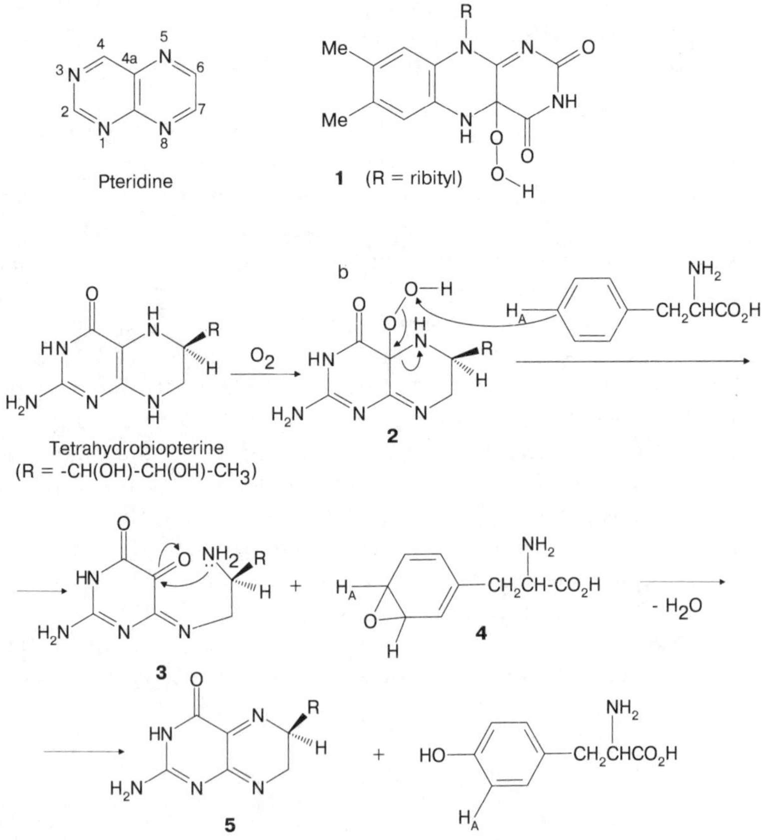

Figure 4.12 Adducts of monooxidases with oxygen and mechanism of the oxidation of phenylalanine to tyrosine with the coenzyme tetrahydrobiopterine.

4.2.2 COENZYMES AS CARRIERS OF MOLECULAR SPECIES

Numerous enzymes are engaged in the transfer of various functional groups, molecules and ions. In some cases, the enzyme acts to transport a molecule; in others, enzymes deliver a group into a receiving molecule during the biosynthesis of a complex natural product. Many transferring enzymes include in their structures a heterocyclic coenzyme which is directly involved in the transfer process.

Hemoglobin and myoglobin as oxygen transporters

Hemoglobin and myoglobin are the most widespread and the best-known molecular transportation vehicles, and their function is to supply tissues with oxygen. Strictly speaking, they are not classified as enzymes, but they have much in common with enzymes as far as their structures and mechanism of action are concerned. Hemoglobin and myoglobin contain heme as the coenzyme. The heme is held within one of the clefts of the polypeptide globule by hydrophobic interactions and by coordination linkages between the Fe^{2+} ion and a histidine residue. Myoglobin molecules contain a polypeptide chain of 153 amino acid residues. Myoglobin is the oxygen-binding protein responsible for the transport of oxygen and also for its storage in muscle tissues. Hemoglobin is more complex than myoglobin. It is a molecular aggregate composed of four intertwined protein chains, of which two are identical α-type protein chains, and two are identical β-type chains. The α-type chains contain 140 residues each and the β-type chains contain 146 amino acid residues each. Each chain binds one heme coenzyme. Consequently, hemoglobin as a whole contains four heme residues.

The natural occurrence of hemoglobin is more extensive than that of myoglobin, the latter being encountered only in the skeletal muscles of whales, sharks and other sea creatures. Hemoglobin is believed to have been the more reliable oxygen-bearing system during biological evolution owing to the presence of several polypeptide chains and several hemes, because the destruction of any one unit could be compensated for by the normal functioning of the others. Genetic disorders and blood diseases induced by such degradation are not uncommon in man and animals, but thanks to natural selection, and perhaps to God's foresight, they do not always prove to be fatal.

A crucial difference between the heme of cytochromes and the heme of globins lies in the fact that in the latter the Fe^{2+} ion is linked with only five ligands: four nitrogens of the porphyrin system and one imidazole nitrogen of a histidine residue in the protein chain. As a result, the Fe^{2+} ion in globins can utilize its vacant sixth coordination site to bind an oxygen molecule

(Figure 4.13).[†] This binding must, of course, be reversible and therefore not too strong. Oxygen becomes linked to heme under the normal partial pressure inherent in lung alveoli and is then released in tissues with an oxygen shortage, where the partial pressure of oxygen is decreased.

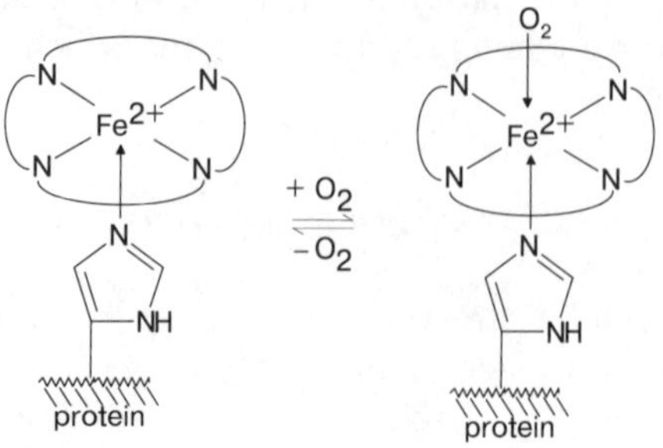

Figure 4.13   The reversible binding of oxygen to the globin heme.

It is instructive to consider what prevents the globin iron from spontaneous oxidation to iron(III), since the rapid oxidation of a cut apple exposed to air is well known. X-Ray crystallography reveals that hemoglobin contains a further histidine residue (in addition to that coordinated to iron) in close proximity to the heme, and that the imidazole ring of this second histidine is protonated. The presence of the positive charge near the Fe^{2+} cation strongly disfavors the loss of an electron, as the electrostatic repulsion between the resulting Fe^{3+} and the imidazolium cation would be great. An example of what can occur when this system does not function correctly is the blood disease called ferrihemoglobinemia. This disorder is characterized by substitution of the second histidine residue (the one not coordinated with iron) by another amino acid. This replacement in the protein chain leads to the facile oxidation of two iron atoms in such a heme to Fe^{3+}. In this state, the iron cations cannot coordinate with oxygen. As a result, the oxygen absorption capacity of the blood in such patients is diminished by a factor of two.

Thus, the chemical reactions of hemes involve not only the iron cation, but also require participation both of the imidazole rings of the polypeptide histidine units and of the porphyrin system as a whole. The role of the highly conjugated porphyrin system is to delocalize and therefore stabilize charge. Such stabilization is especially important for the hemes, as the lifetime of the intermediate oxidized and reduced states must be long enough for it to exercise the corresponding biological effects.

[†]The sixth valence of iron(II) in globin hemes is considered by some authors to be occupied by a water molecule. Under appropriate conditions this molecule of water is readily displaced by oxygen.

Coenzyme A as an acyl group carrier

Coenzyme A is an extended chain molecule containing the following sequence of units: adenine, ribose, pyrophosphate, pantothenic acid and 2-aminoethanethiol (Figure 4.14). Although the heterocyclic moieties of this coenzyme do not participate directly in the enzymatic reactions, consideration of its specific biochemical properties helps in a better understanding of the functions of many other coenzymes, including those based on heterocycles.

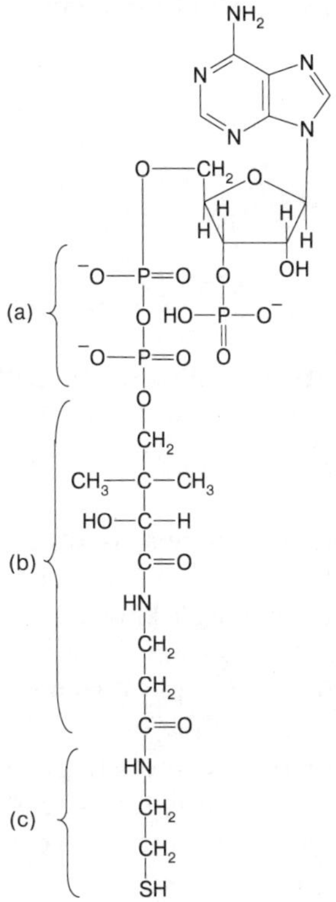

Figure 4.14 Coenzyme A: (a) the pyrophosphate moiety, (b) the pantothenic acid residue and the (c) 2-aminoethanethiol moiety.

Coenzyme A (designated also as CoA or CoA-SH) transports acyl groups (RC=O) and also activates fatty acids in various bioreactions. The critical functionality in CoA is the terminal thiol which converts a carboxyl function into the corresponding thioester

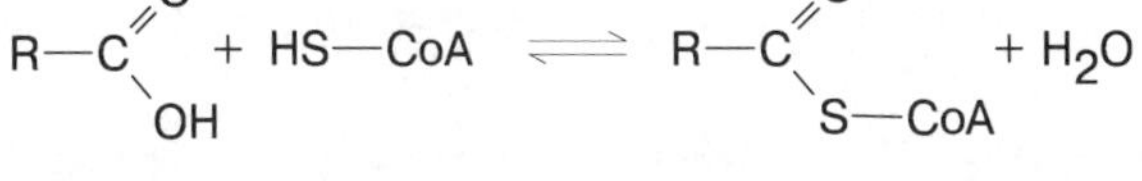

$$R-C\!\!\underset{OH}{\overset{O}{\diagup}} + HS-CoA \rightleftharpoons R-C\!\!\underset{S-CoA}{\overset{O}{\diagup}} + H_2O$$

Thioesters of carboxylic acids have two characteristics important for the properties of coenzyme A. The first is that the energy-rich C–S bond in a thioester is labile. Under the influence of various nucleophilic agents (water, alcohols, amines, etc.) this C–S bond is readily broken to liberate CoA with simultaneous acylation of the nucleophilic agent (Figure 4.15a). The lability of the C–S bond is controlled by several factors, the main being the high polarization of the sulfur valence electrons. A second characteristic property of thioesters important for the biological activity of coenzyme A is that the C–H bonds α to the carbonyl group are strongly activated. Even comparatively weak bases accept a proton from such C–H bonds to form a carbanion, as shown in Figure 4.15b.

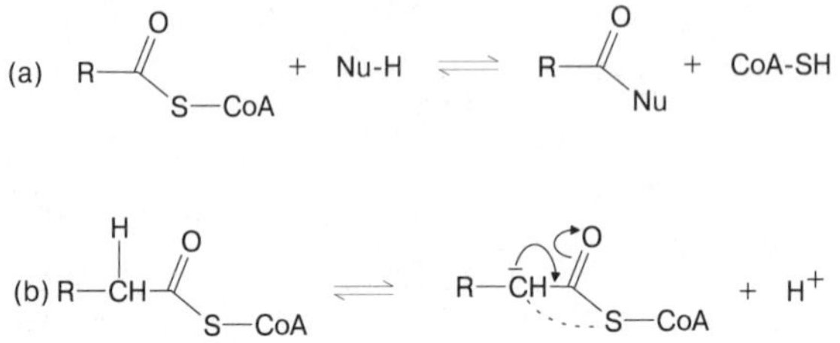

Figure 4.15 Chemical reactions of acylated coenzyme A: (a) acyl group transfer to a nucleophilic agent and (b) ionization of the C–H bond with the formation of a carbanion.

In organic chemistry, the facile generation of carbanions by proton loss from the α-position of aldehydes, ketones and other carbonyl-containing compounds is well known. The negative charge is stabilized by conjugation with the C=O group (indicated by arrows in Figure 4.15b). In thioesters the C–H bond activation is significantly greater than in ordinary esters, owing to the unfilled sulfur 3*d*-orbitals. Their relatively large size allows these 3*d*-orbitals partially to overlap the unshared electron pair of the α-carbon atom (a dashed line in Figure 4.15b), which imparts additional stabilization to the carbanion and facilitates its generation.

$$\text{a)}\ \text{R-}\overset{\gamma}{\text{CH}_2}\overset{\beta}{\text{CH}_2}\overset{\alpha}{\text{CH}_2}\text{-COOH} + \text{HS-CoA} \rightleftharpoons \text{R-CH}_2\text{CH}_2\text{CH}_2\text{-CO-S-CoA} + \text{H}_2\text{O}$$

$$\text{b)}\ \text{R-CH}_2\text{CH}_2\text{CH}_2\text{-CO-S-CoA} + \text{FAD} \rightleftharpoons \text{R-CH}_2\text{CH=CH-CO-S-CoA} + \text{FAD-H}_2$$

$$\text{c)}\ \text{R-CH}_2\text{CH=CH-CO-S-CoA} + \text{H}_2\text{O} \rightleftharpoons \text{R-CH}_2\underset{\text{OH}}{\text{C}}\text{HCH}_2\text{-CO-S-CoA}$$

$$\text{d)}\ \text{R-CH}_2\underset{\text{OH}}{\text{C}}\text{HCH}_2\text{-CO-S-CoA} + \text{NAD}^+ \rightleftharpoons \text{R-CH}_2\underset{\text{O}}{\underset{\|}{\text{C}}}\text{CH}_2\text{-CO-S-CoA} + \text{NAD-H} + \text{H}^+$$

$$\text{e)}\ \text{R-CH}_2\underset{\text{O}}{\underset{\|}{\text{C}}}\text{CH}_2\text{-CO-S-CoA} + \text{HS-CoA} \rightleftharpoons \text{R-CH}_2\text{-CO-S-CoA} + \text{CH}_3\text{-CO-S-CoA}$$

Figure 4.16 Oxidative breakdown of fatty acids.

A combination of these factors is involved in the oxidative decomposition of fatty acids in organisms (Figure 4.16). This multistep process requires the participation of a large number of enzymes including the flavin- and pyridine-dependent dehydrogenases FAD and NAD$^+$. In the first stage (Figure 4.16a) coenzyme A is 'loaded' with an acid to form a thioester in which the activation of the α-C–H bond enables dehydrogenation of the acid by FAD in the second stage of the process (Figure 4.16b). This conversion probably proceeds stepwise in two fast successive steps entailing the loss of a proton to generate a carbanion which then undergoes hydride ion loss

$$\overset{|}{\underset{H}{C}}-\overset{|}{\underset{H}{C}} + FAD \;\rightleftharpoons\; \overset{|}{\underset{H^-}{C}}-\overset{|}{C} + FAD\cdot H^+ \;\rightleftharpoons\; C=C + FAD\cdot H_2$$

The third stage (Figure 4.16c) is the hydration of the C=C bond in the unsaturated thioester thus formed. The orientation of this addition is consistent with electron delocalization theory: a hydroxide anion is added at the β-carbon atom (which has a partial positive charge owing to conjugation with the C=O group) followed by a proton at the α-carbon atom. In the fourth stage (Figure 4.16d) the hydroxy group is oxidized to a carbonyl group by an NAD$^+$-containing dehydrogenase, thus forming a β-keto acid (in the thioester form)

$$CH-OH + NAD^+ \;\rightleftharpoons\; \overset{+}{C}-OH \;\longleftrightarrow\; C=\overset{+}{O}H \;\rightleftharpoons\; C=O + H^+ + NAD-H$$

The C–C bonds situated between the two carbonyl groups of β-keto esters are susceptible to facile cleavage by various reagents. Such a cleavage occurs during the final stage (Figure 4.16e) of the overall process when another enzyme containing CoA converts the β-keto ester into an acetyl CoA and an acyl CoA. The original fatty acid carbon chain is thus shortened by two carbon atoms. The same process is repeated with the newly produced acyl CoA until the initial acid is completely converted into two-carbon acetyl fragments.

The acetyl products of degradation are then subjected to a versatile range of subsequent conversions within the organism. It should be noted that the enzymes containing FAD, NAD and coenzyme A participate together in the form of a single cellular ensemble in the degradation of fatty acids.

Thiamine pyrophosphate as an acetaldehyde activator

Many C–H bonds in organic compounds possess low acidity and the substitution of a hydrogen atom with a metal frequently requires a strong base such as *n*-butyllithium. By such treatment, thiazole is converted into 2-thiazolyllithium, while the *n*-butyllithium is transformed into *n*-butane (Figure 4.17a).

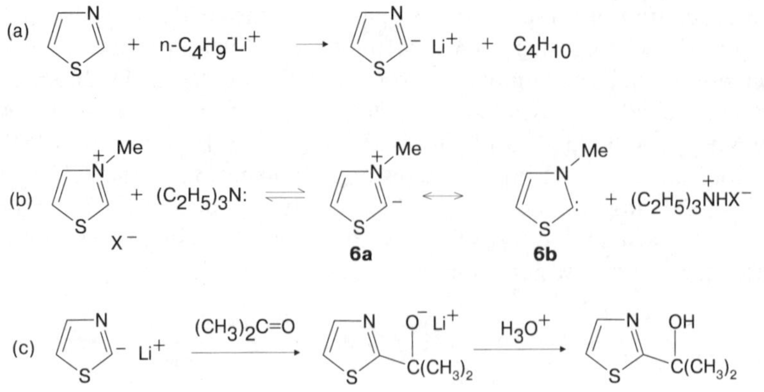

Figure 4.17 The C-2–H bond ionization in (a) thiazole and (b) the 3-methylthiazolium cation. (c) Reaction of 2-lithiothiazole with acetone.

Various structural changes can induce a pronounced increase in CH acidity. One, which we have already discussed, is the introduction of a sulfur atom into the α-position relative to the C–H bond, as in the case of coenzyme A. A large increase in acidity can be induced by cation formation in heterocyclic compounds. Both factors are valid in the case of 3-alkylthiazolium salts which are able to ionize the C-2–H bond even under the action of such a comparatively weak base as triethylamine (Figure 4.17b). The resultant species (**6a**) possesses carbanionic and cationic centers in close proximity and is called an ylide. Ylides can often be represented in an uncharged carbenoid form (**6b**).[†] Carbanions are potent nucleophiles widely used in synthesis. For instance, 2-lithiothiazole reacts with acetone to yield the corresponding alcohol (Figure 4.17c).

At the end of the 1950s it was found that similar chemical reactions occur naturally, though in a more specific manner. One of the most important coenzymes, thiamine pyrophosphate, participates in such transformations (Figure 4.18). Not surprisingly, this compound is a derivative of the thiazole cation which thus plays a leading role in biochemical reactions.

Thiamine pyrophosphate mediates three main types of enzymatic reactions. (i) It decarboxylates α-keto acids; pyruvic acid, a key intermediate in carbohydrate and amino acid metabolism, is thus converted into acetaldehyde. (ii) Thiamine pyrophosphate catalyzes the formation of acyloins (α-hydroxyalkyl ketones). (iii) The oxidative decarboxylation of pyruvic acid to form acetic acid also occurs with the help of this coenzyme. We now consider the general characteristics of these three reactions.

[†]Carbenes are compounds with divalent carbon of the general formula $R^1R^2\ddot{C}$.

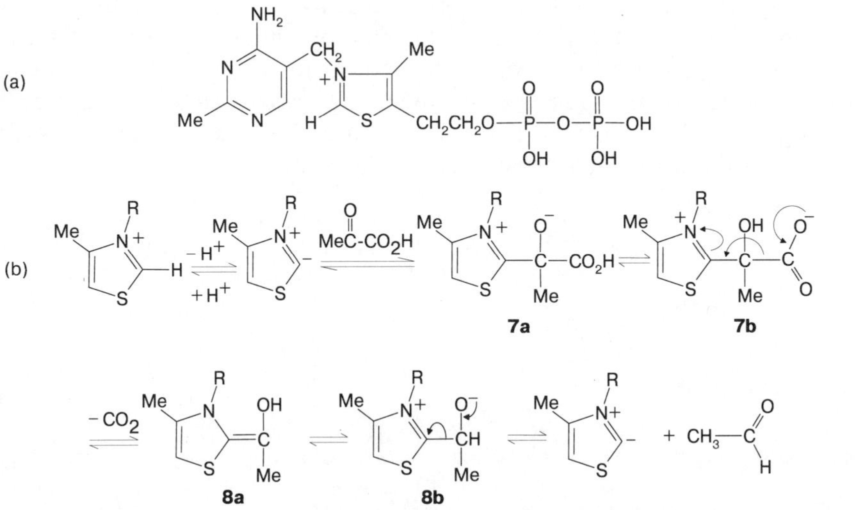

Figure 4.18 (a) Thiamine pyrophosphate and (b) the assisted decarboxylation of pyruvic acid (only the thiazole portion of the coenzyme is shown).

The C-2–H bond in the thiamine pyrophosphate thiazole ring is partially dissociated at physiological pH. The keto group of a pyruvic acid molecule adds to the thiamine pyrophosphate ylide (Figure 4.18b). The addition product (**7a**) undergoes tautomerism into (**7b**), and then decarboxylation to give compound (**8**) which may be considered as 'activated acetaldehyde'. The intermediate can also exist in the two tautomeric forms (**8a**) and (**8b**), and this gives rise to two alternative further transformations. (i) Decomposition into acetaldehyde and the original ylide completes the decarboxylation reaction of pyruvic acid (Figure 4.18b). (ii) The 'activated acetaldehyde' in tautomeric form (**8a**) can attack a second molecule of acetaldehyde (Figure 4.19). The addition product (**9**) thus formed is cleaved with the formation of acyloin and the ylide, enabling the second type of enzymatic reaction which is effectively an acyloin condensation.

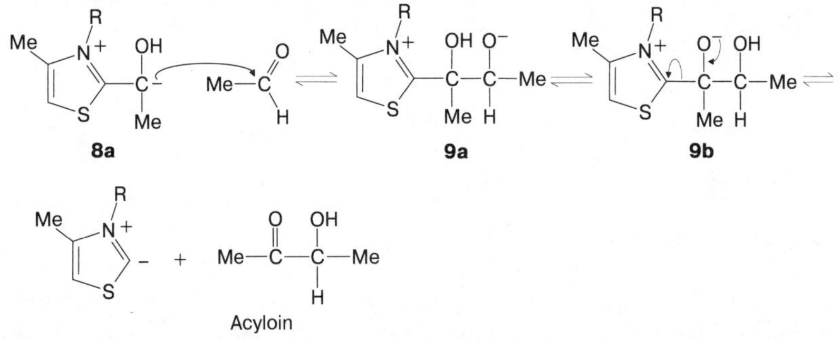

Figure 4.19 Acyloin condensation catalyzed by thiamine pyrophosphate.

The oxidative decarboxylation of pyruvic acid is a more complex process involving other enzymes apart from thiamine pyrophosphate. One of these other enzymes contains as a coenzyme lipoic acid (Figure 4.20), which can exist in two forms: an oxidized state containing a cyclic disulfide bridge or a reduced noncyclic form with two thiol groups. The two forms are readily interconvertible during oxidation–reduction processes.

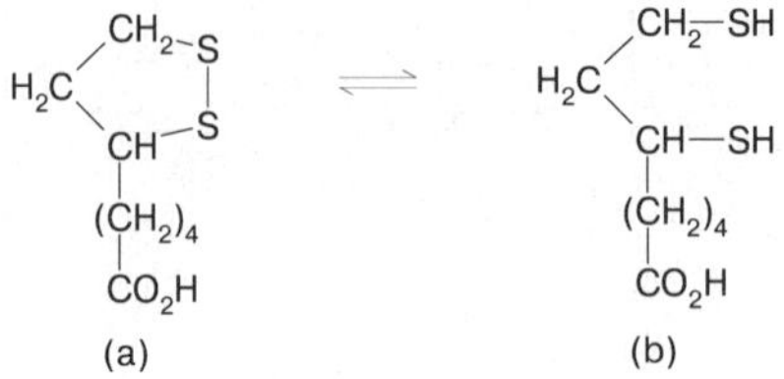

Figure 4.20   Lipoic acid: (a) oxidized form and (b) reduced form.

The S–S disulfide bridge is known to cleave even when treated with rather mild reductants. The 'activated acetaldehyde' serves as such a reducing agent and is represented by form (**8b**) shown in Figure 4.21a. Tautomer **8b** donates a hydride ion to lipoic acid and is oxidized to acetic acid. In the course of the reaction the S–S bond is ruptured, and a hydride ion is added to one sulfur atom while an acetyl group is added to the other.

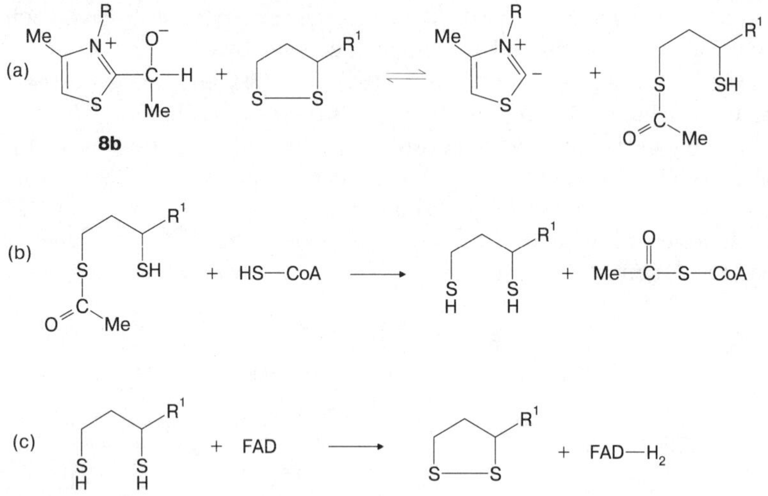

Figure 4.21 Oxidative decarboxylation.

In metabolic processes, acetyl groups are transported via coenzyme A. In the present case, S-acetyllipoic acid reacts with one of the CoA-containing enzymes with subsequent acyl group transfer to the coenzyme (Figure 4.21b).

In the process lipoic acid is converted into the dithiol form. The reaction sequence ends with the oxidation of the reduced form of lipoic acid to the cyclic form by means of the FAD coenzyme (Figure 4.21c).

Pyridoxal phosphate as an amino group carrier

Normal metabolism in man and other animals requires 20 amino acids. Vertebrates are able to synthesize half this number, while the remaining essential α-amino acids must be supplied by the diet. A protein diet is the main source of these essential amino acids. Proteins are hydrolyzed by the digestive system of the organism to amino acids which then undergo complex chemical transformations monitored by dozens of different enzymes. Transaminases are the most important of these enzymes; they function as delivery systems by transferring amino groups and thus interconverting α-amino acids and α-keto acids:

$$
\underset{\underset{NH_2}{|}}{R-CH-COOH} + \underset{\underset{O}{\|}}{R'-C-COOH} \underset{}{\overset{enzyme}{\rightleftharpoons}} \underset{\underset{O}{\|}}{R-C-COOH} + \underset{\underset{NH_2}{|}}{R'-CH-COOH}
$$

Pyridoxal phosphate, a derivative of pyridine-4-aldehyde, serves as a coenzyme for all of the transaminases (Figure 4.22). Enzymatic reactions occur with the direct participation of the aldehyde group which is transformed into an azomethine group (CH=N) by interaction with α-amino acids. The ease of this reaction is due to the interaction between the aldehyde group and the pyridine nitrogen. The electron-withdrawing effect of the nitrogen generates a partial positive charge at the 4-position of the pyridine ring (Figure 2.1), which in turn increases the positive charge on the carbon atom of the aldehyde group. This transfer of electron density substantially facilitates nucleophilic attack by the α-amino acid nitrogen atom at the aldehyde carbon.

A characteristic of the azomethine (**10a**) formed in the first step is the dramatic increase in C–H bond acidity at the α-position of the amino acid residue over the corresponding free amino acid. The reason for this is that the carbanion produced by C–H bond ionization is greatly stabilized by the delocalization of negative charge over the whole π-system of the molecule.

The carbanionic structure may be represented as a set of resonance structures (Figure 4.23). The pyridine ring, azomethine group and amino acid residue all lie in the same plane to maximize conjugation. This arrangement is achieved by the formation of a six-membered ring stabilized by intramolecular hydrogen bonding between the 3-position hydroxy group and the azomethine nitrogen. Pyridoxal phosphate is able to catalyze transamination only with the participation of this hydroxy group.

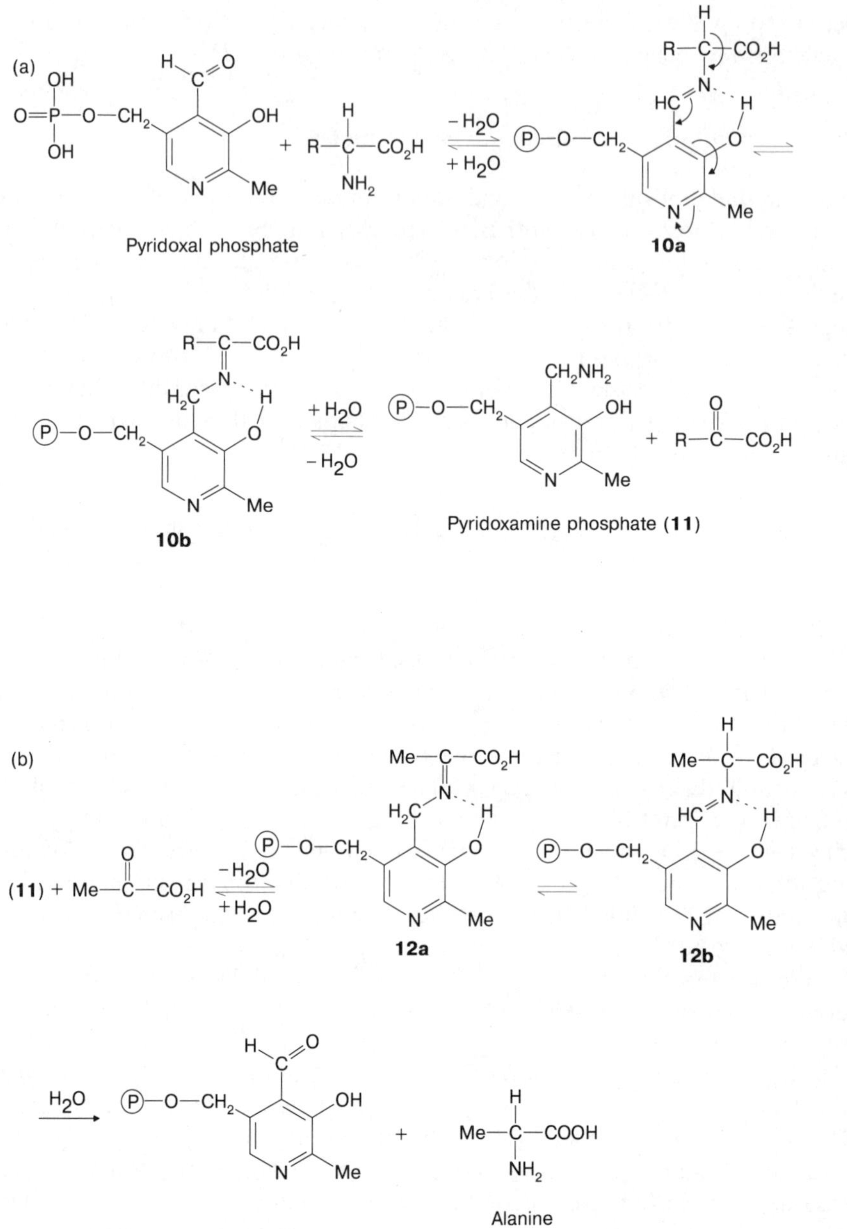

Figure 4.22 (a) Transamination with pyridoxal phosphate participation ($P = H_3PO_2$). (b) Amino group transfer from α-amino acid to pyruvic acid with the formation of alanine.

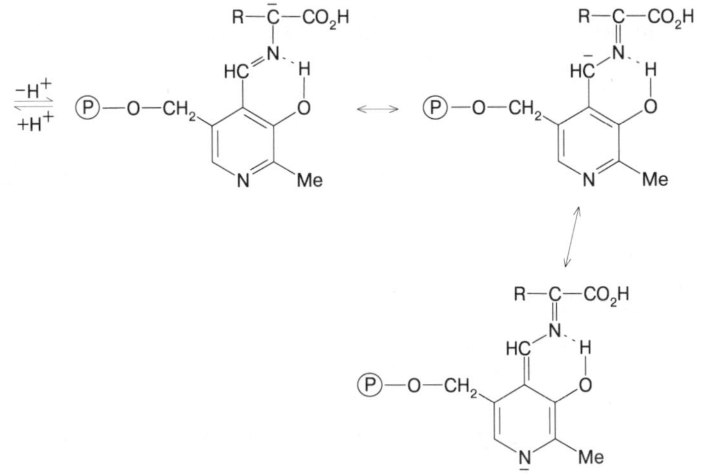

Figure 4.23 Delocalization of the negative charge in the carbanion produced from compound (**10a**).

The higher acidity of the C–H bond in compound (**10a**) favors proton migration to the azomethine carbon and the consequent formation of tautomer (**10b**) in which the azomethine double bond occupies a new position between the nitrogen and carbon of the amino acid residue. Subsequent hydrolysis of structure (**10b**) yields pyridoxamine phosphate (**11**) and a keto acid, the product of oxidative deamination of the parent α-amino acid (Figure 4.22a). Pyridoxamine phosphate (**11**) further reacts with pyruvic acid to produce a new azomethine (**12a**) which tautomerizes to (**12b**). Hydrolysis of (**12b**) leads to the regeneration of pyridoxal phosphate and the formation of alanine (Figure 4.22b).

The amino groups of all the essential α-amino acids are likewise transformed, with simultaneous conversion of pyruvic acid into alanine, each amino acid being controlled by its own transaminase. The alanine itself also undergoes deamination to re-form pyruvic acid; in this case, α-ketoglutaric acid serves as the amino group acceptor (Figure 4.24a) and is converted into glutamic acid. Glutamic acid is then oxidatively deaminated in a process that regenerates α-ketoglutaric acid, this process being mediated by glutamate dehydrogenase and its coenzyme NAD$^+$ (Figure 4.24b).

NAD$^+$ abstracts a hydride ion from the α-carbon of glutamic acid, thus converting it into the corresponding immonium salt which, like all immonium salts, is readily hydrolyzed to liberate α-ketoglutaric acid and ammonia (Figure 4.25). Therefore, the metabolism of all amino acids involves the liberation of nitrogen as ammonia.

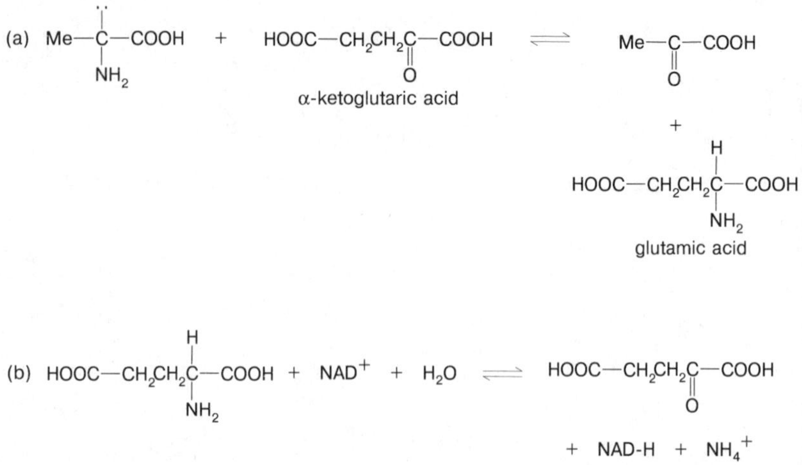

(a) $Me-\overset{\cdot\cdot}{\underset{NH_2}{C}}-COOH$ + $HOOC-CH_2CH_2\overset{\parallel}{\underset{O}{C}}-COOH$ $\rightleftharpoons$ $Me-\overset{\parallel}{\underset{O}{C}}-COOH$

α-ketoglutaric acid

+

$HOOC-CH_2CH_2\overset{H}{\underset{NH_2}{C}}-COOH$

glutamic acid

(b) $HOOC-CH_2CH_2\overset{H}{\underset{NH_2}{C}}-COOH$ + NAD^+ + H_2O $\rightleftharpoons$ $HOOC-CH_2CH_2\overset{\parallel}{\underset{O}{C}}-COOH$

+ NAD-H + NH_4^+

Figure 4.24   (a) Alanine deamination. (b) Oxidative deamination of glutamic acid.

It follows that the oxidation pathways of all α-amino acids include a 'narrow gate' formed by the alanine–glutamic acid pair. This 'narrow gate' is due to the presence within the organism of an active and highly specific enzyme which targets glutamic acid only.

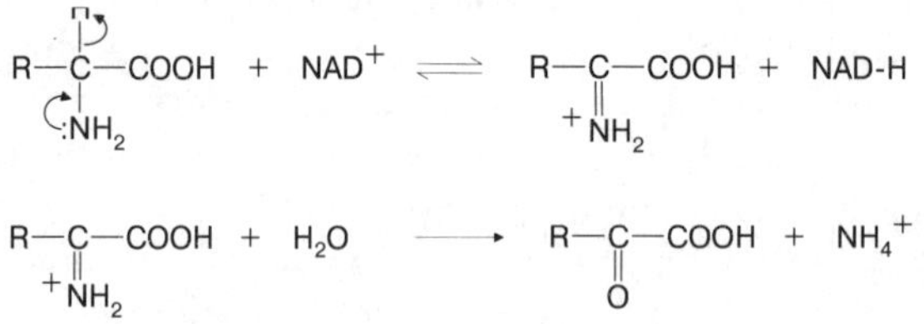

$R-\overset{\cdot\cdot}{\underset{\overset{\cdot\cdot}{NH_2}}{C}}-COOH$ + NAD^+ $\rightleftharpoons$ $R-\overset{}{\underset{\overset{+}{NH_2}}{C}}-COOH$ + NAD-H

$R-\overset{}{\underset{\overset{+}{NH_2}}{C}}-COOH$ + H_2O $\longrightarrow$ $R-\overset{\parallel}{\underset{O}{C}}-COOH$ + NH_4^+

Figure 4.25   The two-step sequence of α-amino acid oxidative deamination.

To complete our discussion of pyridoxal phosphate, we note that pyridoxal phosphate also catalyzes other transformations of α-amino acids, specifically decarboxylations, racemizations and aldol condensations. All of these conversions commence with azomethine formation.

A coenzyme from spinach leaves

Heterocyclization reactions in which heterocycles are formed from acyclic compounds, and ring interconversions in which one heterocyclic ring is formed from another or from a carbocycle, are some of the most interesting and specific reactions of heterocyclic chemistry. So far, we have not mentioned these and have paid attention mainly to reactions which do not involve the heterocyclic skeleton.

There are numerous methods for heterocycle formation from acyclic compounds. The preparations of benzimidazoline (2,3-dihydrobenzimidazole) and benzimidazole from *o*-phenylenediamine on treatment with formaldehyde and formic acid, respectively, are simple examples of heterocyclization (Figure 4.26). Analogous processes are usually accompanied by the formation of intermediate compounds which are often not isolable (generally depicted in square brackets in reaction schemes). Thus, in the benzimidazoline reaction the imine base (**13**) is an intermediate, and in the benzimidazole reaction *N*-formyl-*o*-phenylenediamine (**14**) is an intermediate.

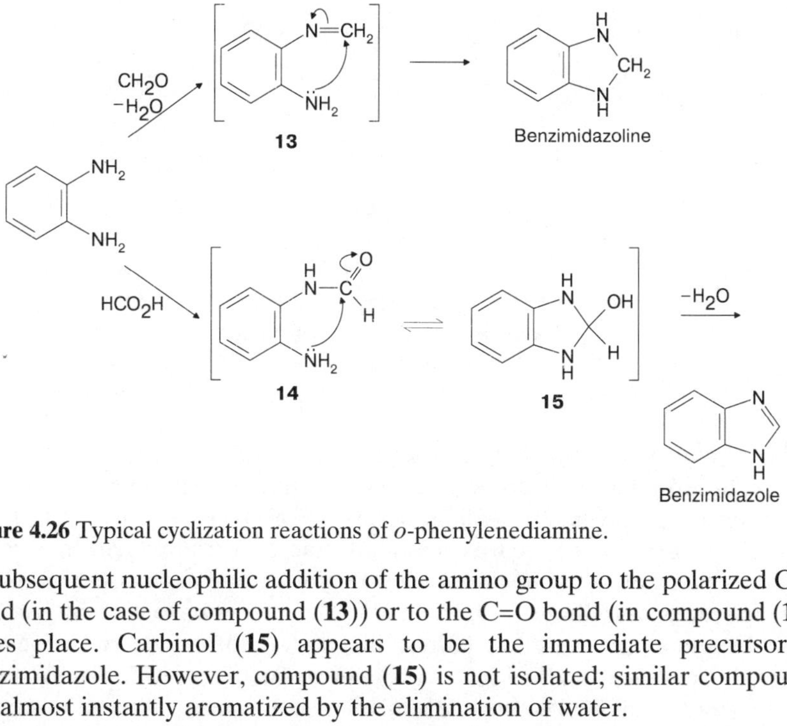

Figure 4.26 Typical cyclization reactions of *o*-phenylenediamine.

Subsequent nucleophilic addition of the amino group to the polarized C=N bond (in the case of compound (**13**)) or to the C=O bond (in compound (**14**)) takes place. Carbinol (**15**) appears to be the immediate precursor of benzimidazole. However, compound (**15**) is not isolated; similar compounds are almost instantly aromatized by the elimination of water.

In contrast to heterocyclization, ring interconversion reactions are accompanied by drastic alterations of the heterorings. Such reactions may occur with enlargement or contraction of the initial ring. Alternatively, the ring size can be preserved but its nature changed. Most ring interconversions are complex multistage processes which involve heteroring opening and subsequent cyclization. A familiar case is the conversion of pyrylium salts into pyridines upon treatment with aqueous ammonia (Figure 4.27a). In the first step one molecule of ammonia attacks the highly electron-deficient α-carbon atom of the pyrylium cation resulting in adduct (16). Ring opening with the formation of acyclic products (17) is characteristic of such covalent adducts. Intermediate (17) then undergoes recyclization with participation of the amino and carbonyl groups.

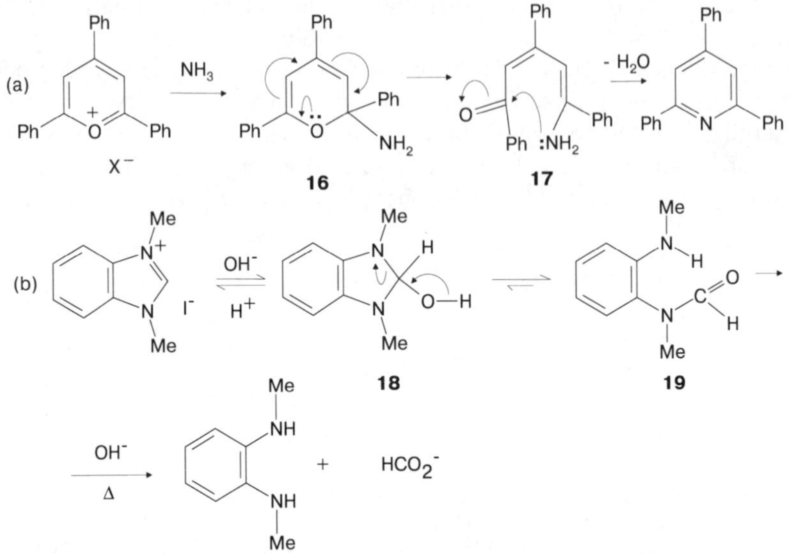

Figure 4.27 Examples of heterocyclic cation transformations: (a) recyclization of the 2,4,6-triphenylpyrylium cation into triphenylpyridine and (b) elimination of the –CH= group from the imidazole ring in the 1,3-dimethylbenzimidazolium cation.

The majority of heteroring interconversions are initiated by nucleophilic attack. In certain circumstances a nucleophile may even eliminate a portion of the heteroring, and this frequently occurs when quaternary salts are treated with alkali. Even at ambient temperature, the 1,3-dimethylbenzimidazolium cation is attacked at the 2-position by hydroxide ion to form carbinol (18), also called a pseudobase (Figure 4.27b). The pseudobase, like adduct (16), exists predominantly in the acyclic tautomeric form (19), which can be isolated. However, on heating, the alkali deformylates compound (18) to produce N,N'-dimethyl-o-phenylenediamine and formic acid. The reaction as a whole represents the formal elimination of a –CH= group from the imidazole ring. Examples of fragment eliminations from neutral heterocycles are also encountered, though they are not so common.

Cyclization and ring transformation reactions are widespread in heterocyclic chemistry and are utilized extensively in synthesis. It is not surprising that transformations of these types are found in a whole series of biochemical reactions in which insertions of one-carbon units such as Me, CH_2OH, CHO or –CH= occur into a host of different molecules. Enzymes which transfer these groups use folic acid derivatives as coenzymes. The acid was named after the Latin word *folium* (leaf), for it was originally isolated from spinach leaves. As can be seen from Figure 4.28, folic acid is a derivative of pteridine (see Figure 4.12) which has a complex substituent combining *p*-aminobenzoic acid and glutamic acid residues at the 6-position.

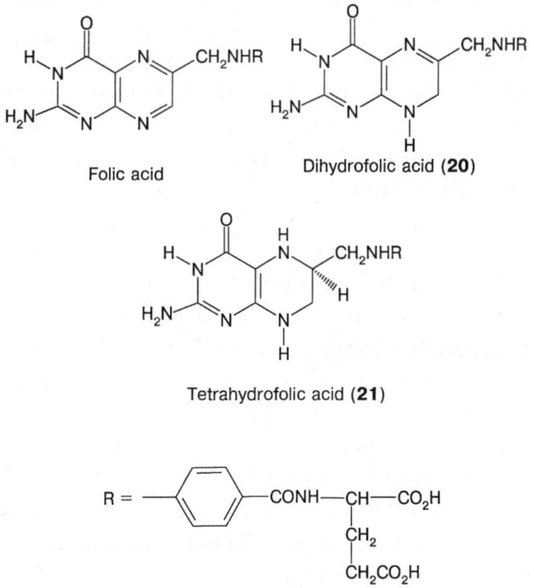

Figure 4.28 Folic acid and its hydro derivatives.

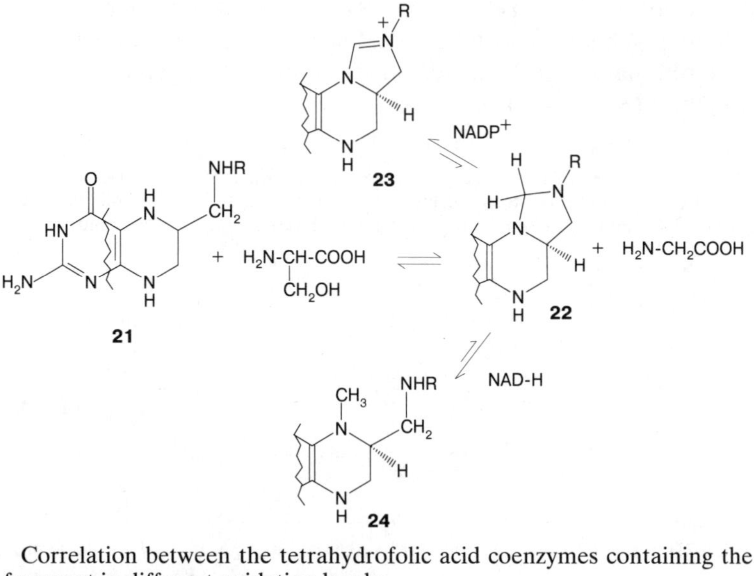

Figure 4.29 Correlation between the tetrahydrofolic acid coenzymes containing the one-carbon fragment in different oxidation levels.

Folic acid itself does not participate in biochemical reactions; the tetrahydro and, less often, the dihydro derivatives are the biologically active forms.

Tetrahydrofolic acid is first 'loaded' with a one-carbon fragment donated by the amino acid serine. In the course of this reaction, enzymatically catalyzed by serine transhydroxymethylase, tetrahydrofolic acid is converted into 5,10-methylenetetrahydrofolic acid (22) and serine is transformed into glycine (Figure 4.29) with the formal loss of a formaldehyde molecule.[†] The cyclization itself resembles the formation of benzimidazoline from *o*-phenylenediamine and formaldehyde (Figure 4.26). Similar to *o*-phenylenediamine, tetrahydrofolic acid contains the NHCCNH fragment of a vicinal diamine.

5,10-Methylenetetrahydrofolic acid (22) undergoes two important transformations within the organism. It can be oxidized to become 5,10-methenyltetrahydrofolic acid (23) by loss of a hydride ion to a dehydrogenase which contains NADP$^+$ as the coenzyme (see Figure 4.29). Alternatively, by reaction with a reductase utilizing the NAD-H coenzyme, the acid is reduced with cleavage of the imidazole ring to form 5-methyltetrahydrofolic acid, i.e. (24). All three of these derivatives of tetrahydrofolic acid, (22)–(24), exhibit coenzymatic action. They differ from each other in the degree of oxidation of the one-carbon unit: in coenzyme 22 it is at the formaldehyde level, whereas in 23 the oxidation level is that of formic acid, and in 24 that of methanol.

We now consider several instances of enzyme-mediated reactions which proceed with the participation of coenzymes (22)–(24). One of the most important of those promoted by 5,10-methylenetetrahydrofolic acid is the conversion of 2′-deoxyuridylic acid (26) by *C*-methylation of the uracil ring into the 2′-deoxythymidylic acid (27) (Figure 4.30a). In the course of this transformation, coenzyme 22 is converted back into dihydrofolic acid (20), and the regeneration of tetrahydrofolic acid (21) necessitates participation of an additional enzyme, dihydrofolate reductase, together with its coenzyme NADP-H (Figure 4.30b).

---

†Pyridoxal phosphate is a coenzyme of serine transhydroxymethylase. In general terms the formaldehyde formation from serine can be represented by the following scheme (Enz = enzyme, B: = a basic site in the enzyme, e.g. the imidazole ring of a histidine)

Anion 25b is then protonated and hydrolyzed to produce glycine and pyridoxal phosphate.

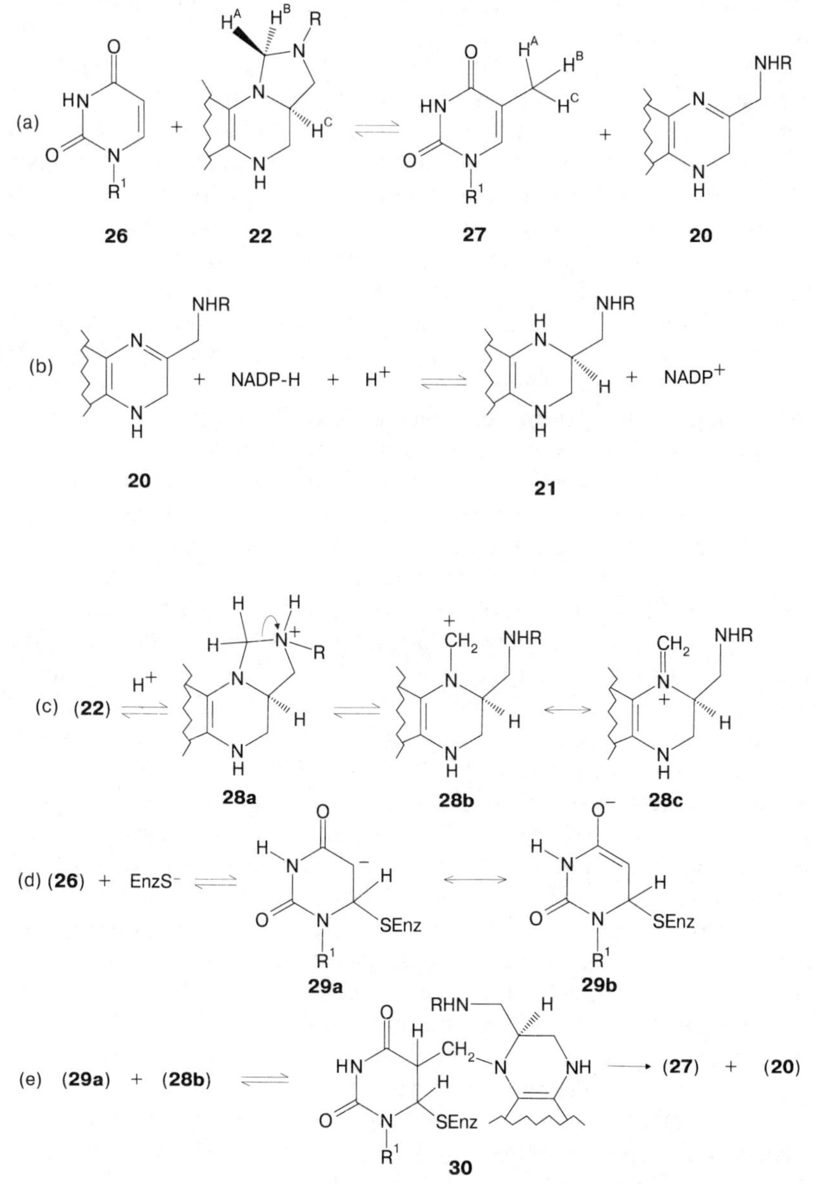

Figure 4.30 Transformation of 2′-deoxyuridylic acid into 2′-deoxythymidylic acid (R¹ = the phosphorylated deoxyribose residue; see Figure 3.2): (a) general overview; (b) regeneration of tetrahydrofolic acid; (c) transformation of coenzyme **22** into its active methylating form **28b**, (d) activation of the uracil ring by enzyme addition; and (e) proposed mechanism of *C*-methylation of the uracil nucleus.

At first sight, coenzyme (22) appears to behave as though it introduces the carbon atom at the methanol oxidation level rather than at that of formaldehyde. This is not the case because the methyl group is created in the uracil ring in stages. Initially, the coenzyme fashions a methylene link containing H^A and H^B atoms (formaldehyde level of oxidation), the H^C atom, first connected to the enzyme C-6 atom, is transferred only at the next step. The details of the process are believed to be as follows. Coenzyme 22 functions in this reaction as the open chain analogue with a positively charged methyleneimmonium group (28) (in which the carbenium resonance structure (28b) participates) rather than as the cyclic imidazolidine (22) or its protonated form (28a) (Figure 4.30c). 2'-Deoxyuridylic acid in its turn is activated at the C-6 atom by a cysteine residue in an ancillary enzyme (Figure 4.30d). The nucleophilicity of the C-5 center in the corresponding intermediate, which can be represented as a resonance hybrid of structures 29a and 29b, is thus greatly enhanced. The methylene group of the coenzyme 28b attacks the uracil ring at position C-5, and the addition product 30 then undergoes stepwise proton and enzyme elimination, C–N bond cleavage, and hydride ion transfer from the C-6 atom of the coenzyme to give finally 2'-deoxythymidylic (27) and dihydrofolic (20) acids (Figure 4.30e). This is, of course, a simplified formulation as the real processes are synchronous.

A major function of coenzyme 23 is its involvement in purine biosynthesis (Figure 4.31). The imidazole and pyrimidine ring closures require the incorporation of the C-8 and C-2 atoms.[†] In the course of these transformations coenzyme 23 undergoes certain preliminary changes. As previously mentioned, treatment of imidazolium salts with alkali produces pseudobases which exist primarily in the acyclic form containing an N-formyl group (Figure 4.27, structure (19)). A similar situation occurs with 5,10-methenyltetrahydrofolic acid (23), which first adds hydroxide ion to yield a carbinol (31a) and then is stabilized by ring opening to one of the acyclic forms (31b) or (31c). The 10-formyl derivative 31c is comparatively the less stable but more active of the pair and is responsible for the transfer of formyl groups to N-ribonucleotides of glycineamide (32) and 5-amino-4-carbamoylimidazole (35). The intermediate N-formyl derivatives (33) and (36) undergo cyclization by dehydration (Figure 4.31b and 4.31c). The cyclized product from 36 is inosine which, in the course of subsequent reactions, provides the organism with all the necessary purine bases.

The integration of a methyl group into bioorganic molecules is carried out by means of 5-methyltetrahydrofolic acid (24), a coenzyme at the methanol oxidation level. A typical example is the synthesis of methionine from homocysteine (Figure 4.32). The enzyme homocysteine methyltransferase

---

[†]The numbering of the atoms in the purine system does not coincide with that in imidazole (Figure 2.16). In particular, the C-8 atom of purine corresponds to the C-2 atom of imidazole.

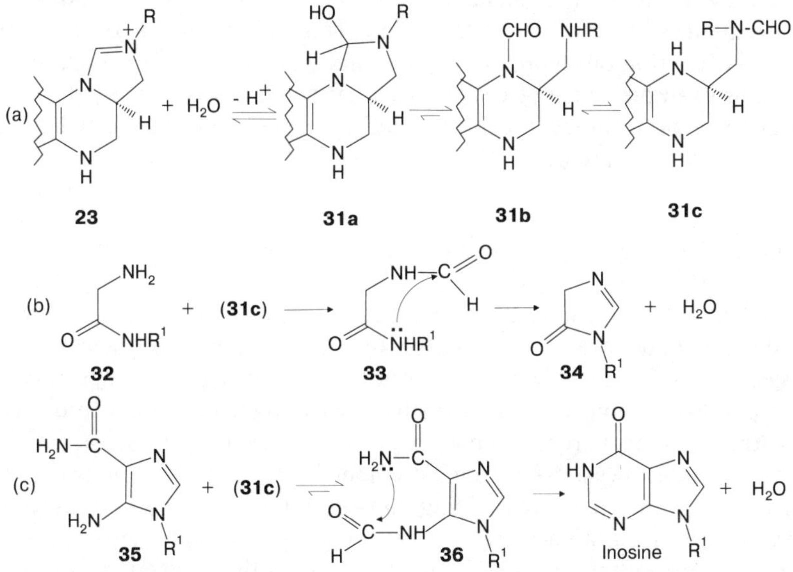

Figure 4.31 Introduction of one-carbon fragments with the assistance of coenzyme 23 in the course of imidazole and pyrimidine ring closures during the biosynthesis of purines (R1 = residue of phosphorylated ribose or deoxyribose; the scheme does not include the biochemical reaction sequence leading to compound **35**).

participates in the reaction alongside the coenzyme methylcobalamin which has a porphyrin-like structure incorporating a trivalent cobalt atom (see Section 4.3). The cobalt coordinates with the four pyrrole ring nitrogens and one of the nitrogen atoms of a 5,6-dimethylbenzimidazole unit. The sixth coordination site is essentially available to other ligands. In methionine formation, the methyl group of 5-methyltetrahydrofolic acid is first

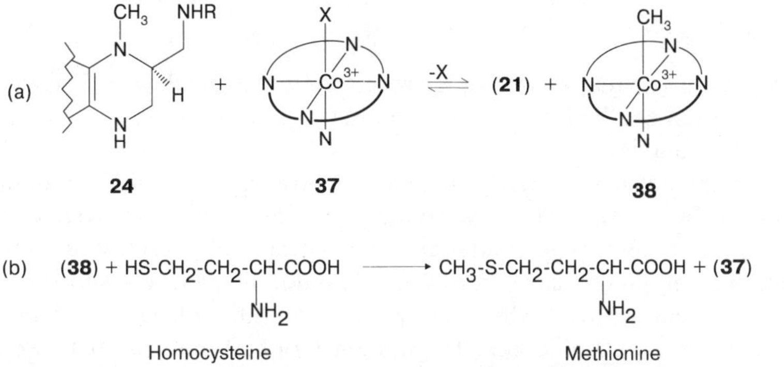

Figure 4.32 Formation of methionine from homocysteine involving methyl group transfer by coenzyme 24 and methylcobalamin.

transferred to the methylcobalamin molecule to give a Co–Me linkage, with displacement of the sixth (mobile) ligand. This Me group is later transferred to the SH function of homocysteine. Many details of this process are still not clear. For example, it is not yet known what promotes N–Me bond cleavage in coenzyme **24**, or in what form the methyl group is transferred (whether as cation, anion or radical).

Biotin as a carboxylic group carrier

Carbon–carbon bond formations are the cornerstone of organic chemistry, and the planning of any synthesis of an organic compound, whether it be simple or complex, must consider carefully the various possibilities of carbon skeleton construction. In living organisms carbon–carbon bonds are formed biosynthetically by a great diversity of constructions in an immense variety of bioorganic substances. As a rule, fragments containing one or two carbon atoms serve as the chemical building units used by a host of enzymes to form new C–C bonds. In this way coenzyme A combines acetate units into chains while synthesizing fatty acids. The mechanisms of these reactions are similar to those already described for the cleavage of fatty acids, but the sequence is now reversed (refer back to Figure 4.16). Thiamine and pyridoxal phosphate mediate C–C bond formation by means of acyloin and aldol condensations, respectively (Figure 4.19).

In the preceding section we considered the effect of the folic acid coenzymes, catalyzing the extension of a carbon skeleton by one atom in the synthesis of thymine from uracil (Figure 4.30). Carboxylation of organometallic compounds, to give the corresponding carboxylic acids, is one of the simplest and most widespread of reactions, both in nature and in the laboratory.

$$R^-M^+ + O{=}C{=}O \xrightarrow{} \underset{\delta+}{} R\text{-}C\text{-}O^-M^+ \xrightarrow{H_3O^+} R\text{-}C\text{-}OH$$

Just as for the majority of other reactions which produce new C–C bonds, carboxylation can be interpreted as a nucleophilic addition of a carbanion to a carbonyl function.

In an organism, the CO_2 molecule is transported and simultaneously activated by biotin. This coenzyme contains a bicyclic system of two condensed heterocyclic moieties, the completely saturated nuclei of thiophene and imidazole (Figure 4.33). The thiophene ring is substituted with a valeric acid residue which allows the biotin molecule to bind to an apoenzyme via an amide link. The influence of biotin is crucial at two stages. In the first stage a bicarbonate anion (equivalent to a CO_2 molecule) binds to the heterocyclic nitrogen. In the second stage this loosely held and activated carboxy group is transferred to the carbanion. In this fashion, the biosynthesis is achieved of oxaloacetic acid from pyruvic acid (for details see Section 5.2.2). Another enzyme, working jointly with pyridoxal phosphate as the coenzyme, facilitates the generation of the pyruvic acid carbanion. As regards the addition of the bicarbonate anion to the cyclic amide nitrogen, it should be noted that carbonic acid, as well as the bicarbonate anion, is a hydrated form of carbon dioxide, and the three species exist in equilibrium. Therefore, in the reaction scheme the HCO_3^- ion may be interchanged with CO_2. Furthermore, all of the reaction centers of an enzyme–substrate complex are usually activated simultaneously; hence, in the transition state, the N–H bond in the amide and the C–OH bond in the HCO_3^- anion are polarized synchronously. This leads to a significant increase in both the nitrogen nucleophilicity, and the carbon electrophilicity and, as a consequence, in the rate of the addition reaction.

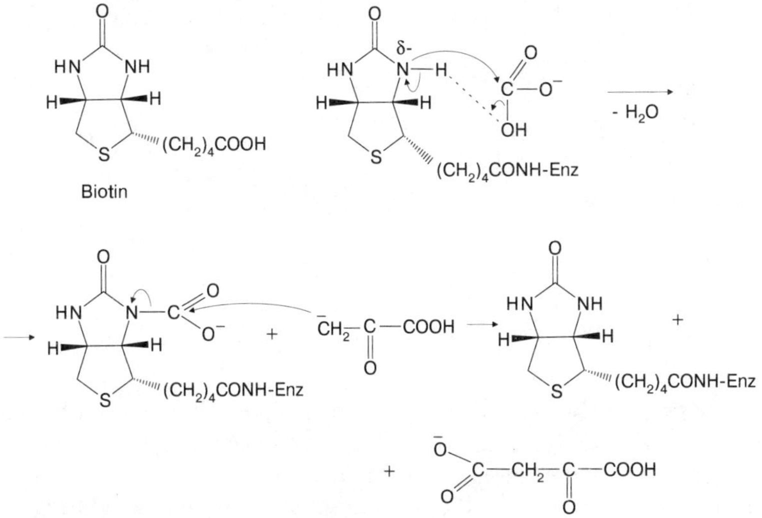

Figure 4.33 The transformation of pyruvic acid into oxaloacetic acid as effected by biotin (Enz = apoenzyme residue).

4.3 Vitamins, the 'Molecules of Health'

We have seen that coenzymes are ubiquitous in their initiation and coordination of the chemistry of living organisms. We now consider the origin of the coenzymes. Major portions of coenzyme structures cannot be synthesized by human or other vertebrate organisms and must therefore be obtained from the diet. It has long been noticed that many of the vital precursors from which organisms are able to synthesize the required coenzymes contain nitrogen and may be classified as amines, more or less.

The Polish scientist Funk named these substances vitamins (Latin: *vita*, life). For example, nicotinic acid and its amide (vitamin B_5) are precursors of the pyridine-dependent dehydrogenases NAD^+ and $NADP^+$. Similarly, pyridoxal phosphate is biosynthesized from nutritionally supplied pyridoxine, pyridoxal and pyridoxamine (i.e. vitamin B_6, Figure 4.34). In some cases, vitamins are the actual coenzymes rather than their precursors: the coenzymes thiamine, riboflavin, biotin and folic acid are indeed vitamins B_1, B_2, B_7 and B_C, respectively.

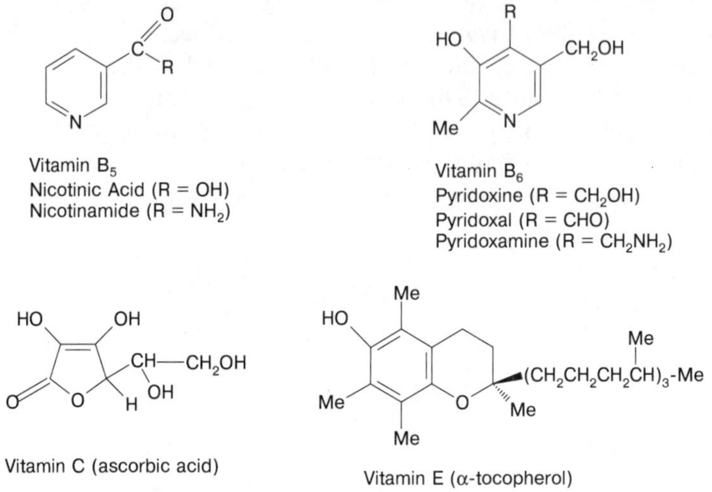

Vitamin B_5
Nicotinic Acid (R = OH)
Nicotinamide (R = NH_2)

Vitamin B_6
Pyridoxine (R = CH_2OH)
Pyridoxal (R = CHO)
Pyridoxamine (R = CH_2NH_2)

Vitamin C (ascorbic acid)

Vitamin E (α-tocopherol)

Figure 4.34 Some heterocyclic vitamins.

The biological activity of vitamins is so high that only a few milligrams of each are required daily. Thus, the daily requirements of nicotinic acid, riboflavin and pyridoxine are 15–20, 1–3 and 1–2 mg, respectively. Vitamin B_{12} shows even higher activity and the daily dose necessary for an adult is as low as 0.1-0.2 mg. This vitamin deserves special attention not only for its high activity but also because of its extraordinarily complicated chemical structure.

The distinguishing feature of vitamin B_{12} (cobalamin) is the ability of the cobalt(III) ion to coordinate with six donor ligands (Figure 4.35). The so-called corrin system serves as the main tetradentate ligand in this vitamin. Corrin resembles the porphyrins but differs in that two of the four pyrrole rings in corrin are directly connected (A and D rings), not through a –CH= bridge as in the case of porphyrins. Moreover, the pyrrole rings in the corrin system are partially reduced. Vitamin B_{12} is produced by anaerobic bacteria located in the alimentary canal of ruminants; therefore, corrin biosynthesis does not include steps which would require oxygen to produce a more highly oxidized aromatic porphyrin-type structure.

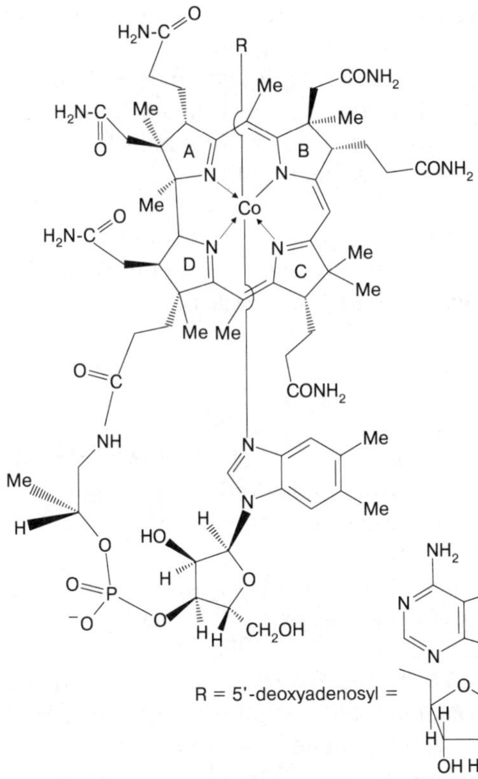

Figure 4.35 Vitamin B$_{12}$ (cobalamin) and its modifications: cyanocobalamin (R = CN), hydroxycobalamin (R = OH), methylcobalamin (R = Me) and adenosylcobalamin (R = 5'-deoxyadenosyl).

The fifth coordination site, oriented almost normal to the corrin plane, binds the cobalt atom with the N-3 atom of a 5,6-dimethylbenzimidazole unit. This unit is attached by a long chain formed by ribophosphate and amino acid units. The sixth ligand (R substituent in Figure 4.35) can vary, giving rise to several vitamin B$_{12}$ modifications including cyanocobalamin, hydroxycobalamin and adenosylcobalamin. All of these analogues are present in organisms, the major component (about 50%) being the 5'-deoxyadenosyl derivative. A remarkable feature of methylcobalamin and adenosylcobalamin is the carbon–cobalt linkage. These coenzymes are rare examples of naturally occurring organometallic compounds.

What are the functions of vitamin B$_{12}$? Its methyl group carrier properties have already been discussed in the synthesis of methionine. The vitamin is believed to have many other functions, among them rearrangement reactions, some of which are not yet clearly understood. However, the most important role involves the generation and management of blood. A lack of vitamin B_{12} causes pernicious anemia, one of the most serious of blood diseases. The main dietary source of this vitamin is meat, especially liver.

Other vitamins with a heterocyclic structure are known. Some of the most familiar are vitamin C (ascorbic acid) and vitamin E (α-tocopherol), which contain tetrahydrofuran and tetrahydrobenzopyran rings, respectively (Figure 4.34).

To sum up this chapter we would like to add to the ancient wisdom '*memento de mortis*' a relatively new biochemical maxim '*memento de vitaminum*', or even '*memento de heterocyclus*'.

4.4   Problems

1. 2,6-Dimethyl-3,5-diethoxycarbonyl-1,4-dihydropyridine does not reduce benzaldehyde or *m*-nitrobenzaldehyde under normal conditions. In contrast, salicylic and especially 5-nitrosalicylic aldehydes are readily reduced. Account for this fact.

2. The reduction of flavins by NAD-H is catalyzed by magnesium(II) and zinc(II) ions. Suggest a mechanism for the catalysis.

3. Xanthine oxidase, a key enzyme which controls the levels of all purines in living organisms, assists the oxidation of hypoxanthine, xanthine and their precursors (adenine and guanine) to uric acid. Write the stepwise mechanism of the hypoxanthine $\longrightarrow$ xanthine reaction (hint: the first step involves hydration of the C=N bond; FAD serves as the cofactor).

4. What role does the 3-hydroxy substituent play in pyridoxal phosphate?

5. Draw a scheme showing the decarboxylation of tryptophan to form tryptamine catalyzed by pyridoxal phosphate.

6. Which heterocyclic coenzymes participate in the formation of new carbon–carbon bonds during biochemical reactions?

7. Which widespread heterocyclic fragment is found in the structures of many coenzymes, vitamins and energy-rich compounds (see also Chapter 5)?

8. Coenzyme PQQ (Figure 4.8b) was first isolated as the product of nucleophilic addition of acetone to one of its carbonyl groups. Similar products are formed by PQQ with many other nucleophiles such as methanol. Which one of the two carbonyl groups in PQQ is more active? Draw the structures of the PQQ adducts with acetone and methanol.

9. The biologically important decarboxylation reaction has been extensively studied using the conversion of benzisoxazole-3-carboxylates (A) into *o*-cyanophenolate anions (B) as a model. These species are convenient model compounds as they are not susceptible to general acid–base catalysis but are sensitive to other media conditions. Predict how the rate of the enzyme-catalyzed decomposition of carboxylates (A) would

change if the protic solvent was replaced by an aprotic one?

A   B

10. The biochemical pathways of fertilization include protection of the embryo via oxidative crosslinking of its protein surface, as demonstrated in sea urchin eggs. The mechanism requires the bioproduction of hydrogen peroxide (which has a deleterious effect on the embryo) as regulated by the enzyme oxidase. H_2O_2 evolution results from an increased oxygen uptake by the fertilized egg (called the 'respiratory burst of fertilization')

$$NADP\text{-}H + H^+ + O_2 \xrightarrow{\text{oxidase}} NADP^+ + H_2O_2$$

The enzyme peroxidase catalyzes the consumption of hydrogen peroxide in conjunction with the intracellular family of antioxidants known as ovothiols (C). (a) To which widespread, naturally occurring compound are the ovothiols related? (b) How can H_2O_2 levels be maintained by the use of ovothiols? Indicate schematically the ovothiol conversion. (c) Provide a mechanism for regeneration of the consumed ovothiols.

$NR_2 = N(CH_3)_2$ (ovothiol A)

$NR_2 = NHCH_3$ (ovothiol B)

C

4.5 Suggested Reading

1. Lehninger, A. L., *Principles of Biochemistry*, 2nd Edn, Worth, New York, 1993.
2. Branden, C. and Tooze, J., *Introduction to Protein Structure*, Garland, New York, 1991.
3. Dordick, J. S. (ed.), *Biocatalysts for Industry*, Plenum Press, New York, 1991.

4. Wyman, J. and Gill, S. J., *Binding and Linkage. Functional Chemistry of Biological Macromolecules*, University Science, Mill Valley, CA, 1990.
5. Stryer, L., *Biochemistry*, 3rd Edn, Freeman, New York, 1988.
6. Ferdinand, W., *The Enzyme Molecule*, Wiley, Chichester, 1987.
7. Jencks, W. P., *Catalysis in Chemistry and Enzymology*, Dover, New York, 1987.
8. Prentis, S., *Biotechnology: A New Industrial Revolution*, Orbis, London, 1984.
9. Godfrey, T. and Reichelt, J., *Industrial Enzymology: The Application of Enzymes in Industry*, Nature Press, New York, 1983.

5 HETEROCYCLES AND BIOENERGETICS

I flow like a river, as lilac I bloom,
Silent as a shadow, I flame as a bonfire,
Blaze as the Sun, and shine as the Moon
And flutter as a colored butterfly.

I. Severyanin

All living organisms need energy for such diverse purposes as nerve impulses, movement, nutrient transport and maintenance of body temperature. Energy is also consumed in the biosynthesis of such complex molecules as nucleic acids, proteins, polysaccharides, phospholipids, steroids and so on. In fact, all living matter on Earth owes its existence to solar energy. However, only plants and certain bacteria can directly assimilate solar energy. The physicochemical processes by which solar radiation is absorbed by plants are collectively called photosynthesis. This phenomenon will be dealt with in Chapter 6. The present chapter is devoted to the problems of energy supply in those living organisms without a photosynthetic system, i.e. the animal kingdom, including mankind.

Animals receive all of their required energy from food. In proteins, fats and carbohydrates—the main constituents of the diet—energy is stored in chemical bonds formed during photosynthesis. The process used by organisms for the release of energy from nutrient molecules is similar to that utilized by man to obtain heat from wood, coal or natural gas. This process is oxidation. The oxidizing agent used by the body is molecular oxygen inhaled during respiration. At some of the intermediate stages of biological oxidation, the role of oxidant may also be performed by organic molecules such as pyruvic acid.

Large quantities of energy are evolved in the oxidation of proteins and especially of fats and carbohydrates. For example, the combustion of one mole of palmitic acid to carbon dioxide and water liberates 2338 kcal, while one mole of glucose liberates 686 kcal:[†]

$$C_6H_{12}O_6 + 6O_2 \rightarrow 6CO_2 + 6H_2O + 686 \text{ kcal}$$

[†]Here and elsewhere, the energies quoted are $\Delta G°$ values. In the cell, of course, the true ΔG will be affected by the relative concentrations of the products and starting materials, according to the Nernst equation. Because in living cells products and starting materials are kept at near equilibrium values, the true ΔG is often significantly larger than the $\Delta G°'$.

It is clear that if such quantities of energy were to be liberated instantly, the organism would neither be able to utilize them completely nor to conserve the energy for the future. Therefore, in the course of evolution, living organisms have developed intricate oxidative–reductive mechanisms which enable slow, adjustable rates of oxidation of food. The basis of the oxidation–reduction system is a set of diverse enzymes, the chief among them being the pyridine- and flavin-dependent dehydrogenases and cytochromes. The heterocyclic compound adenosine triphosphate (ATP) is the main substance responsible for the accumulation and supply of energy for all cellular processes.

5.1   ATP as the Universal Currency of Energy

When we buy something, we usually pay with small change and bills of different values. Similarly, organisms pay for their expenditures of energy, small and large, with a type of 'molecular currency'. Such a system is required because physiological processes are based on chemical reactions with activation energies usually in the range of $7–15\,kcal\,mol^{-1}$. Thus, units of energy of the same order of magnitude need to be available.

The main molecular carrier of energy, the unit of the biological 'energy coinage', is the tetraanion of adenosine triphosphoric acid, namely adenosine triphosphate (ATP) (at physiological pH the acidic protons of adenosine triphosphoric acid are almost completely ionized). The mode of action of ATP is based on the hydrolytic elimination of one or both of the two terminal phosphate groups to form adenosine diphosphate (ADP) or adenosine monophosphate (AMP), respectively (Figure 5.1). These processes each release approximately $9\,kcal\,mol^{-1}$ of energy.

In some biochemical reactions, energy transfer is carried out by guanosine triphosphate (GTP), uridine triphosphate (UTP) or cytidine triphosphate (CTP). Though there is no evidence yet for the direct participation of the heterocyclic moieties of any of these compounds in the energy transfer, the presence of a purine unit in ATP suggests some function: the heterocyclic bases probably bind the ATP, GTP, UTP and CTP molecules (by nonbonding interactions) to the enzymes which transport them.

A number of effects facilitate the liberation of energy by hydrolysis of the pyrophosphate bonds in ATP†. Firstly, the negatively charged oxygen atoms mutually repel each other, and when the length of the pyrophosphate chain decreases, the repulsion between the remaining oxygen atoms becomes less

---

†The actual amount of energy liberated by hydrolysis of the pyrophosphate bonds in ATP is considerably influenced by the solvation energies. Because of the strong solvation of phosphate anion, the energy liberated is higher in a phase of low dielectric constant.

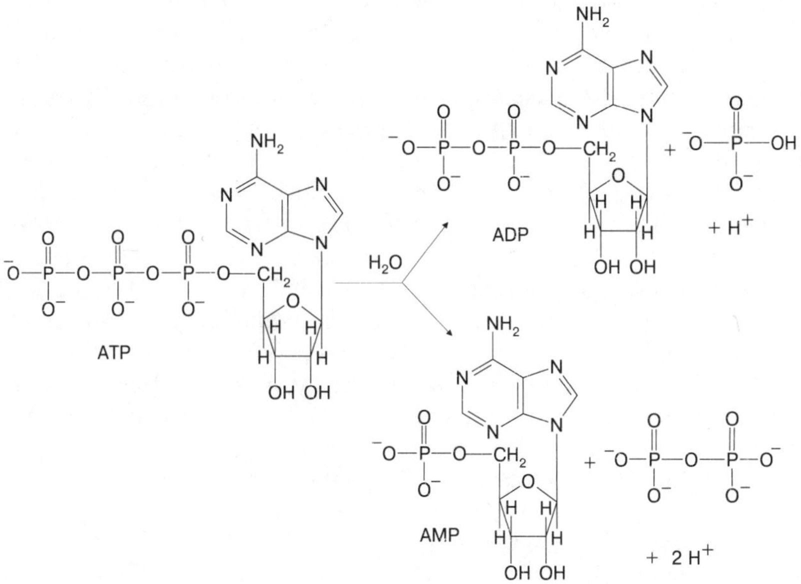

Figure 5.1 Hydrolytic cleavage of the phosphate groups of ATP with the formation of ADP and AMP.

pronounced. Moreover, the ATP hydrolysis provides one additional negatively charged oxygen, increasing the possibility of resonance stabilization in the resulting adenosine diphosphate and phosphate anion compared with the initial molecule of ATP. Overlapping of the unshared electron pairs of such oxygen atoms with the vacant $3d$-orbitals of the phosphorus atom gives rise to a substantial gain in energy:

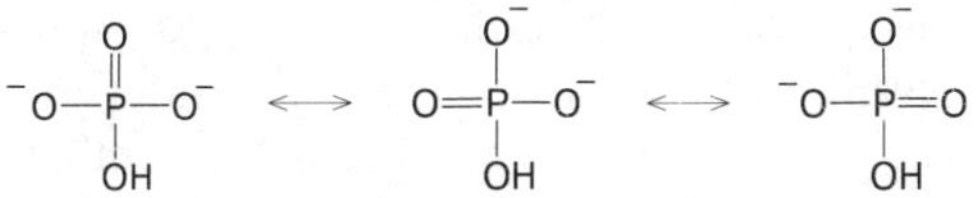

In biochemistry the energy balance of a reaction is preferably expressed not in terms of the heat evolved ΔH but as the free energy change $\Delta G°$. The $\Delta G°$ is connected to the equilibrium constant K by the expression:

$$\Delta G° = -RT \ln K$$

where R is the gas constant and T is the absolute temperature. When $\Delta G°$ is negative, the products of the reaction predominate in the equilibrium mixture. By contrast, the equilibrium is shifted in favor of the reactants when $\Delta G°$ is positive. Obviously, if $\Delta G° = 0$, then $K = 1$; that is, the concentrations of the products and the reactants are equal in the equilibrium mixture. Many biochemical reactions are unfavorable from the thermodynamic point of view. This means that under standard conditions their $\Delta G°$ values are positive. To shift the equilibrium toward product formation requires energy. The key problem becomes the mechanism of energy transfer from ATP to the reactants.

A constant temperature must be maintained for the normal functioning of the living cells of many organisms. Therefore, direct utilization of the heat from the hydrolysis of ATP for activation of the biochemical reactions in the cell is impossible. The only biochemical procedure to transfer energy is by the so-called coupling of separate reactions by means of a common intermediate. Such is the case, for example, in the esterification of carboxylic acids by alcohols. Under conventional conditions the equilibrium for this reaction is largely shifted toward the reactants. However, the introduction of ATP molecules into the reaction sequence allows the process to be modified to give more of the ester. The process may be represented as shown in Figure 5.2.

$$R—COOH \ + \ ATP \ \rightleftharpoons \ R—COO—AMP \ + \ PP, \quad \Delta G^0 = 0 \ kcal/mol$$

$$R—COO—AMP \ + \ R'—OH \ \rightleftharpoons \ R—COO—R' \ + \ AMP, \quad \Delta G^0 = -2 \ kcal/mol$$

Figure 5.2 Coupling of two reactions during the esterification of carboxylic acids in the presence of ATP.

The initial step represents an activation of the carboxylic acid. In the course of this activation a pyrophosphate (PP) unit separates from the ATP molecule and a mixed anhydride of the carboxylic acid and adenosine monophosphoric acid is formed. The energy previously stored in the ATP molecule is now conserved in a new P–O bond between the acyl group of the carboxylic acid and AMP. This bond is called an energy-rich bond and can be compared with a compressed spring ready to be released at any moment. The release of this 'spring' takes place in the second step when the mixed anhydride reacts with an alcohol to produce the ester and AMP. This step has a $\Delta G°$ value of $-2\,kcal\,mol^{-1}$ which shifts the equilibrium to the right-hand side of the equation, toward esterification. The following analogy may help. If you push your car (with a dead battery) up a steep incline by the shortest route, you must expend some effort. However, you will be compensated if you choose to go downhill by a longer route. Such is the general principle of ATP action as the energy source.

To conserve energy in ATP, ATP is synthesized by the addition of orthophosphate to ADP or pyrophosphate to AMP. Clearly, these reactions necessitate energy expenditures. The organic derivatives of phosphoric acid such as phosphoenolpyruvate, 1,3-bisphosphoglycerate, creatine phosphate and acetyl phosphate are all also energy-rich compounds. Indeed, they are all richer in energy than ATP, and can therefore operate as phosphate transfer agents on AMP and ADP (Figure 5.3 and Table 5.1). In muscle tissues, where energy consumption is very high, ATP molecules are replenished by means of phosphate group transfers from creatine phosphate to ADP.[†] Table 5.1 reveals that phosphoenolpyruvate (an ester of phosphoric acid and pyruvic acid in the enol form) possesses the most energy-rich phosphate group of all the carriers. The additional energy capacity results from the lability of the enol form, which instantly isomerizes to the more stable keto form following donation of the phosphate anion.

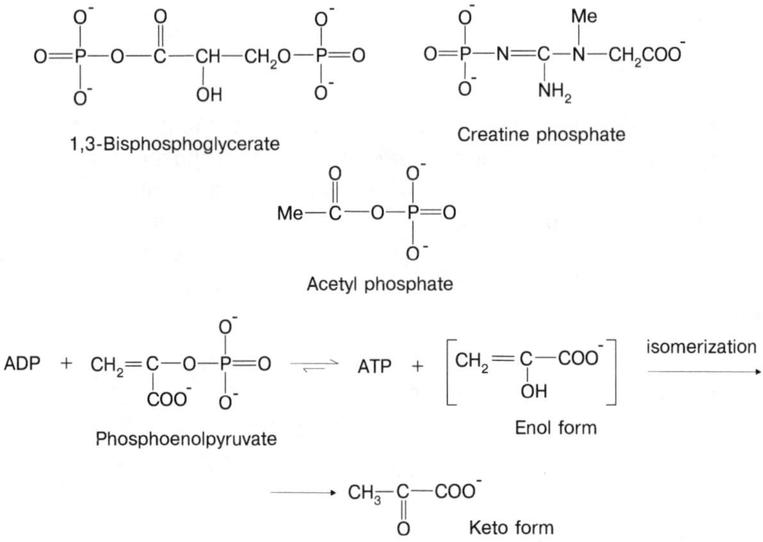

1,3-Bisphosphoglycerate

Creatine phosphate

Acetyl phosphate

Phosphoenolpyruvate

Enol form

Keto form

Figure 5.3 Natural energy-rich phosphates and phosphate group transfer from phosphoenolpyruvate to ADP.

Table 5.1 Standard free energies of hydrolysis ($\Delta G°$) for some biologically important phosphates (adapted from Lehninger, A. L., *Biochemistry*, Worth, New York, 1970, Chap. 14, p. 302, Table 14-4, with permission).

Compound	$\Delta G°$ (kcal mol^{-1})	Compound	$\Delta G°$ (kcal mol^{-1})
Phosphoenolpyruvate	−14.8	Glucose 1-phosphate	−5.0
1,3-Bisphosphoglycerate	−11.8	Fructose 6-phosphate	−3.8
Creatine phosphate	−10.3	Glucose 6-phosphate	−3.3
Acetyl phosphate	−10.1	Glycerol 1-phosphate	−2.2
ATP (to ADP)	−7.3		

[†]Creatine phosphate is particularly suitable as the transfer of a phosphate group from it to ADP is almost energy neutral.

Thus, the role of the ATP–ADP system in an organism is to transfer phosphate groups from the high energy phosphates to the low energy acceptors, such as glucose. By inspecting Table 5.1 one can conclude that this role of ATP is facilitated by its median position in the thermodynamic scale, because it readily receives phosphate groups from the high energy phosphates and can also easily transfer them to the low energy acceptors. As we shall see later, in Section 5.2.3, there exists a further method of chemical energy transfer in living cells, namely electron transfer.

5.2 Breathing

We now consider the main energetic principles of the chemical processes of breathing. Breathing is a method of extracting energy from fuel in the cell in a regulated manner by oxidation with oxygen. Respiration is a very complex chemical process comprising numerous reactions in which dozens of enzymes take part. To achieve a better understanding of the essence of respiration, we can advantageously divide the whole process into three stages, as illustrated in Figure 5.4.

$$(a) \quad C_6H_{12}O_6 \longrightarrow 2\ CH_3{-}\underset{\underset{O}{\|}}{C}{-}COOH \overset{2\ H_2O}{\longrightarrow}$$

$$\longrightarrow 2\ CH_3COOH + 2\ CO_2 + 4\ H^+ + 4\ e^-$$

$$(b) \quad CH_3COOH + 2\ H_2O \longrightarrow 2\ CO_2 + 8\ H^+ + 8\ e^-$$

$$(c) \quad 8\ H^+ + 8\ e^- + 2\ O_2 \longrightarrow 4\ H_2O$$

Figure 5.4 Generalized chemistry of the three principal stages of the respiratory process: (a) glycolysis and the link reaction; (b) the tricarboxylic acid cycle, or Krebs cycle; and (c) the respiratory chain.

For example, in the oxidation of the carbohydrate glucose, the first stage, called glycolysis, involves breakdown of the sugar molecule into two three-carbon pyruvic acid residues which then undergo oxidative decarboxylation to give acetic acid in the form of acetyl CoA, MeCOSCoA. In the second stage, called the tricarboxylic acid cycle or Krebs cycle, the molecules of acetic acid are degraded with the participation of water and special electron acceptors. The degradation liberates hydrogen ions, electrons and CO_2. The third stage of the respiratory process is named the respiratory chain and concerns the transfer of the electrons, produced in the second stage, to oxygen by means of a complex series of electron carriers. Each stage of respiration produces energy for the organism from the processed raw materials, energy which is then stored in ATP molecules.

Although the first stage in the biochemical processing of fats and proteins differs slightly from glycolysis, acetic acid (in the form of MeCOSCoA) is, nevertheless, always the crucial intermediate product at the end of the first stage. The second and third stages are identical regardless of the source of the cellular fuel. We now consider each stage of the respiratory process in detail.

5.2.1   GLYCOLYSIS

Glycolysis commences with the phosphorylation of a glucose molecule by ATP to produce D-glucose 6-phosphate. The glucose molecule is thus activated for a subsequent conversion; moreover, in the phosphate form it can be more readily and more strongly linked to the appropriate enzyme. D-Glucose 6-phosphate is then isomerized into D-fructose 6-phosphate which in turn is phosphorylated by another ATP molecule to yield D-fructose 1,6-bisphosphate (Figure 5.5).

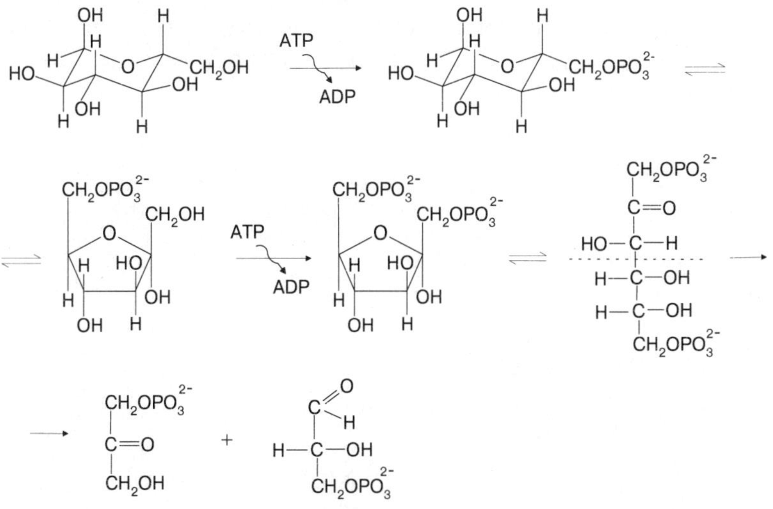

Figure 5.5   Cleavage of a glucose molecule into two three-carbon fragments in the first steps of glycolysis.

As for all monosaccharides, D-fructose 1,6-bisphosphate exists as an equilibrium mixture of cyclic and acyclic forms. The next step in glycolysis involves cleavage of the open chain form of D-fructose 1,6-bisphosphate into two three-carbon fragments: dihydroxyacetone phosphate and

D-glyceraldehyde 3-phosphate (Figure 5.5). This transformation, a retroaldol condensation, demands considerable energy (equilibrium $\Delta G° = 5.7\,kcal\,mol^{-1}$) and is normally the rate-determining step in glycolysis. Dihydroxyacetone phosphate is isomerized into D-glyceraldehyde 3-phosphate by the action of an enzyme. As a result, one molecule of glucose gives two molecules of glyceraldehyde 3-phosphate.

From the point of view of heterocyclic chemistry, the next stage is of special significance. Oxidation of the aldehyde group in D-glyceraldehyde 3-phosphate to a carboxylic acid is followed by phosphorylation with phosphoric acid to form 1,3-bisphosphoglycerate (Figure 5.6a). The oxidation is mediated by the enzyme glyceraldehyde 3-phosphate dehydrogenase with NAD^+ as the coenzyme. A thiol (SH) group linked to the protein chain exercises an important catalytic function in the action of this enzyme. In the enzyme–substrate complex, the thiol group is believed to attach itself to the C=O group of the aldehyde to produce a hemithioacetal structure, which is able to donate a hydride ion. The hydride ion is accepted by the adjacent coenzyme, NAD^+, which is thus transformed into the corresponding reduced form NAD-H. The aldehyde group is simultaneously converted to the S-acylated form, i.e. the thioester. The C–S bond in such thioesters increases their reactivity toward nucleophiles. In the given case this bond is cleaved under the action of a very weak nucleophilic agent, inorganic phosphate, which thus gives a phosphorylated carboxy group in the product (Figure 5.6b).

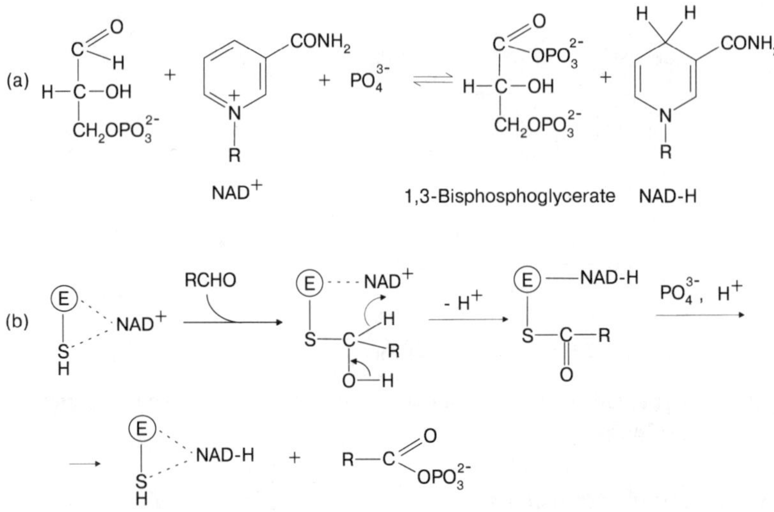

Figure 5.6 D-Glyceraldehyde 3-phosphate conversion into 1,3-bisphosphoglycerate: (a) stoichiometry of the reaction and (b) proposed mechanism.

The energetics of 1,3-bisphosphoglycerate are of interest. Phosphorylation of a carboxylic group requires a great deal of energy ($\Delta G° = 11.8\,\text{kcal}\,\text{mol}^{-1}$), and it is not immediately clear how such a reaction might occur. Again we encounter here conjugative reactions which possess a common intermediate. In the preceding stage of aldehyde oxidation, the quantity of energy ($\Delta G° = -10.3\,\text{kcal}\,\text{mol}^{-1}$) made available is typical of that required for the majority of oxidative processes.

In the next stage of glycolysis, 1,3-bisphosphoglycerate phosphorylates adenosine diphosphate (ADP) to form ATP and 3-phosphoglycerate (Figure 5.7a). The direction of this step is determined by the greater free energy of hydrolysis of a phosphate bond in 1,3-bisphosphoglycerate than in ATP (see Table 5.1). Some free energy is evolved in the reaction owing to the difference in the $\Delta G°$ values of the two compounds, and this enables completion of the previous stage of 1,3-bisphosphoglycerate formation.

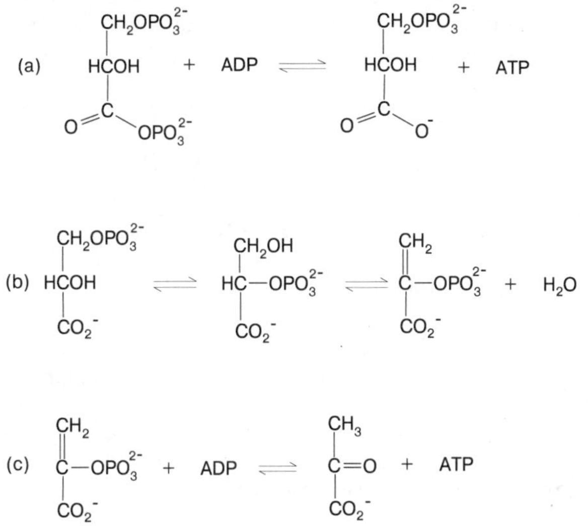

Figure 5.7 Formation of ATP: (a) 1,3-bisphosphoglycerate conversion into 3-phosphoglycerate and ATP; (b) isomerization and dehydration of 3-phosphoglycerate; and (c) transformation of phosphoenolpyruvate into pyruvate and ATP.

The 3-phosphoglycerate produced is isomerized by 3-phosphoglycerate mutase to 2-phosphoglycerate, which is then dehydrogenated to give phosphoenolpyruvate (Figure 5.7b). The function of phosphoenolpyruvate is (as discussed in Section 5.1) to phosphorylate ADP to produce ATP and the anion of pyruvic acid (Figure 5.7c).

Though pyruvate formation is not the final step of glycolysis, the overall energetics can now be calculated. The arithmetic of the energy balance is straightforward. Two ATP molecules are lost in the initial stage (in the phosphorylation of glucose and D-fructose 6-phosphate). Bearing in mind that the cleavage of one glucose molecule releases two molecules of 1,3-bisphosphoglycerate, four new ATP molecules are formed in the final stage. Thus, glycolysis results in the liberation of about 18 kcal mol^{-1}, this energy being stored in two molecules of ATP.

What happens next to the pyruvate, varies according to the type of living organism and the conditions inside the cell. Under anaerobic conditions (the absence of oxygen) pyruvate is reduced to lactate (the anion of lactic acid) (Figure 5.8a and 5.8b) by the NAD-H which was formed during the oxidation of glyceraldehyde 3-phosphate (see also Figure 5.6a). We conclude from the above that the main purpose of the coenzyme pair NAD$^+$-NAD-H in glycolysis is to transfer two electrons and a hydrogen ion from glyceraldehyde 3-phosphate. The lactate formation is accompanied by an appreciable decrease in free energy ($\Delta G° = -6.0$ kcal mol^{-1}) which represents a significant shift of the equilibrium toward the product. When a lack of oxygen occurs in the cells during intense muscular activity, lactic acid is also produced.

(a) $CH_3\text{-}\underset{\underset{O}{\|}}{C}\text{-}COO^- + NAD\text{-}H + H^+ \rightleftharpoons CH_3\text{-}\underset{\underset{OH}{|}}{CH}\text{-}COO^- + NAD^+$

(b) $C_6H_{12}O_6 + 2 PO_4^{3-} + 2 H^+ + 2 ADP \rightleftharpoons$

$\rightleftharpoons 2 CH_3CH(OH)COO^- + 2 H_2O + 2 ATP$

(c) $CH_3\text{-}\underset{\underset{O}{\|}}{C}\text{-}COO^- + H^+ \overset{-CO_2}{\rightleftharpoons} CH_3\text{-}CHO \overset{NAD\text{-}H + H^+}{\rightleftharpoons} CH_3CH_2OH + NAD^+$

(d) $C_6H_{12}O_6 + 2 PO_4^{3-} + 2 H^+ + 2 ADP \rightleftharpoons$

$\rightleftharpoons 2 CH_3CH_2OH + 2 CO_2 + 2 H_2O + 2 ATP$

Figure 5.8 Glycolysis under anaerobic conditions: (a) pyruvate conversion into lactate by lactic acid fermentation; (b) stoichiometric equation of glycolysis under lactic acid fermentation; (c) pyruvate transformation into ethanol by alcohol fermentation; and (d) stoichiometric equation of glycolysis in alcohol fermentation.

During the fermentation of alcohol (e.g. by beer yeast), pyruvate is decarboxylated to form acetaldehyde which is subsequently reduced to ethanol (Figure 5.8c and 5.8d). The first reaction is already familiar to us owing to the participation of the coenzyme thiamine pyrophosphate (Section 4.2.2). The reduction of acetaldehyde is carried out by another coenzyme, alcohol dehydrogenase, containing the coenzyme NAD-H.[†]

Lactic acid and ethanol are the end products of anaerobic glycolysis; in other words, they are the waste materials of these bioprocesses. Such a route for the supply of energy is extremely inefficient: the complete oxidation of one mole of glucose liberates 686 kcal of energy, but the yield during anaerobic glycolysis is but 6.8% of this total. The greater portion of the energy remains conserved in the chemical bonds of lactic acid and ethanol, which are discarded by the anaerobic organism as waste. Aerobic cells, by contrast, are able fully to extract energy from the cellular fuel in the presence of oxygen. Under aerobic conditions, the process begins with oxidative decarboxylation of pyruvic acid (Section 4.2.2) resulting in acetyl coenzyme A formation

$$Me-\underset{\underset{O}{\|}}{C}-COO^- + NAD^+ + HS\text{-}CoA \rightleftharpoons Me-\underset{\underset{O}{\|}}{C}-S-CoA + CO_2 + NAD\text{-}H$$

This reaction proceeds with a dramatic decrease in free energy ($\Delta G^\circ = -8.0$ kcal mol^{-1}), causing a massive shift of the equilibrium to the right. Lipids and proteins, in addition to carbohydrates, also generate acetyl CoA under biological oxidation. Acetyl CoA is a raw material of primary importance for the next stage of the respiratory process, the tricarboxylic acid cycle.

5.2.2 THE KREBS CYCLE, OR THE 'MOLECULAR MERRY-GO-ROUND'

The generalized transformations depicted in Figure 5.9 detail the basic chemistry of the Krebs cycle, or tricarboxylic acid cycle. Readers with imagination may recognize it as an analogue of a merry-go-round (carousel). Indeed, as in the well-known attraction, some molecules in this biochemical merry-go-round are in permanent circular gallop to produce compounds and chemical energy at a great speed. Obviously, it is not the mechanical energy of movement, but the energy stored in different chemical reactions. The only organic chemical to enter the cycle is acetic acid, in the form of the acetyl group, with the assistance of coenzyme A. Coenzyme A, in its S-acetyl form, transfers the acetyl group to oxaloacetic acid. This operation results in the formation of citric acid (citrate) and the regeneration of the coenzyme. The chemistry of this process, similar to an aldol condensation, is schematically depicted in Figure 5.10.

---

†In humans and other animals, alcohol dehydrogenase fulfills the opposite role to dehydrogenate ethanol to acetaldehyde. The human organism obtains ethanol not only from beer and wine but also from numerous food products such as yogurt, fruits preserved in sugar, jams, etc.

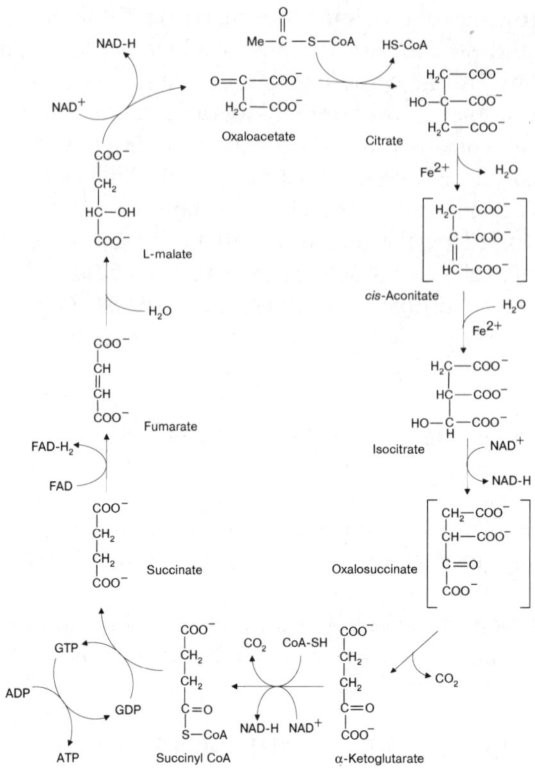

Figure 5.9 The tricarboxylic acid or Krebs cycle (acids are represented by their anions).

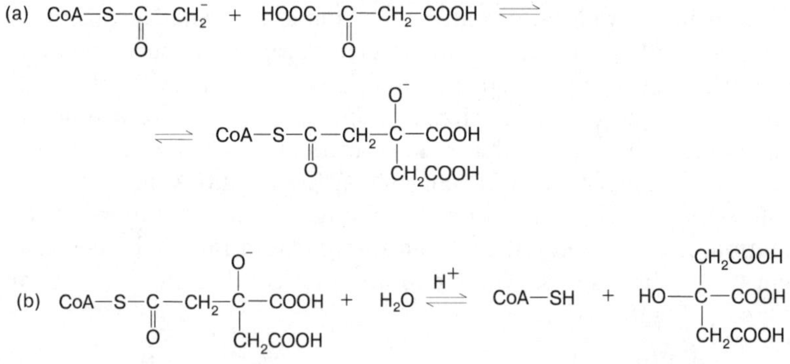

Figure 5.10 Citric acid (citrate) formation in the Krebs cycle: (a) aldol condensation of acetyl CoA in the anionic form (see also Figure 4.15b) with oxaloacetic acid and (b) citryl CoA hydrolysis.

According to the Krebs cycle, citric acid isomerizes into the isocitric form (isocitrate). The process is presumed to proceed via a dehydration–hydration sequence involving the intermediate *cis*-aconitic acid. The enzymatic mediator requires the presence of iron(II) ions for operation. Isocitrate is further oxidized and decarboxylated to produce α-ketoglutaric acid. The oxidation is carried out by the NAD-dependent enzyme isocitrate dehydrogenase with the participation of magnesium(II) or manganese(II) ions. Special attention should be paid to the transformation of isocitric acid into α-ketoglutaric acid. Firstly, this is the overall rate-determining step of the Krebs cycle. Secondly, isocitrate dehydrogenase is inhibited by ATP and NAD-H, but conversely activated by ADP. This situation has the following consequences. When energy expenditures are intense in a cell, the amount of ATP decreases and that of ADP accordingly increases. As a result, the oxidation of isocitric acid and, consequently, the cyclic process as a whole are automatically accelerated. The intensity of electron transfer along the respiratory chain is therefore itself enhanced, causing in turn an increase in the rates both of oxidative phosphorylation and of conversion of ADP into ATP (see also Section 5.2.3). As the cellular stockpile of ATP is regenerated, isocitrate dehydrogenase inhibition begins to occur and the metabolic processes once again slow down to reach their optimal levels.

Figure 5.9 illustrates the two-step conversion of α-ketoglutaric acid into succinic acid (succinate). The first step leads to the formation of succinyl CoA and is, in principle, analogous to the oxidative decarboxylation of pyruvic acid (Section 4.2.2). A well-known group of heterocyclic coenzymes, namely, thiamine pyrophosphate, CoA and NAD⁺, all participate. Succinyl CoA reacts subsequently with inorganic phosphate under the control of guanosine diphosphate (GDP). This pathway is a further example of energy transfer by means of reactions coupled via a common intermediate. Here, the critical intermediate is apparently the mixed anhydride of succinic and

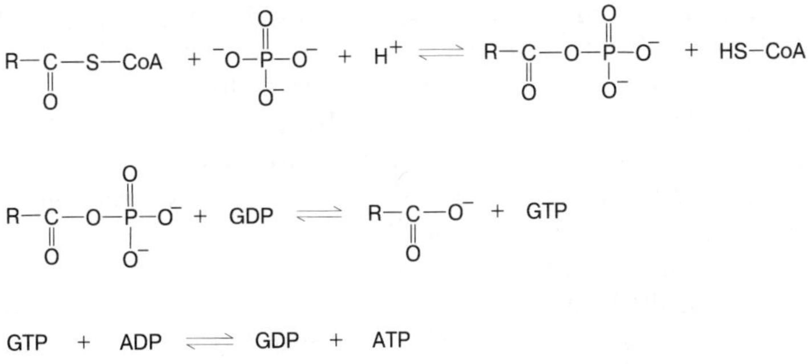

Figure 5.11 Coupled reactions during the conversion of succinyl CoA to succinic acid (R = HOOCCH₂CH₂).

phosphoric acids. This anhydride possesses great phosphorylating power and serves as a means of transporting a phosphate group to GDP to form guanosine triphosphate (GTP). The latter phosphorylates ADP, thus connecting the α-ketoglutaric acid conversion into succinic acid with the formation of one ATP molecule (Figure 5.11).

In a subsequent stage of the Krebs cycle, succinic acid is oxidized to give fumaric acid with participation of the enzyme succinate dehydrogenase and the coenzyme flavin adenine dinucleotide (FAD) (Figure 4.6). The coenzyme accepts two hydrogen atoms and is thus transformed into the reduced compound FAD·H_2. Fumaric acid then undergoes hydration, initiated enzymatically by fumarase, to yield the L-stereoisomer of malic acid (L-malate). The final stage in the tricarboxylic acid cycle involves the oxidation of malic acid, monitored by the NAD-dependent enzyme malate dehydrogenase, to oxaloacetic acid. This returns us to the starting point of the overall process. Thus, the merry-go-round has come full circle and is ready to continue in a nonstop fashion.

Two carbon atoms per turnover are exported from the Krebs cycle in the form of CO_2 and two carbon atoms are imported as an acetyl group attached to CoA. The scheme also demonstrates that the carbon atoms entering and leaving the cycle are in different forms. A most important point is that eight hydrogen atoms are liberated as four pairs per full turn. Three pairs are used in the reduction of NAD^+ and one pair in the hydrogenation of FAD. The stoichiometric chemical equation embracing the overall process of the Krebs cycle was described earlier (Figure 5.4). The bioenergetics of the Krebs cycle will be calculated later when we consider the overall energy balance of the respiratory process.

5.2.3 THE RESPIRATORY CHAIN

As we now know, all of the electrons and hydrogen atoms released in glycolysis and in the tricarboxylic acid cycle are retrieved by the coenzymes NAD^+ and FAD which are thus reduced to NAD-H and FAD·H_2, respectively. The subsequent path taken by these electrons to their terminal destination, i.e. to the oxygen which finally accepts them, is called the respiratory chain. A long series of electron carriers constitutes this chain (Figure 5.12).

The first reaction in the respiratory chain is the oxidation of an NAD-H molecule by a flavin mononucleotide (FMN) unit of the enzyme NAD-H dehydrogenase. The coenzyme FMN·H_2 formed is then oxidized during the next stage by an iron(III) ion of a protein enzyme. Such iron is called nonheme iron to distinguish it from the heme iron which plays a decisive role in later stages of the respiratory chain. The reduced enzyme with the

1) $NAD\text{-}H + H^+ + E_1\text{-}FMN \rightleftharpoons NAD^+ + E_1\text{-}FMN\cdot H_2$

2) $E_1\text{-}FMN\cdot H_2 + 2\,E_2\text{-}Fe^{3+} \rightleftharpoons E_1\text{-}FMN + 2\,E_2\text{-}Fe^{2+} + 2\,H^+$

3) $2\,E_2\text{-}Fe^{2+} + 2\,H^+ + CoQ \rightleftharpoons 2\,E_2\text{-}Fe^{3+} + CoQ\cdot H_2$

4) $CoQ\cdot H_2 + 2\,Cyt.b(Fe^{3+}) \rightleftharpoons CoQ + 2\,H^+ + 2\,Cyt.b(Fe^{2+})$

5) $2\,Cyt.b(Fe^{2+}) + 2\,Cyt.c(Fe^{3+}) \rightleftharpoons 2\,Cyt.b(Fe^{3+}) + 2\,Cyt.c(Fe^{2+})$

6) $2\,Cyt.c(Fe^{2+}) + 2\,Cyt.a(Fe^{3+}) \rightleftharpoons 2\,Cyt.c(Fe^{3+}) + 2\,Cyt.a(Fe^{2+})$

7) $2\,Cyt.a(Fe^{2+}) + 2\,Cyt.a_3(Fe^{3+}) \rightleftharpoons 2\,Cyt.a(Fe^{3+}) + 2\,Cyt.a_3(Fe^{2+})$

8) $2\,Cyt.a_3(Fe^{2+}) + 1/2\,O_2 + 2\,H^+ \rightleftharpoons 2\,Cyt.a_3(Fe^{3+}) + H_2O$

Stoichiometric equation: $NAD\text{-}H + H^+ + 1/2\,O_2 \rightleftharpoons NAD^+ + H_2O$

Figure 5.12 Respiratory chain reactions (by convention, E_1 and E_2 designate the corresponding apoenzymes) (adapted from Lehninger, A. L., *Biochemistry*, Worth, New York, 1970, Chap. 17, p. 380, Table 17-5, with permission).

nonheme iron(II) component subsequently reduces coenzyme Q (CoQ), a *p*-benzoquinone derivative with a long isoprenoid side chain (Figure 5.13), in the presence of protons.

Oxidation of the reduced form of CoQ, and all of the subsequent reactions, proceeds under the control of cytochrome proteins containing heme iron (see Section 4.2.1). It is clear that electrons can be transferred from one carrier to another only when they are in the correct location and relative orientation. Attempts to gain insight into the organization of the respiratory chain have indeed shown that the different components are very close to one another and form respiratory ensembles integrated into the lipid matrix, located on the inner side of the mitochondrial membrane. The components are

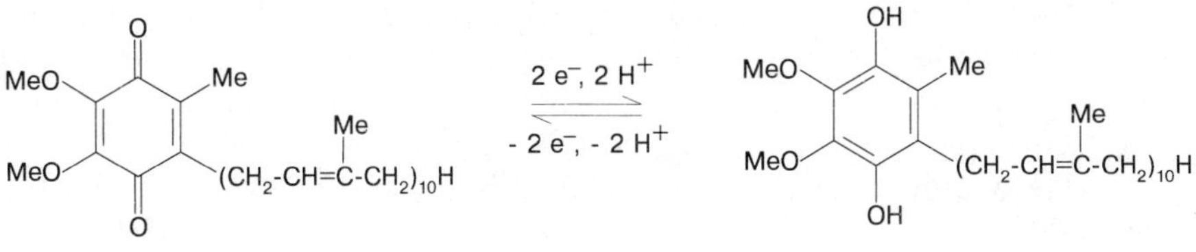

(a)   (b)

Figure 5.13   Coenzyme Q: (a) oxidized form and (b) reduced form.

therefore difficult to isolate and investigate. Cytochromes a and a_3, for instance, are aggregated in one complex called cytochrome oxidase which transports the electrons directly to oxygen. The migration of electrons along the respiratory chain is accompanied by a loss of energy, and at the lowest energy level the electrons combine with oxygen. The energy evolved in this process might be compared with that of river rapids or waterfalls. The energy of electrons proceeding to a lower level is determined by the energy gap between the initial and final levels, similar to the dependence of the energy of water flow on the difference in height between the higher and lower levels. In the case of oxidation–reduction processes the energy liberated may be calculated from the redox potentials of the substances engaged in the electron transfer (Table 5.2).

Table 5.2   Standard redox potentials (E'_0) for some pairs of compounds (two-electron transfer at pH 7.0 and 25–37 °C) (adapted from Lehninger, A. L., *Biochemistry*, Worth, New York, 1970, Chap. 17, p. 366, Table 17-1, with permission).

Reductant	Oxidant	E'_0 (V)
Acetaldehyde	Acetate	−0.60
H_2	$2H^+$	−0.42
Isocitrate	α-Ketoglutarate + CO_2	−0.38
NAD-H	$NAD^+ + H^+$	−0.32
FAD·H_2[a]	FAD[a]	−0.11
Coenzyme Q (reduced form)	Coenzyme Q (oxidized form)	−0.05
Cytochrome b (Fe^{2+})	Cytochrome b (Fe^{3+})	0.00
Cytochrome c (Fe^{2+})	Cytochrome c (Fe^{3+})	0.26
Cytochromes a and a_3 (Fe^{2+})	Cytochromes a and a_3 (Fe^{3+})	0.285
H_2O	$\frac{1}{2}O_2$	0.82

[a]In association with NAD-H dehydrogenase.

Each reductant on loss of its electrons becomes an oxidant. Similarly, an oxidant gains electrons and is thus transformed into a reductant. Together they form a coupled redox pair. The more negative the value of the redox potential (E'_0), the stronger the reductant and, correspondingly, the weaker the oxidant. On the contrary, the more positive the potential, the stronger the oxidant, and the weaker the reductant. The strongest reductant listed in Table 5.2 is acetaldehyde, and the strongest oxidant is oxygen. Electron transfer in the respiratory chain occurs between agents from top to bottom of Table 5.2 in strict conformity with their redox potentials.

The free energy change for the interaction of two redox pairs can be calculated from

$$\Delta G^{\circ\prime} = -nF\Delta E'_0 \qquad (5.1)$$

where n is the number of electrons transferred, F is Faraday's constant

(23.06 kcal) and $\Delta E'_0$ is the difference between the redox potentials of the electron donor and acceptor. As an example, for the redox pair NAD-H–NAD$^+$ and H_2O–$\frac{1}{2}O_2$, E'_0 is equal to 1.14 V. Hence, the free energy $\Delta G'_0$ is lowered by -52.6 kcal mol^{-1} in the transfer of two electrons from NAD-H to oxygen. Theoretically, this is enough to form approximately six molecules of ATP. However, experimentally the oxidation of each mole of NAD-H was found to produce three moles of ATP, and only two moles of ATP in the case of FAD·H$_2$. Since the energy of a phosphate bond in ATP is assumed to be -9.0 kcal mol^{-1}, the efficiency of the energy conservation in NAD-H oxidation seems to be around 50%.

In a similar way, the energetics of each separate reaction can be estimated so that the probable stage of synthesis of ATP in the respiratory chain might be elucidated. The likelihood of synthesis can be estimated at three sites, namely the locations of the following transformations:

$$\text{NAD-H} \rightleftharpoons \text{FAD} \cdot \text{H}_2 \qquad (\Delta G^{0'} = -9.7 \text{ kcal/mol})$$

$$\text{cytochrome } b \rightleftharpoons \text{cytochrome } c \quad (\Delta G^{0'} = -12.0 \text{ kcal/mol})$$

$$\text{cytochromes } (a + a_3) \rightleftharpoons \text{oxygen} \quad (\Delta G^{0'} = -24.7 \text{ kcal/mol})$$

Table 5.3 Biosynthesis of NAD-H, FAD·H$_2$ and ATP during the oxidation of one mole of glucose.

Reactions	Number of moles formed		
	NAD-H	FAD·H$_2$	ATP[a]
Glycolysis			
1,3-Bisphosphoglycerate formation from D-glyceraldehyde 3-phosphate	2	0	0
Phosphoenolpyruvate reaction with ADP	0	0	2
Oxidative decarboxylation of pyruvate	2	0	0
Krebs cycle			
α-Ketoglutarate formation from isocitrate	2	0	0
Succinyl CoA formation from α-ketoglutarate	2	0	0
Succinic acid (succinate) formation from succinyl CoA	0	0	2
Fumaric acid formation from succinate	0	2	0
Oxaloacetic acid (oxaloacetate) formation from L-malate	2	0	0
Respiratory chain			
Oxidation of 10 moles of NAD-H by oxygen	0	0	30
Oxidation of two moles of FAD·H$_2$ by oxygen	0	0	4

[a] Two moles of ATP, produced by the phosphorylation of ADP with the assistance of 1,3-bisphosphoglycerate, are not taken into consideration since they are compensated for by the expenditure of two moles of ATP at the start of glycolysis.

The energy liberated is sufficient to combine ADP with inorganic phosphate to form ATP. The mechanism of energy conservation in the respiratory chain, called oxidative phosphorylation, is very complicated and will not be discussed here.

We now examine the overall energy balance of glucose oxidation in the respiratory process. Calculations can be made on the basis of the formation of three moles of ATP from each mole of NAD-H oxidized by oxygen

$$NAD\text{-}H + H^+ + 3ADP + 3P_i + \tfrac{1}{2}O_2 \longrightarrow NAD^+ + H_2O + 3ATP$$

The data in Table 5.3 imply that in glycolysis and in the Krebs cycle the overall yield of NAD-H is 10 molecules and that of FAD·H$_2$ is two molecules. Their oxidation in the respiratory chain thus provides 34 molecules of ATP. Four more molecules of ATP are formed in the phosphorylation of ADP by phosphoenolpyruvate (during glycolysis) and by guanosine triphosphate (in the Krebs cycle). Overall, the oxidation of one mole of glucose gives 38 moles of ATP

$$C_6H_{12}O_6 + 6O_2 + 38P_i + 38ADP \longrightarrow 6CO_2 + 38ATP + 44H_2O$$

Since the free energy of hydrolysis of ATP to ADP is assumed to be $-9\,\text{kcal}\,\text{mol}^{-1}$, we can estimate that the oxidation of one mole of glucose liberates only approximately 342 kcal. Therefore, about half of the energy which could have theoretically been produced by the combustion of one mole of glucose is utilized.

In summary, heterocyclic molecules play an important role in living organisms as the main carriers of electrons and hydrogens during the extraction of energy from cellular fuel. In some cases (coenzymes NAD$^+$, FAD and their reduced forms) the structures of the heterocyclic nuclei change during the course of the reactions, whereas in others (ATP, cytochromes) the heterocyclic functions play a more passive role.

5.3 Problems

1. What are energy-rich bonds? Indicate which bonds are energy-rich in the following compounds. Provide equations showing the products of hydrolysis.

A   B   C

D   E

2. Explain the expression 'coupling of reactions'. Indicate its significance in biochemical processes and give two examples.
3. Aside from ATP, indicate which other compounds are utilized by living organisms in energy transfer. Draw their structures.
4. Discuss the circumstances in which energy is released by the compounds in Problem 3.
5. Name the three main biochemical stages of respiration and give their overall chemical equations.
6. Which of the compounds listed in Table 5.1 are capable of phosphorylating ADP? Which will be phosphorylated by ATP?
7. Calculate the standard free energy change for the following reaction if the equilibrium constant K is 7.2×10^{-9} at 25 °C and pH 7

$$(CH_3)_2CHOH \;+\; NAD^+ \;\rightleftharpoons\; (CH_3)_2C{=}O \;+\; NAD\text{-}H \;+\; H^+$$

8. How many molecules of NAD-H are formed per one turn of the Krebs cycle? How many ATP molecules does this provide for the respiratory chain?
9. Ethanol is not only a poisonous narcotic but also a source of energy. Estimate the amount of energy stored in a human organism after an intake of 100 ml of vodka.

5.4 Suggested Reading.

1. Lehninger, A. L., *Principles of Biochemistry*, 2nd Edn, Worth, New York, 1993.
2. Moore, G. R. and Pettigrew, G. W., *Cytochromes c. Evolutionary, Structural and Physicochemical Aspects*, Springer, Berlin, 1990.
3. Stryer, L., *Biochemistry*, 3rd Edn, Freeman, New York, 1988.
4. Racker, E., *A New Look at Mechanisms in Bioenergetics*, Academic Press, New York, 1976.
5. Lynen, F., in *Nobel Lectures: Physiology and Medicine (1963–1970)*,

Elsevier, New York, 1973, p. 103.
6. Newsholme, E. A. and Start, C., *Regulation in Metabolism*, Wiley, London, 1973.
7. Krebs, H. A., *Perspect. Biol. Med.*, 1970, **14**, 154.

6 HETEROCYCLES AND PHOTOSYNTHESIS

Like hymns to the luminous Sun
Breathe the lotus' delightsome
O'er the mirror-like space of the lake.

K. Bal'mont

Photosynthesis is that combination of chemical reactions and physical processes which occurs in the green parts of plants and cyanobacteria, resulting in the reduction of carbon dioxide to carbohydrates with the utilization of solar energy. Water acts as the reducing agent; it gives up its hydrogen atoms and is consequently transformed into oxygen. Oxygen, the by-product of photosynthesis, has promoted the development of life on Earth. The overall equation of photosynthesis is well known

$$6CO_2 + 6H_2O + light \longrightarrow C_6H_{12}O_6 + 6O_2$$

Photosynthesis also occurs in algae, phytoplankton and certain bacteria. It is of interest that some bacteria use reductants other than water. For example, green and purple sulfur bacteria reduce carbon dioxide with hydrogen sulfide. Elemental sulfur is a by-product of this reaction

$$6CO_2 + 12 H_2S + light \longrightarrow C_6H_{12}O_6 + 6H_2O + 12S$$

During photosynthesis the energy of sunlight is transformed into energy stored in the chemical bonds of the reduced compounds. The primary products of photosynthesis are sugars. Biosynthesis of fats and proteins can also take place in plants. The process of photosynthesis can be divided into two stages. Reactions in the first stage proceed only in the presence of light. The subsequent conversions in the second stage do not directly require energy from light. The main result of the first stage is the formation of the heterocyclic carriers of energy (i) ATP and (ii) nicotinamide adenine dinucleotide phosphate (NADP-H) in its reduced form. In the second stage, CO_2 is reduced by NADP-H in association with various enzymes. A subsequent sequence of complex chemical reactions leads to the formation of carbohydrates.

It is evident that, in contrast to animals, 'the lotus' delightsome' of the vegetable kingdom breathe during the day not with oxygen but with carbon

119

dioxide. From a chemical point of view, photosynthesis is the reverse of the process of breathing. Such interdependence maintains a biological equilibrium and recycles carbon in nature (Figure 6.1). Plants (as phototrophs) synthesize reduced organic substances and oxygen utilizing the sun's energy, water and carbon dioxide as raw materials. Animals (as heterotrophs) consume plants and oxygen as raw materials to obtain energy, metabolizing them into water and carbon dioxide.

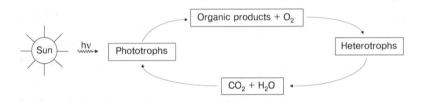

Figure 6.1　The carbon cycle in the biosphere (Bering, C. L., *J. Chem. Educ.*, 1985, **62**, 659).

6.1　Chlorophyll: Sunlight-receiving Antenna and Energy Carrier

Chlorophyll is the green pigment of plants which is the basis of the photosynthetic complex. Like heme in the hemoglobin of blood, the structure of chlorophyll has a porphyrin core which contains a coordinated magnesium(II) ion in its central cavity instead of iron(II) as in heme. A peculiar feature of the chlorophyll structure is the presence of a long hydrocarbon side chain formed by the alcohol phytol. It is by means of this phytyl group that the chlorophyll molecule associates with the nonpolar lipids of cellular membranes through hydrophobic interactions. Various modifications of chlorophyll are known (Figure 6.2). The most abundant analog is chlorophyll *a*. Chlorophyll *b* differs from chlorophyll *a* by the presence of an aldehyde group instead of a methyl group in pyrrole ring II. The chlorophyll found in photosynthetic bacteria has a somewhat different structure and is called bacteriochlorophyll; a C–C bond in the pyrrole ring II is now saturated.

The role of chlorophyll is to capture the energy of visible light with the subsequent production of excited state electrons. A second function is to transfer these excited electrons to the corresponding electron acceptors. These functions are performed by different types of chlorophyll. The chlorophyll which transfers an electron directly to the acceptor is called the reaction center. Another type of chlorophyll acts solely to trap light energy. There is only one such reaction center per 400 molecules of light-fixing chlorophyll. 'Antenna chlorophyll', composed of 75% chlorophyll *a* and 25% chlorophyll *b,* are the

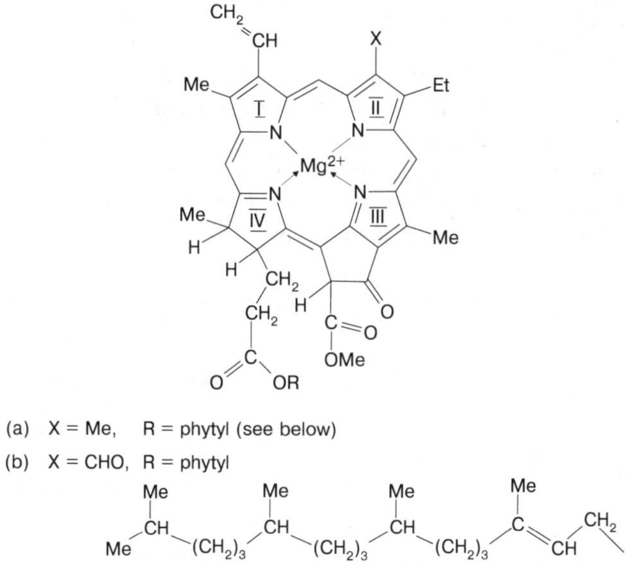

Figure 6.2   Chlorophyll structures: (a) chlorophyll *a* and (b) chlorophyll *b*.

molecules responsible for light fixation. To understand better the coordination between the functions of the antenna chlorophyll and the reaction center, we consider the physical fundamentals of energy absorption by these molecules.

It is well established that the electrons in conjugated molecules are distributed in discrete energy levels called molecular orbitals (MOs). In the ground state the paired electrons occupy the lowest energy levels called bonding MOs. Absorption of energy allows an electron to move from an occupied bonding MO to one of the unoccupied (nonbonding) orbitals. This migration results in excitation of the molecule. The first excited state is readily reached by transfer of an electron from a high energy occupied MO to a low energy unoccupied MO. When sufficient energy is supplied, the migration of electrons to higher orbitals becomes possible. Such energy is found in X-rays and ultraviolet radiation.[†] However, the energy of the radiation reaching the Earth's surface as visible light is considerably lower. Such radiation induces electron migration to the first excited state in only those molecules which have rather narrow energy gaps between the HOMO (highest occupied MO) and LUMO (lowest unoccupied MO). Small

---

[†]X-rays, γ-rays and near-ultraviolet radiation (λ< 300 nm) are destructive to living cells because they remove electrons from purine and pyrimidine bases which have long wavelength absorption maxima around 270 nm. Radiation of λ< 300 nm destroys nucleic acids and all enzyme systems. Thankfully, most of the damaging ultraviolet solar radiation is absorbed by the ozone layer of our planet. We are shielded from X-rays and γ-rays by the ionosphere and magnetosphere.

HOMO–LUMO gaps are characteristic of substances with an extensive system of conjugated bonds, which therefore have many π-electrons. Practically all organic pigments, including chlorophyll, belong to this category. The absorption spectrum of chlorophyll in the visible range (Figure 6.3) helps us to understand the origin of its green color. Chlorophyll a has two intense absorption bands with maxima at 450 and 660 nm. Therefore, this compound readily absorbs blue and red light but is transparent to green light (in the 450-650 nm range no absorption is observed).

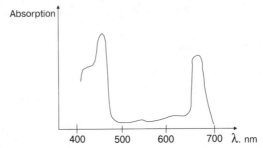

Figure 6.3   Visible spectrum of chlorophyll a (in diethyl ether solution) (Bering, C. L., *J. Chem. Educ.*, 1985, **62**, 659).

Since chlorophyll absorbs only a narrow range of visible light, its effectiveness as the antenna pigment could be questionable. However, three factors increase its efficiency. Firstly, the antenna complexes of green plants have additional pigments which readily absorb light in the very range where the chlorophyll molecules are transparent. These compounds are the carotenoids and phycobilins (Figure 6.4). The former are isoprenoid hydrocarbons containing long chains of conjugated bonds, whereas the latter are linear tetrapyrrole analogues of chlorophyll devoid of the Mg^{2+} ion. The carotenoids absorb light below 550 nm, and the phycobilins absorb in the range 570–650 nm. Incidentally, the beautiful yellow and orange colors of autumn leaves are largely due to the carotenoids, which are unmasked after the chlorophyll is destroyed.

Secondly, chlorophyll molecules in a cell are closely surrounded by proteins, lipids and pigments. These pigments generally extend the range of absorption by the chlorophyll molecule. For example, purified bacteriochlorophyll a has a long wavelength absorption maximum at 770 nm, while the same compound absorbs anywhere between 800–890 nm when involved in nonbonding interactions with various cellular components.

The third reason for the increased efficiency of chlorophyll as an antenna pigment is connected with the special properties of the molecules which distinguish them from a mere assembly of atoms. An electron transfer from one energy level (orbital) of an atom to another level takes place only with the

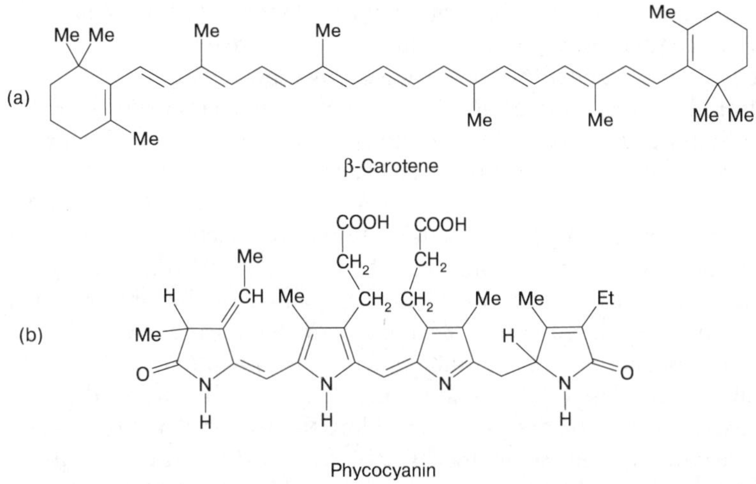

(a)

β-Carotene

(b)

Phycocyanin

Figure 6.4 Structures of (a) a typical carotenoid: β-carotene and (b) a typical phycobilin: phycocyanin.

absorption of a strictly determined quantum of energy which must correspond to the energy gap between the levels. In cases where the energy of the photon is greater or smaller, the photon is simply not absorbed. An atom in the excited state has a very short lifetime (about 10^{-12}s), after which it returns to the ground state and a quantum of energy of the same frequency is released (Figure 6.5a and 6.5b). By contrast, the restrictions on the quantum energy are not so rigorous when a molecule is excited. Each electronic state of a molecule has numerous substates related to its vibrational and rotational motions. The laws of quantum mechanics allow the transfer between many sublevels of the

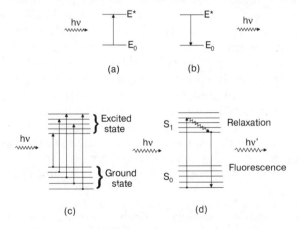

Figure 6.5 Absorption and emission of light by an atom (a, b) and by a molecule (c, d).

ground and excited states. Therefore, light is absorbed by molecules not at a single wavelength only but over a certain range (Figure 6.5c).

Deactivation of excited molecules has its own peculiarities. As a rule, light absorption drives an electron to one of the central excited substates. During its short lifetime (10^{-12} s) the excited molecule succeeds in transferring part of its energy to vibration or rotation or to the environment as a result of thermal equilibration. Thus, the molecule is transformed into one of the lower excited substates by a process called relaxation. When the electron returns from this lower excited state to the ground state, the molecule will emit light of a lower frequency, i.e. with a longer wavelength than that of the photon originally absorbed to excite the molecule. Such emission is known as fluorescence (Fig 6.5d).[†] Fluorescence is an undesirable phenomenon in photosynthesis because of the superfluous loss of absorbed energy. Evidently, it is not without reason that chemical evolution has selected chlorophyll as the main pigment and energy-carrying system in photosynthesis. In contrast with other pigments, such as heme, chlorophyll displays extremely feeble fluorescence.

The second method of deactivation from the photoexcited state involves energy transfer to another molecule. The excited molecule E* transfers its energy to a second molecule M, resulting in deactivation of the former (E) and excitation of the latter (M*). Such transfer does not mean that E* emits energy initially and M then captures it. The process occurs in a more specific way, resembling the mechanical resonance of two linked pendulums in which the energy of motion is directly transferred from one pendulum to the other. In the case of molecules, similar resonance energy transfer is possible only when the absorption of one molecule overlaps that of the other. This requirement is met in the case of different pigments such as carotenoids and chlorophyll, and also in the case of two identical molecules of chlorophyll. In association with other pigments, antenna chlorophyll absorbs quanta of incident light energy and transfers them from one molecule to another until they reach the reaction site. The energy transfer should be fast and efficient and only occurs when there is very tight packing of the chlorophyll molecules.[‡] The antenna chlorophyll molecules form gigantic supermolecular associations, and delocalization of an excited electron between the fragments results in the transfer of the energy from light excitation.

The third path of deactivation results from the ability of the molecule, once it is excited, to undergo a photochemical reaction, e.g. a redox conversion. Thus, if a suitable acceptor A is in the vicinity of the excited molecule E*, an electron from the latter may migrate to the acceptor to give anion A^- and cation E^+. As E^+ has now become an oxidant, it is capable of accepting an

---

[†]Additional information on fluorescence is given in Section 9.3.1.

[‡]The efficiency of the excitational energy transfer is inversely proportional to r^6, r being the distance between E* and M (i.e., if the distance increases by a factor of two, the efficiency is decreased 64-fold).

electron from a donor D. This possibility may lead to the formation of the triad D^+EA^- with separated charges

$$DEA \xrightarrow{\text{hv}} D\overset{*}{E}A \longrightarrow DE^+A^- \longrightarrow D^+EA^-$$

The molecule E itself remains apparently unchanged. Photochemists say that this substance serves as an energy carrier or a photosensitizer, making the reaction sensitive to light. Chlorophyll functions in plants in the same manner: water molecules are the initial donors of electrons and CO_2 molecules are the final acceptors. The intermediate processes are discussed in detail in the following section.

6.2 What Daylight can Achieve

At the end of the 1950s it was established that green plants and algae possessed two different reaction sites where electron transfers from the photoexcited chlorophyll to the acceptor molecules occurred. These reaction centers differ from each other in their light absorption characteristics. One center is designated as P680 since its chlorophyll absorbs at 680 nm, the other is P700 with an absorption maximum at 700 nm. The difference in absorption is caused by different environments of the chlorophyll a molecules in the two centers, both of which are constructed from protein–pigment complexes incorporated into a lipid matrix.[†]

Each reaction site performs its specific functions in photosynthesis but their actions are coordinated in a very efficient manner. The photosynthetic complex composed of pigment P680, antenna pigments, electron carriers, protein molecules and a number of other components all working in cooperation is called photosystem II. The analogous complex based on pigment P700 is named photosystem I.

The discovery of the two types of photosystems led to the elaboration, in 1960, of the presently accepted scheme of photochemical reactions known as scheme Z because of its resemblance to the letter Z (Figure 6.6). Incident light excites both reaction centers of a plant. Each photoexcited pigment thus becomes capable of reducing the acceptors in its photosystem. Photoexcited pigment P700 in photosystem I transfers its electron to the oxidized form of the iron-containing protein ferredoxin. The electron is then abstracted by a flavin-containing protein and is transferred to coenzyme $NADP^+$, the terminal acceptor during the light-dependent photosynthetic reactions; the $NADP^+$ is thus converted to NADP-H.

[†]In plants, the entire photosynthetic complex (which includes two photosystems and the associated enzymes) is located either on the surface of or inside lipid-protein membranes. These membranes are components of the cellular organelles called chloroplasts.

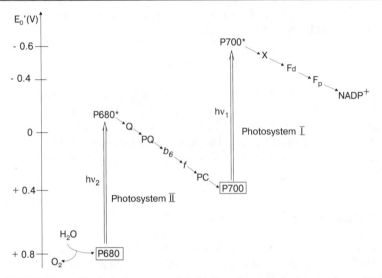

Figure 6.6   Scheme Z of electron transfer during photosynthetic reactions (Q is the initial acceptor composed of plastoquinone and nonheme iron; PQ is the pool of plastoquinones; b_6 and f are cytochromes; PC is plastocyanin; X is tightly bound ferredoxin; Fd is ferredoxin; and F_p is a flavoprotein) (Bering, C. L., *J. Chem. Educ.*, 1985, **62**, 659).

Photosystem II has its own chain of electron transporters. The excited pigment P680 first reduces plastoquinone complexed with nonheme iron. The electron is further transferred to an extensive pool of other unbound plastoquinones which serves as a reservoir of electrons during the oxidation of water. Cytochromes b and f and the copper-containing protein plastocyanin are the subsequent electron carriers in photosystem II. The oxidized pigment $P700^+$ is located at the end of this chain. Pigment $P700^+$ accepts an electron from photosystem II and reverts to the original state (P700) ready to absorb the next quantum of light. The oxidized pigment $P680^+$ also obviously has to be converted back into the initial state. Water acts as the reductant, though there are reasons to believe that between water and P680 one further set of electron transporters may exist, in particular, a manganese-containing protein.[†] Water is the original and inexhaustible source of all of the electrons involved in photosynthesis; these electrons are then employed in the reduction of plastoquinones, flavins and $NADP^+$ (see Figures 4.5, 4.7 and 5.13). Oxygen and protons are also liberated simultaneously.

To understand the necessity for plants to have two discrete photosystems, we examine the energetics of electron transport. We use two formulas in the

---

[†]It is possible that not all of the electron carriers of scheme Z have yet been uncovered.

calculations: equation (5.1) that we encountered while investigating the dependence of redox potentials on the free energy changes of reactants, and the Planck equation (6.1) for calculating the energy of photons

$$E = h\nu = hc/\lambda \tag{6.1}$$

where ν is the frequency, λ is the light wavelength, h is the Planck constant $(1.58 \times 10^{-34} \, \text{cal s})$ and c is the speed of light $(3 \times 10^8 \, \text{m s}^{-1})$.

Consider Figure 6.6 in which the redox potentials of the main participants of electron transfer are shown (see also Table 5.2). Water is situated at the bottom of the redox scale $(E'_0 = 0.82 \, \text{V})$ in accordance with the high stability of water molecules and the observation that removing an electron from water is difficult. The fact that the oxidation of water occurs under mild circumstances is a very remarkable feature of photosynthesis. As water is oxidized, the photosynthetic complex of plants must contain an oxidant with a greater positive potential than oxygen: in photosystem II this is the oxidized pigment P680$^+$. The estimated value of E'_0 for the redox couple P680–P680$^+$ falls in the range 0.8–0.9 V. Such a high value is rationalized by the tight packing of pigment P680 in the hydrophobic matrix surrounded by nonaqueous media. Water has no direct contact with P680$^+$ and therefore transfers electrons to this complex via other hydrophobic carriers such as the manganese-containing protein mentioned above.

We emphasize once again that the photochemical reactions of photosynthesis generally culminate in the reduction of coenzyme NADP$^+$ by water to form NADP-H. As the difference between the redox potentials of the two pairs H$_2$O–½O$_2$ and NADP$^+$–NADP-H $(E'_0 = -0.32 \, \text{V})$ amounts to $-1.14 \, \text{V}$, the free energy increase calculated by equation (5.1) is approximately 52 kcal mol^{-1}, taking into account the two electrons involved in the reduction process. A single quantum of light energy with a wavelength of 680 nm is insufficient to meet this substantial energy requirement as we can see by using equation (6.1)

$$E = \frac{(1.58 \times 10^{-34} \, \text{cal s}) \, (3 \times 10^8 \, \text{m s}^{-1})}{680 \times 10^{-9} \, \text{m}} = 6.97 \times 10^{-20} \, \text{cal}$$

The value obtained is the quantum of energy taken up by one molecule. Since all of the calculations are given for one mole of compound, this value must be multiplied by the Avogadro constant

$$E = (6.97 \times 10^{-20} \, \text{cal}) \times (6.02 \times 10^{23} \, \text{mol}^{-1}) = 42.2 \, \text{kcal mol}^{-1}$$

This value is well below 52 kcal mol^{-1}, and therefore one quantum of light with a wavelength of 680 nm is not sufficient to reduce NADP$^+$ to NADP-H. The necessary energy can be supplied by two quanta, but the problem is how to incorporate the energy at one time since the reaction center of chlorophyll transfers electrons one by one. In other words, the reductive energy of the photoexcited pigment P680$^+$ is insufficient to transport an electron to the lowest unoccupied MO of NADP$^+$. This is clearly shown in Figure 6.6, where the higher reductive potential of NADP-H can be compared with that of P680* and reduced plastoquinone Q. These considerations demonstrate the need for a second photosystem, of which the reaction center P700 is characterized by a lower energy of excitation and a less positive redox potential $(E'_0 = 0.4$–$0.5\,V)$. This potential is not sufficiently high for the oxidized form of pigment P700* to remove an electron from water. The unoxidized excited pigment P700*, however, can reduce the NADP$^+$ molecule, which implies that it is a stronger reductant than P680*. The calculations involving equations (5.1) and (6.1) indicate that the energy of one quantum of light is more than enough to transfer an electron from P700* to NADP$^+$ (Table 6.1).

Table 6.1 Energy characteristics of some light processes in photosynthesis.

Process	Energy (kcal mol^{-1})
Free energy change during the two-electron reduction of NADP$^+$ by water $(\Delta E'_0 = -1.14\,V)$	52.6
Free energy change during the one-electron reduction of plastoquinone $(E'_0 = -0.1\,V)$ by pigment P680 $(E'_0 = 0.83\,V)^a$ $(\Delta E'_0 = -0.93\,V)$	21.4
Free energy change during the two-electron reduction of NADP$^+$ coenzyme by pigment P700 $(\Delta E'_0 = -0.72\,V)$	33.2
Free energy change during the one-electron reduction of bound ferredoxin $(E'_0 = -0.55\,V)$ by pigment P700 $(E'_0 = 0.4\,V)^a$ $(\Delta E'_0 = -0.95\,V)$	21.9
Quantity of energy in one quantum of light with a wavelength of 680 nm	42.2
Quantity of energy in one quantum of light with a wavelength of 700 nm	40.8

^{a}Approximate values.

A general picture of the energetic processes occurring during the photochemical reactions of photosynthesis may be gained from the following considerations. Initially, two independent quanta of light are absorbed by pigments P680 and P700. The excited form P700* carries an electron to coenzyme NADP$^+$. The oxidized form P700$^+$ thus generated is immediately reduced by the excited pigment P680* and is thus available to absorb the next quantum of light. The oxidized form P680$^+$ is, in turn, reduced by water. The next two quanta of light trapped by both photosystems promote a second electron transfer to NADP$^+$ (to be precise, to radical NADP·), resulting in the formation of an NADP-H molecule by the simultaneous addition of a proton.

Thus, pigment P680 is a powerful oxidant in the oxidized form and a relatively weak reductant when photoexcited. On the contrary, pigment P700 is a weak oxidant in the oxidized state but a strong reductant when excited. Since photosynthesis requires the simultaneous availability of both a strong oxidant and a powerful reductant in the same cell, no single pigment can carry out photosynthesis. Only the joint action of both photosystems achieves the required result.

Between 8 and 10 quanta seem experimentally necessary to form one molecule of oxygen. This is consistent with the following simple considerations. Water is the only source of molecular oxygen in photosynthesis, one molecule of oxygen being derived from two molecules of water. This process involves the release of four protons and four electrons from two molecules of water. Since the oxidized pigment $P680^+$ functions as an electron acceptor, the transfer must involve four steps, and therefore necessitates four quanta of light. Four further quanta will, in parallel, be absorbed by pigment P700 in photosystem I in preparation for these four electrons. So, in total, eight quanta of light should be absorbed. The numbers of electrons and protons evolved are then sufficient for the formation of two molecules of NADP-H. This process is illustrated in Figure 6.7.

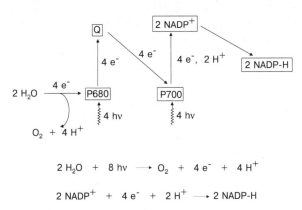

Figure 6.7   Light quanta and electron balance during the formation of one molecule of oxygen in the course of photosynthesis.

NADP-H is not the only molecule in which the energy of the light quanta is stored. ATP plays a similar role. A mere glance at the Z scheme reveals two sites where a stream of electrons can flow spontaneously like water down a waterfall. Nature does not waste the energy released by this process. However, one of the sites, namely that situated between the reduced form of bound ferredoxin and $NADP^+$, provides a relatively small gain in energy ($\Delta G^\circ = -5.3\,\mathrm{kcal\,mol^{-1}}$) which is almost impossible to utilize. By contrast, the quantity of energy liberated at the second site, located between the reduced plastoquinone and the oxidized pigment P700$^+$, during the passage of even one electron ($\Delta G^\circ = -11.5\,\text{kcal mol}^{-1}$) is quite sufficient for the formation of one molecule of ATP.[†] Experimental evidence shows that three molecules of ATP are produced from organic phosphate (P$_i$) during the transfer of four electrons through the site. The summary equation of the light reactions is

$$2\text{H}_2\text{O} + 8h\nu + 2\text{NADP}^+ + 3\text{ADP} + 3\text{P}_i \longrightarrow \text{O}_2 + 2\text{NADP-H} + 3\text{ATP} + 2\,\text{H}^+$$

The present discussion would not be complete without mention of the specificity of photochemical reactions in photosynthetic bacteria. The reaction center in such species contains bacteriochlorophyll a which absorbs light in the range 800–890 nm (the near-infrared region). Less energy, therefore, is needed to photoexcite bacteriochlorophyll a compared with chlorophyll a. The energy requirement equals 35.9 kcal mol^{-1} in the case of light with a wavelength of 850 nm. The energy of two such quanta is adequate, in principle, to transfer one electron from water to NAD$^+$.[‡] However, the oxidized form of bacteriochlorophyll a has a positive potential ($E'_0 = 0.4\,\text{V}$) which renders it unable to extract electrons from water. This is why none of the known photosynthetic bacteria can utilize water as a substrate for photosynthesis, and therefore why they do not produce oxygen. Alternatively, more easily oxidized materials such as hydrogen sulfide, alcohols, organic acids and so on are selected as substrates. This fact reflects one more remarkable feature inherent to photosynthetic bacteria: they possess a single photosystem with electron transport (Figure 6.8) occurring in a cyclic fashion.[§]

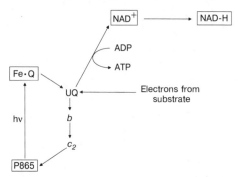

Figure 6.8 Scheme of electron transport in photosynthetic nonsulfur purple bacteria (UQ is the ubiquinone pool and b and c_2 are cytochromes).

[†]Discussion of the mechanism of energy conversion from the free electrons to the chemical bonds of ATP is beyond the scope of this book. Interested readers may find further information in the publications listed in Section 6.5.
[‡]In contrast to green plants, the coenzyme NAD$^+$ is the final acceptor of electrons in bacteria.
[§]We remind readers that green plants have two photosystems complexed in a linear manner (See Figures 6.6 and 6.7).

In these bacteria, an electron is transferred from the photoexcited pigment P865 to the initial acceptor, the ubiquinone[†] complex containing nonheme iron (FeQ). The electron is further passed to the oxidized pigment P865$^+$ via a series of carriers including the ubiquinone pool and cytochromes b and c_2. The energy liberated during this process is utilized in ATP synthesis. A portion of the ubiquinone pool is reduced by ethanol or some other substrate. The electrons donated by the substrate are finally accepted by NAD$^+$ which thus converts to NAD-H. The specificity of this process results from the redox potential of NAD-H being substantially more negative than that of the reduced ubiquinone. The process, consequently, cannot occur spontaneously. The energy required is provided by ATP molecules which are formed in the course of the cyclic transport of electrons.

It should be mentioned that cyclic electron transport also occurs in plants. Photosystem II uses this type of operation as an auxiliary or reserve. When insufficient CO_2 is available excess NADP-H accumulates in plant cells. In such cases the further transfer of electrons to NADP$^+$ is blocked via regulatory mechanisms. The electrons then begin to flow from reduced ferredoxin back to the plastoquinone pool. This process is accompanied by the storage of energy in ATP molecules to be used for biosynthesis.

6.3   Photosynthesis without Light

The photochemical reactions provide the photosynthetic cells with the energy to carry out their main function, the subsequent reduction of CO_2. This energy, as already mentioned, is stored in two forms: as the energy of the phosphate bond in ATP and in the form of the 'reductive potential' of NADP-H coenzyme molecules. Thus, carbon dioxide fixation does not need light initiation and may even occur in darkness[‡] according to the following general equation

$$6CO_2 + 12NADP\text{-}H + 12H^+ + 18ATP \longrightarrow$$
$$C_6H_{12}O_6 + 12NADP^+ + 6H_2O + 18ADP + 18P_i$$

Adding this equation to the stoichiometric equation of the light-dependent reactions (see Section 6.2) and balancing all of the coefficients, we get the simple stoichiometric equation of photosynthesis presented at the beginning

---

[†]Ubiquinones are structurally similar to coenzyme Q (Figure 5.13), the main difference being the length of the alkyl side chain. Plastoquinones differ from ubiquinones by the presence of methyl groups in the benzene ring in place of methoxy substituents.

[‡]As a matter of course, periods of darkness (night) alternate with light (day). Plants normally respire at night. However, a long sojourn in the dark prevents a plant from receiving light energy. In this case it has to switch on a photorespiratory mechanism to support its living processes. As a result, carbohydrates amassed by photosynthesis become reoxidized by air to give carbon dioxide and water. An extended period of darkness eventually brings about plant exhaustion and death.

of the chapter. In addition, the exact number of light quanta necessary to produce one mole of glucose can now be calculated (48).

The dark reactions of glucose photosynthesis are not of special interest from a heterocyclic viewpoint. These reactions involve a long and complicated chain of enzymatic conversions in which a number of C_3-C_7 sugars and glycols participate. Nevertheless, we do wish to 'shed some light' on the dark reactions to understand the overall picture of photosynthetic chemistry.

The assimilation of a CO_2 molecule commences with its integration into ribulose 1,5-bisphosphate (RUBP) (Figure 6.9). The carboxylation is accompanied by cleavage of an intermediate C_6 compound resulting in the formation of two molecules of 3-phosphoglycerate (PG). This RUBP carboxylase catalyzed reaction has no simple analogues in classical organic chemistry.

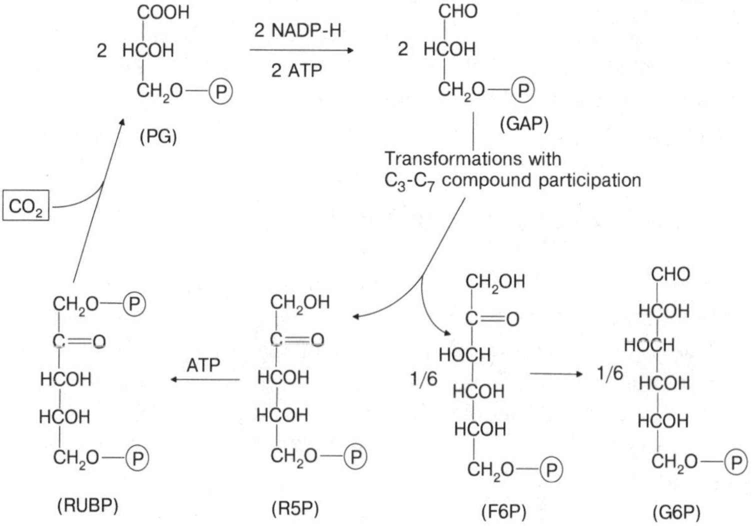

Figure 6.9 Simplified scheme of CO_2 fixation in the reductive pentose phosphate cycle (Calvin cycle).

Both molecules of 3-phosphoglycerate are further reduced by the coenzyme NADP-H to give glyceraldehyde 3-phosphate (GAP). As it may be recalled, the same reaction occurs during respiration, but in the opposite direction (Figure 5.6). The reversibility of this reaction allows the equilibrium to be shifted toward aldehyde formation owing to the participation of ATP and other regulatory mechanisms. The GAP molecules are subsequently subjected to complex conversions including a number of isomerizations and condensations. In this transformation pathway, two GAP molecules combine five of their carbon atoms to assemble one molecule of ribulose 5-phosphate

(R5P), to which one additional carbon atom is added to yield fructose 6-phosphate (F6P). The fructose derivative is finally transformed into glucose 6-phosphate (G6P). It is important to note (Figure 6.9) that one of the carbon atoms incorporated into glucose is taken out of the transformation cycle, whereas R5P is converted to ribulose 1,5-bisphosphate (RUBP) and is recycled in the CO_2 assimilation scheme. This process, known as the Calvin cycle, was discovered in the 1950s. The assimilation of one molecule of CO_2 thus requires two molecules of NADP-H and three molecules of ATP. In total, the formation of one glucose molecule necessitates the uptake of six molecules of CO_2 in the Calvin cycle.

We have seen that almost identical heterocyclic structures are engaged in the chemical processes of both photosynthesis and respiration: the pyridine and flavin coenzymes, adenosine triphosphate, the cytochromes and a series of other tetrapyrrole compounds. However, the central function in photosynthesis is carried out by chlorophyll, which is not found in animals. In the evolution of plants there seem to be several reasons why nature chose this class of compound as the photocatalyst. Firstly, the porphyrin system is highly aromatic and, hence, very stable chemically and thermodynamically. This is a crucial factor since each molecule of chlorophyll must undergo several thousand electron transfers without degradation. Secondly, the elongated π-electronic system of chlorophyll is well adapted to absorb light not only in the blue region of the spectrum but in the red and near-infrared regions also. Thirdly, the π-electronic system is very flexible regarding the matrix, making chlorophyll a functionally versatile catalyst. Finally, chlorophyll does not lose absorbed energy by nonirradiative deactivation processes, in contrast to heme and other tetrapyrrole structures.

For several years the intriguing problems of creating artificial photosynthesis and, therefore, chlorophyll-like compounds have been the focus of attention. Chapter 10 covers the latest developments in this field.

6.4 Problems

1. List some of the naturally occurring modifications of chlorophyll. How do they differ from each other? What are their functions?
2. Which pigments constitute the antenna complex of green plants? Outline their specific functions.
3. How are the photoexcited molecules of chlorophyll deactivated?
4. Which compound is the terminal acceptor of electrons in the light reactions of photosynthesis? Write an equation detailing its formation. How many light quanta are required to form one molecule of this compound?
5. Why does the photosynthetic apparatus of green plants utilize two

reaction centers and two photosystems? Draw the Z scheme of electron transfer and characterize in general terms the principles of its operation.

6. What is the specificity of light reactions in photosynthetic bacteria?

7. Calculate the maximum numbers of ATP molecules which can be synthesized from ADP and inorganic phosphate under standard conditions by absorption of a quantum of light of wavelength (a) 400, (b) 550 and (c) 700 nm, respectively (assume the existence of the appropriate mechanisms).

8. Determine the standard free energy change that occurs during the transfer of one pair of electrons in photosystem I from bound ferredoxin $(E'_0 = -0.55\,\text{V})$ to $NADP^+$ $(E'_0 = -0.32\,\text{V})$.

6.5   Suggested Reading

1. Lehninger, A. L., *Principles of Biochemistry*, 2nd Edn, Worth, New York, 1993.

2. Gregory, R. P. F., *Biochemistry of Photosynthesis*, 3rd Edn, Wiley, New York, 1989.

3. Clayton, R. K., *Photosynthesis: Physical Mechanisms and Chemical Patterns*, Cambridge University Press, Cambridge, 1980.

7 HETEROCYCLES AND HEALTH

There're four hundred and four disease'.
Some hundred and one of these
by drugs are treated with ease.
Another one hundred and one
are cured by doctor and charm.
The same number of illnesses
pass harmlessly by themselves.
Yet I'm with anxiety ill
that the rest are incurable still.

L. Martynov

7.1 Medicines from a Natural Storehouse

For thousands of years the sick were cured worldwide by remedies obtained from 'nature's own drugstore', comprising such sources as leaves, fruits, barks and herbs. Until recently, it was not realized that successful traditional treatments of this type were frequently triggered by the presence of various heterocyclic compounds in extracts derived from plants, animals and insects.

Perhaps no other naturally occurring compound has received as much attention as quinine, both in the scientific literature and in fiction. As children reading Jules Verne's novel *The Mysterious Island*, we were all concerned for the life of Herbert when he became stricken with malaria. The youth seemed to be doomed, but a miracle occurred. On the table appeared a small box bearing the inscription 'Quinine Sulfate'. The box had been secretly left by Captain Nemo. It is further written in the novel

> It [the box] contained nearly two hundred grains of a white powder, a few particles of which he carried to his lips. The extreme bitterness of the substance precluded all doubt; it was certainly the precious extract of quinine, that preeminent antifebrile.

Within a few days Herbert was on his way to recovery.

Quinine is a representative of the alkaloids. The alkaloids comprise numerous families of nitrogen-containing organic compounds which occur widely in the plant kingdom. Alkaloids are often considered to be 'waste products' of the vital processes in plants because they are accumulated in the

135

easily detachable parts: the bark, leaves and fruits. Cinchona trees, whose bark contains quinine and approximately 20 other alkaloids, grow in the jungles of South America. Various legends exist about ancient Indians who prepared infusions from cinchona bark for the prevention of malarial fever.

Almost all alkaloids are structurally derived from either aromatic or hydrogenated heterocycles. This relationship has been used as the basis for a classification system which includes, for example, the isoquinoline, pyridine, purine and quinazoline series of alkaloids. Quinine is a rather complex derivative of quinoline (Figure 7.1).

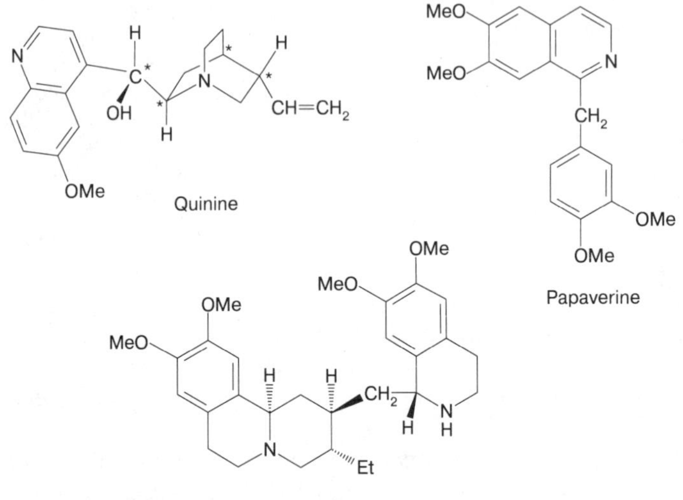

Figure 7.1   Representative alkaloids of the quinoline and isoquinoline series.

The quinine molecule has several asymmetric carbon (C*) atoms and can exist as a number of different stereoisomers. A dextrorotatory diastereomer of quinine, called quinidine, is used as a powerful antiarrhythmic agent in the treatment of tachycardia and ciliary arrhythmia. The isoquinoline alkaloid papaverine has found applications in therapy as a spasmolytic and vasodilator. Another isoquinoline alkaloid called emetine, extracted from the roots of the ipecacuanha plant, was for long an efficient remedy against amoebic dysentery until resistant strains emerged.

Few of us do not indulge in at least one cup of tea, coffee or cocoa daily. These drinks, having a pleasant taste and aroma, have served from time immemorial as mild tonics. Their effects are derived from caffeine, theobromine and theophylline, alkaloids of the purine group which are present in tea leaves and the beans of coffee and cacao (Figure 7.2). All of these substances are stimulants of the central nervous system (CNS) and can

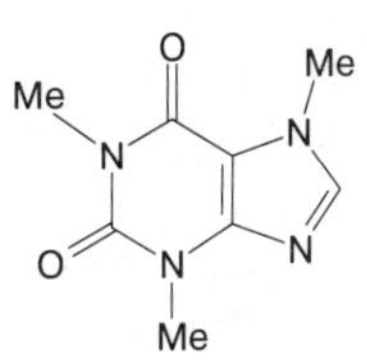

Caffeine

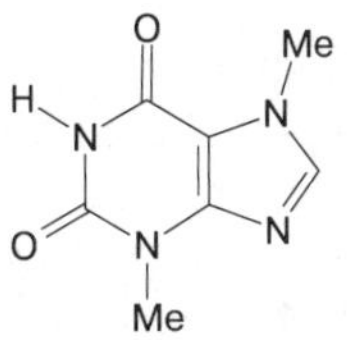

Theobromine

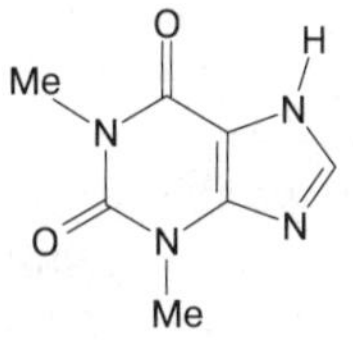

Theophylline

Figure 7.2 Purine alkaloids.

exert a diversity of effects on other organ systems.

Theobromine and theophylline possess vasodilator and diuretic properties. An inherent property of many alkaloids, which makes them medically useful, is their ability to act on the central or peripheral nervous system. For example, the piperidine alkaloid morphine is well known as a pain killer and is sometimes used as an adjunct in cancer chemotherapy (Figure 7.3). These substances suppress the sensitivity of nerve endings, inhibit the conduction of impulses through nerve fibers and induce a weak hypnotic effect. *O*-Methyl morphinate, otherwise known as codeine, is used as an antispasmodic and expectorant in cough treatments. Morphine and codeine are components of opium which is obtained as a milk from unripe somniferous poppy seeds. Cocaine is present in the leaves of the coca shrub which grows in South America, South Eastern Asia and elsewhere.

Atropine is another alkaloid of the piperidine series found in belladonna,

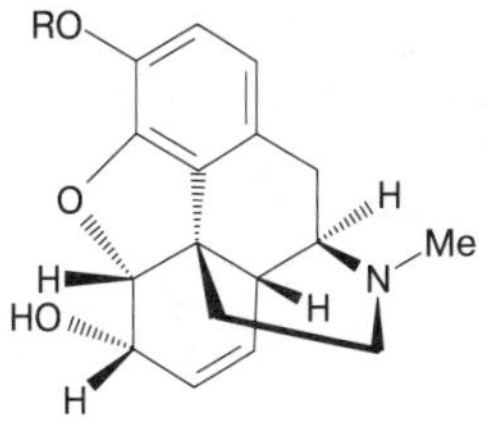

Morphine (R = H)
Codeine (R = CH_3)

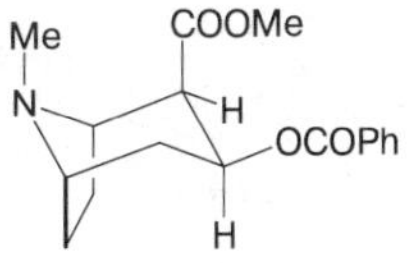

Cocaine

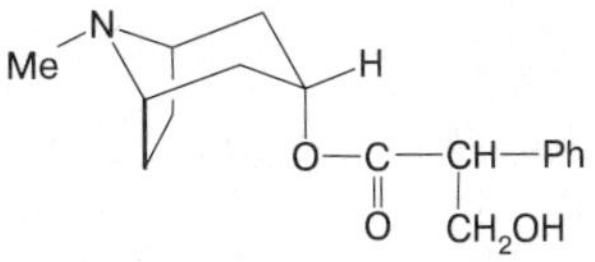

Atropine

Figure 7.3 Piperidine alkaloids.

henbane, Jamestown weed and other plants of the nightshade family, and was at one time widely used by doctors in ophthalmic practice for diagnosis and treatment. Atropine is an antispasmodic muscle relaxant. When applied to the eyeball, it induces pupil dilation. The use of atropine as an antidote in cases of intoxication with narcotics, hypnotics or strong toxins such as muscarine and pilocarpine is of great interest. Atropine is assumed to replace these poisons at the corresponding biological receptor sites, thus eliminating their toxic effect.

The shrub *Rauvolfia serpentina*, native to South and South Eastern Asia, contains the indole alkaloid reserpine (Figure 7.4) which is used as a tranquilizing agent. Moreover, reserpine decreases arterial blood pressure and is useful in the treatment of hypertension.

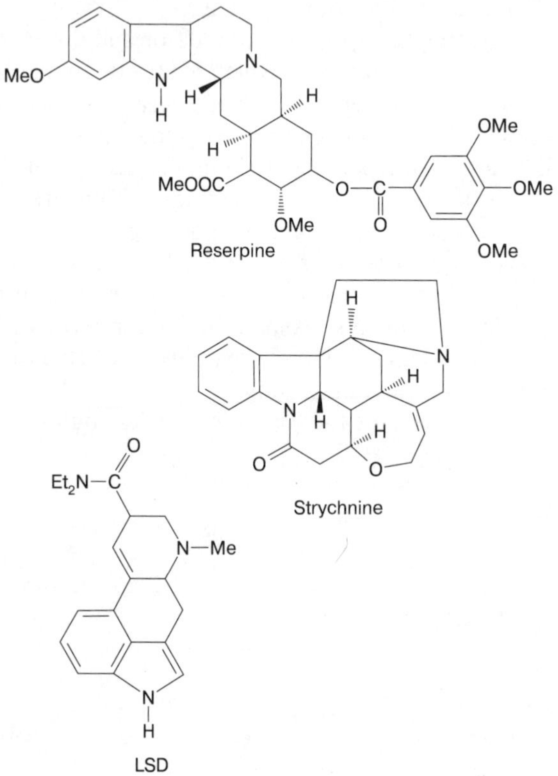

Figure 7.4   Indole alkaloids.

We should emphasize that the biological activity of many alkaloids depends dramatically on the dose level, as can occur with all biologically active compounds. Natural substances can be positive or curing ('angelic'), but at a different dose the same compounds can also display a 'negative' or toxic ('evil') face. For example, the alkaloid strychnine in small doses acts as a cardiac stimulant. In larger doses, however, the same strychnine acts as a convulsive poison and can cause respiratory paralysis and rapid death. The powerful narcotic effects of morphine and cocaine are well known. Repeated infusion of these alkaloids causes drug addiction (morphinism, cocainism). The diethylamide of lysergic acid (the indole alkaloid with the abbreviation LSD) is a notorious drug which may induce hallucinations as a side effect. Figuratively, the ancient warning by Homer seems timely

Trojans, be cautious of gifts offered you by the Greeks!

Many attempts have been made to eliminate the unfavorable physiological properties of the above-mentioned alkaloids. An efficient way to achieve this aim seemed to be to test various substituted derivatives. However, random variation of structure is at best expensive and may lead to the synthesis of even more dangerous compounds, a notorious example of which is synthetic morphine O,O'-diacetate, the narcotic heroin. A more effective strategy would entail determining which fragments of the molecular structure are responsible for the useful biological activity. Thus, the analgesic effects of morphine, cocaine and a series of other local anesthetics have been shown to be induced by so-called anesthesiophorous groups, such as

$$-N-(C)_n-X-\underset{\underset{O}{\|}}{C}-Ar$$

In the case of cocaine $X = O$ and $n = 3$. For morphine $X = O$ and $n = 4$, and the arylcarbonyl group is replaced by an aryl group. In both alkaloids, however, the distance between the tertiary nitrogen atom and the aryl group is approximately the same. A number of synthetic local anesthetics have been developed using this fragment as a basis, and the necessity for cocaine and morphine application has been reduced. Promedol, alphaprodine and procaine (Figure 7.5) are examples of synthetic analgesics which are used extensively. It can be seen that anesthetic molecules may or may not be heterocyclic.

Despite its rather high effectiveness, quinine is not an ideal antimalarial preparation because of its marked toxicity. A further drawback is that cinchona trees still cannot be cultivated worldwide. Malaria remains a serious health problem: according to the data of the World Health Organization, the number of cases worldwide each year is assessed to be about 200×10^6, 3×10^6 of which are fatal. The first approaches to obtaining analogues of quinine were directed at modifying the structure of quinine itself. In Russia this approach led to the discovery of some new efficient preparations, namely chloroquine, plasmocid and others (Figure 7.5). Similar programs involving dozens of universities and companies were developed in the USA during World War II and the Vietnam war. More than 30 000 compounds, mainly quinoline derivatives, were synthesized and tested for antimalarial activity.

Almost all alkaloids have a three-dimensional structure with an intricately

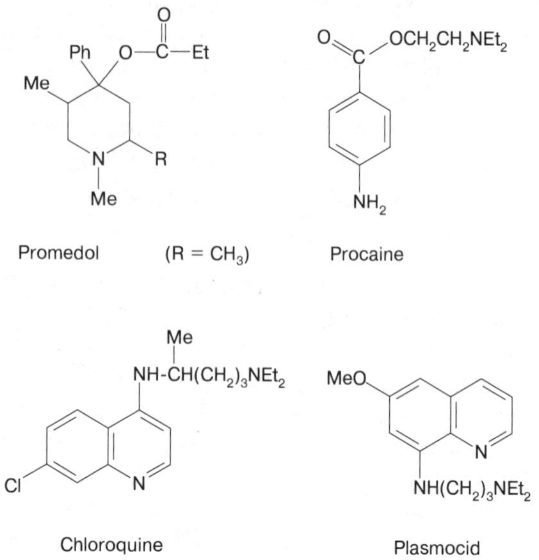

Figure 7.5   Examples of synthetic alkaloids.

developed substitution pattern. Such molecules generally include asymmetric centers and a variety of functional groups capable of forming hydrogen bonds. Pharmacologists and organic chemists describe many of the synthetic compounds as 'alkaloid-like structures'. This indicates that the substance is likely to display prominent biological activity. The specific molecular structures of alkaloids are assumed to consolidate their attachment to the biological receptors responsible for the appearance of a certain function.

7.2   Heterocycles versus Infectious Microbes

7.2.1   IN SEARCH OF 'MAGIC BULLETS'

The twentieth century can be considered as the age of the great drug revolution. Medicinal preparations synthesized in the last 60 years have brought about a decrease in the mortality rate of numerous diseases, and provided relief for many ailments. Widespread success was achieved first and foremost with infectious diseases such as pulmonary inflammation, tuberculosis, cholera and various purulent infections. For thousands of years, these afflictions were a scourge of mankind.

The drug revolution was preceded by a number of earlier discoveries. The most important of these was made by Louis Pasteur in the 1860s. He established that the source of infectious diseases was invisible pathogenic microbes. Later, the German scientist Koch elaborated procedures for growing pure bacterial cultures. It thus became possible to study the action of chemical substances on different species of bacteria.[†] For the initiator of chemotherapy, Erlich, the objective was to find a 'magic bullet' among the compounds tested which would be specific for the target pathogenic microorganism, and would therefore be absolutely harmless to the patient. This approach is the core of the selective toxicity principle. At present many heterocyclic compounds have been found to possess 'magic bullet'-like properties. At the beginning of this century antibacterial activity was discovered in some heterocyclic cationic dyes. In particular, acridinium salts such as proflavine and ethacridine (Figure 7.6) were used with great success during World War I as antiseptics for the disinfection of wounds. Another well-known antiseptic is the phenothiazine dye methylene blue (Figure 7.6).

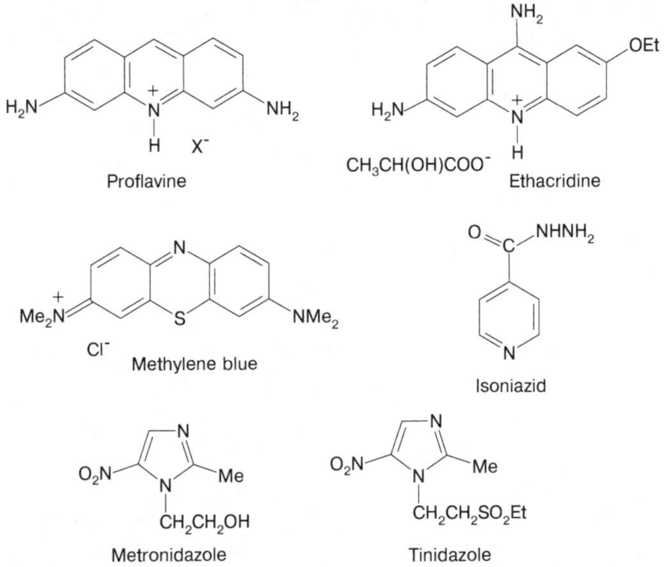

Figure 7.6 Typical heterocyclic pharmaceuticals used in the treatment of infectious diseases.

The hydrazide of isonicotinic acid (isoniazid) played a principal role in the treatment of tuberculosis, and around 1980 gave rise to hope that this disease had been conquered. Unfortunately, resistant forms have since emerged. From the mid-1960s nitroimidazole drugs such as metronidazole and tinidazole became available (Figure 7.6). Such preparations radically changed the treatment of *Trichomonas* infections. These heterocycles proved to be

---

[†]Viruses, pathogenic fungi and protozoa (the simplest one-cell animals such as the cholera germ, *Plasmodium malariae*, etc.) can also serve as infectious disease inducing pathogens. Information concerning antiviral treatments can be found in Section 7.3.

'magic bullets' because of their high potency in relation to the parasites and surprisingly low toxicity toward humans. It is of interest that metronidazole also assists in the cure of alcoholism.

A true revolution in the struggle against infectious diseases was brought about by two classes of medical preparations: the sulfanilamides and antibiotics. Their discovery and widespread use almost coincided with one of the most dreadful events in human history, the beginning of World War II. As a result of these treatments millions of lives were saved.

7.2.2 SULFANILAMIDES AND HETEROCYCLES

Red sulfanilamide and the less toxic white sulfanilamide (second generation) were the first sulfa drugs (Figure 7.7), and these contained no heterocyclic fragments. However, the intensive research work that followed their discovery

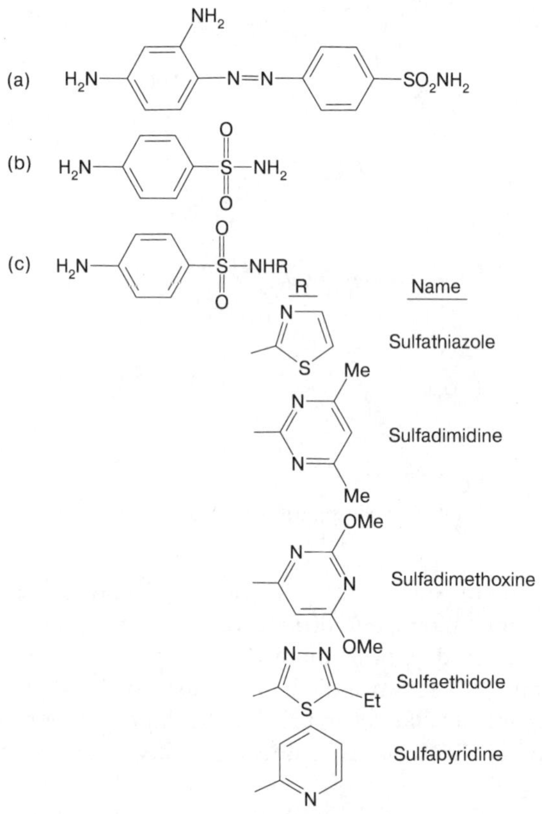

Figure 7.7 Typical sulfa drugs: (a) red sulfanilamide, (b) white sulfanilamide and (c) heterocyclic derivatives.

demonstrated that modification of the p–aminobenzenesulfonamide structure by the introduction of heterocyclic substituents into the amide markedly enhanced the biological activity. More than 30 derivatives of this type, including the well-known sulfathiazole, sulfadimidine, sulfadimethoxine, sulfaethidole and others, were gradually introduced into clinical treatment. The first sulfanilamide medicine synthesized in Russia in the 1930s by Postovsky contained an α-pyridyl substituent at the amide nitrogen (the structure of sulfapyridine is shown in Figure 7.7).

Sulfa drugs are highly efficient against many bacterial species and against some protozoa. Catarrhal illnesses, gastrointestinal infections, meningitis, scarlet fever, tuberculosis and bubonic plague have been successfully treated by such preparations. Simple changes in the heterocycle substitution pattern enable the formation of drugs with either short-lived or prolonged action. With the passage of time, however, the increasing evidence of clinical toxicity of these drugs has led to a diminution in their use, and they have been replaced to a great extent by penicillins, cephalosporins and, more recently, quinolone drugs (see Sections 7.2.3 and 7.2.4).

For sulfanilamides, the biological target and mechanism of action are well established. The normal growth and development of bacteria requires p-aminobenzoic acid for the synthesis of folic acid, which in turn regulates the production of purines and pyrimidines (see Section 4.2.2). The basic fragment of all sulfanilamide drugs is structurally very similar to p-aminobenzoic acid. Therefore, the enzyme which controls the attachment of p-aminobenzoic acid to the pteridine cycle mistakenly binds p-aminobenzenesulfonamide. This brings about an abrupt deceleration of folic acid biosynthesis, and consequently delays the biosynthesis of nucleic acid and bacterial cell proteins. The final result of the process is disastrous for the bacteria.[†] The reason sulfanilamides do not exert a similar negative effect on human cells is that human beings cannot produce folic acid, and therefore have no enzymes which manipulate its biosynthesis. Man and other animals must obtain this vitally important substance (vitamin) from their diet.

The role of the heterocyclic radicals in sulfanilamides is not yet precisely known. However, all the heterocyclic sulfa drugs contain a pyridine-like nitrogen and the heterocycles are rather strong electron-accepting moieties. This supports the assumption that the heterocyclic fragment increases the acidity of the sulfamide N–H linkage making it close to that of p-aminobenzoic acid. The anion formed by dissociation of the N–H bond (Figure 7.8a) is likely to be delivered to the target infection more quickly owing to its increased solubility in blood compared with the neutral molecule. Moreover, the anion might be more readily attached to the active

---

†The creation of medicines which act as antagonists of the natural substrates (metabolites) of specific enzymes is the essence of the antimetabolite concept. We deal later with further applications of this concept.

site of the enzyme. There may, of course, be more than one explanation. The presence of a heterocyclic substituent, together with the enhanced NH acidity, favors the conversion of sulfanilamide into its tautomeric imino form (Figure 7.8b). This tautomer is stabilized by intramolecular hydrogen bonding and may succeed in binding to and thus inhibiting the enzyme.

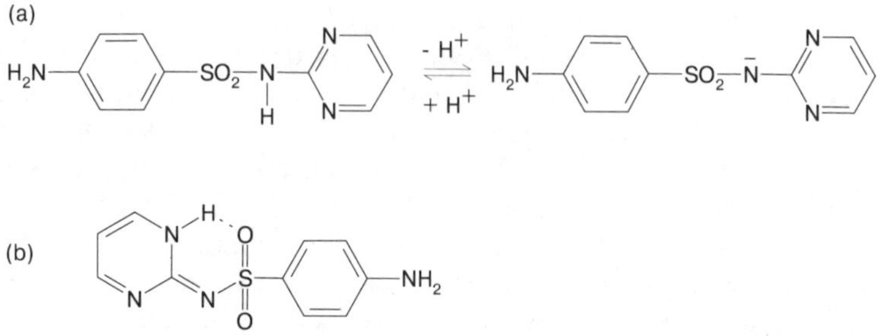

Figure 7.8 (a) Acidic ionization of sulfadimidine and (b) tautomeric form of sulfadimidine.

7.2.3 ANTIBIOTICS

Antibiotics are compounds with rather complicated structures that are produced by certain species of microorganisms supposedly as 'chemical weapons' in the battle against hostile microbes. However, this assumption has not been proven conclusively and the role of antibiotics in nature is still in question. Man has managed to 'domesticate' some species of microorganisms and use their antibiotics for protection against pathogenic bacteria, fungi and even some viruses.

The history of antibiotics began in 1929 when the Scottish scientist Fleming discovered that a *Staphylococcus* culture, contaminated with green mold, had been killed. In 1940 Florey and Chain separated the active agent from the mold of *Penicillium notatum* and named it penicillin. Just one year later, the use of penicillin had spread far and wide, scoring resounding triumphs over the omnipresent microenemies of the human race. Interestingly, the structure of this efficient and low toxicity antibacterial drug was elucidated only in 1945 with the assistance of X-ray structural analysis. The molecular structure of the antibiotic is a condensed heterocyclic system composed of a five-membered thiazole ring and a four-membered azetidine nucleus in the form of a β-lactam, the single cyclic nitrogen atom being common to both rings (Figure 7.9).

The structure of penicillin was surprising to many chemists because of the well-known instability of β-lactam rings, especially toward hydrolysis. Therefore, the natural occurrence of similar structures seemed unlikely at the

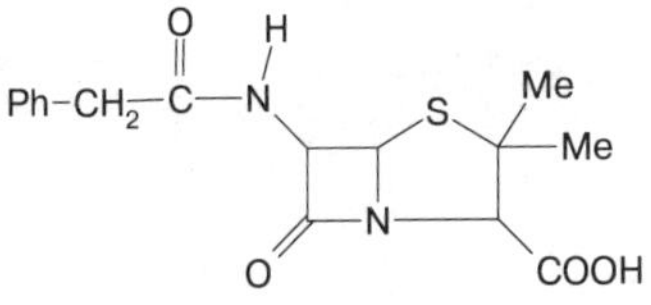

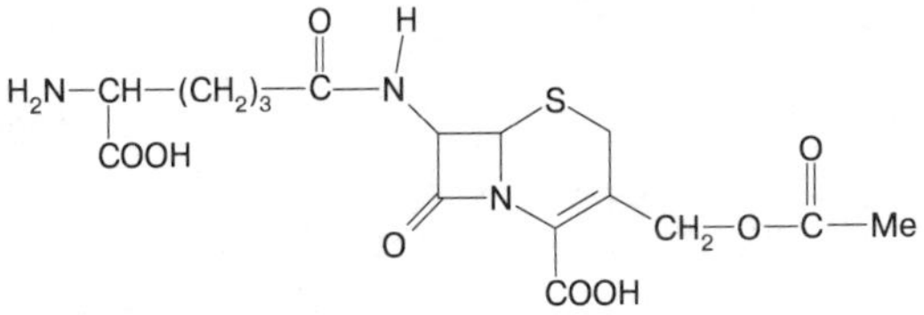

Penicillin G

Cephalosporin C

Figure 7.9 β-Lactam antibiotics.

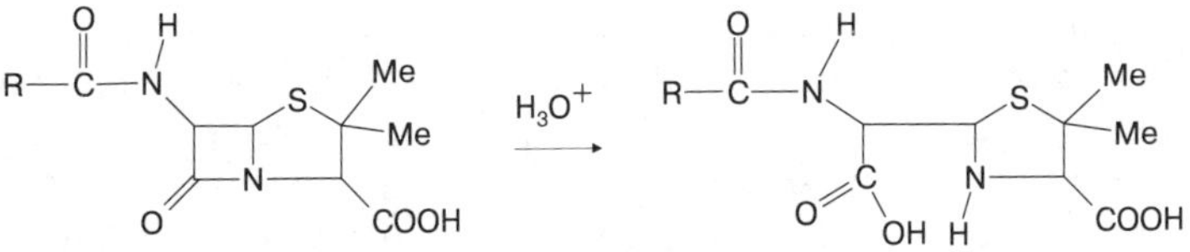

Figure 7.10 Hydrolytic ring opening of penicillin.

time. The β-lactam fragment is indeed a critical feature of the penicillin molecule, responsible for both its frailty and for its bioactivity. Once the heterocycle is ring opened (for instance, by acidic hydrolysis), the antibacterial activity is destroyed because the penicilloic acid formed is not active (Figure 7.10).

The first medicinally important penicillin contained an amino group acylated by phenylacetic acid at the 6-position. This compound, however, had the significant disadvantage of exhibiting activity against Gram-positive bacteria only. Gram-negative species including *Escherichia coli* remained completely unaffected. Gram-negative bacteria possess a highly effective protective mechanism involving a β-lactamase enzyme (penicillinase) which cleaves the lactam OC–N bond with great selectivity, thus deactivating the antibiotic.

Attempts were made to change the structure of the original penicillin G to render it resistant toward hydrolysis. However, the only type of modification to be successful without the loss of biological activity involved changing the substitution pattern at the aminocarbonyl group in penicillin G. This involves changing the acid residue attached to the amino nitrogen. Many new penicillins are prepared biosynthetically from mixtures of water, carboxylic acid and other components using a diversity of mold fungi. In the penultimate stage of biosynthesis 6-aminopenicillanic acid is formed. Subsequent acylation by the chosen carboxylic acid gives the desired penicillin derivative (Figure 7.11).

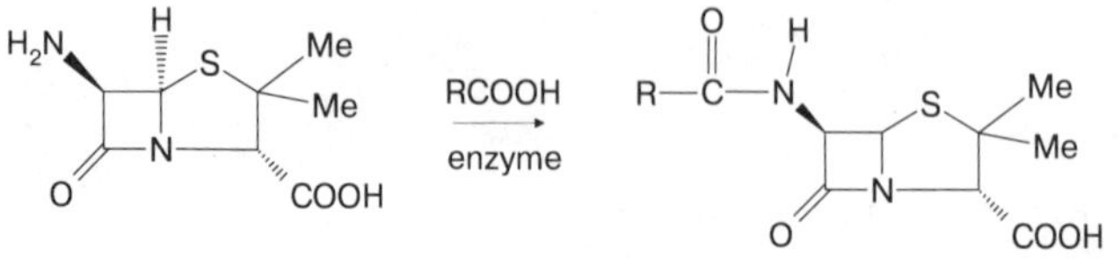

6-Aminopenicillanic
acid

e.g.,
Phenoxymethyl penicillin (R = PhOCH$_2$)
Ampicillin (R = PhCH(NH$_2$)-)

Figure 7.11　Final stage of the preparation of semisynthetic penicillins.

Among the thousands of semisynthetic penicillins prepared, a number exceed penicillin G in terms of hydrolytic stability and demonstrate an appreciable effect on Gram-negative bacteria. Such was the case for phenoxymethylpenicillin and ampicillin (Figure 7.11). The situation is further complicated by the fact that the increased activity against Gram-negative bacteria is associated with a decreased potency toward Gram-positive species. Nevertheless, antibiotics of this type have a wide spectrum of action and are therefore useful in medical therapy.

Because of the greater stability of the lactam ring, cephalosporin antibiotics show greater effectiveness against Gram-negative bacteria than the penicillins. Moreover, the structure of cephalosporins may be modified not only at the amido nitrogen (7-position) but also at the 3-position. Numerous semisynthetic first-, second- and third-generation cephalosporins have been obtained, including cefatrizine, cefuroxime and cefotaxime (Figure 7.12). Each new generation of cephalosporins is characterized by an increase in activity. Thus, cefotaxime (a third-generation antibiotic) is highly effective against resistant forms of Gram-negative bacteria.

The high activity of β-lactam antibiotics suggests that they, like the sulfanilamides, are antimetabolites of a specific bacterial metabolite. Although this metabolite has yet to be found, it is known that penicillins and cephalosporins block the activity of an enzyme called transpeptidase which is responsible for the crosslinking of bacterial cell walls. Thus, in the presence of antibiotics, the cell walls become weaker and are ruptured by osmotic pressure, ultimately killing the bacteria.

Unfortunately, there will never be a drug without some disadvantage.

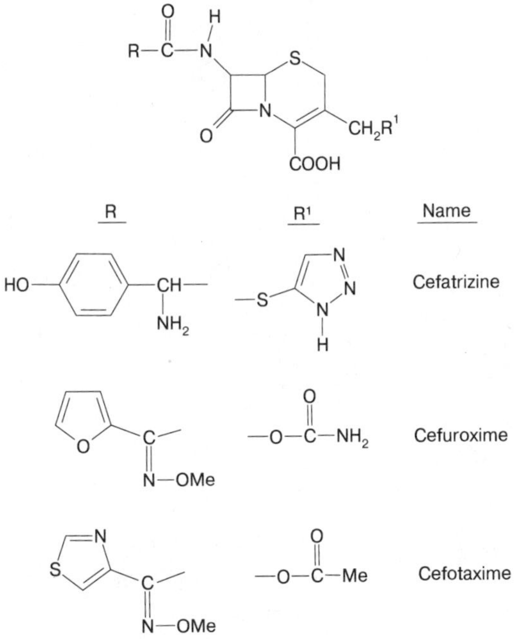

Figure 7.12   Several generations of cephalosporins.

Therefore, it is no surprise that antibiotics are not devoid of defects. Bacterial species that are exposed for what are long periods, on the time scale of bacterial reproduction cycles, to a particular antibiotic develop resistant strains. All other antibacterial drugs also fall victim to this ability of bacteria. Consequently, there is a constant need for the development of new antibiotics.

7.2.4   COMPETITORS ARE WANTED

In the last 40 years, much effort has been devoted to the discovery of new classes of antimicrobial drugs that could compete with the effectiveness of sulfa drugs and antibiotics for human health protection. Heterocyclic compounds make up the majority of successes in this endeavor. Thus, in the 1950s nitrofuran drugs possessing high potency toward both Gram-positive and Gram-negative bacteria were introduced into clinical practice. Furazolidone, for example (Figure 7.13), is used in the treatment of dysentery, and enteric and paratyphoid fevers. Nitrofurans often moderate the development of microorganisms which are resistant to sulfa drugs and antibiotics. However, in recent years the use of nitrofurans has been phased out owing to their tendency to show mutagenic activity.

Great hopes are now held for a relatively new group of antibacterial drugs

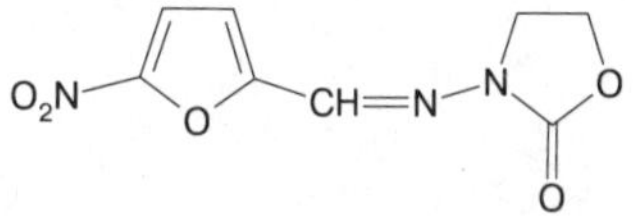

Furazolidone

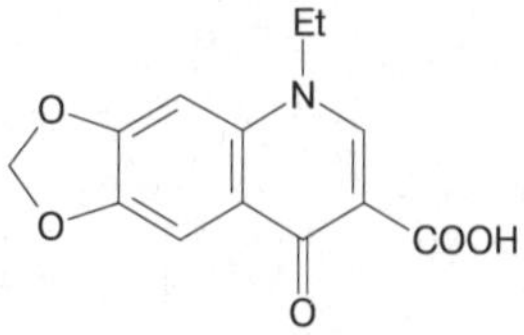

Oxolinic acid

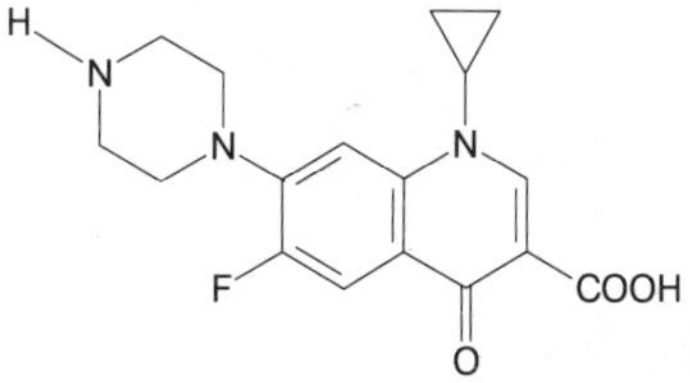

Ciprofloxacin

Figure 7.13   Antibacterial drugs of the nitrofuran and 4-quinolone-3-carboxylic acid series.

which are derivatives of 4-quinolone-3-carboxylic acid. A typical representative is oxolinic acid (Figure 7.13). Like other compounds of the series, oxolinic acid displays a wide spectrum of action, especially towards Gram-negative bacteria. These quinolone derivatives are active against bacterial strains which are resistant to sulfanilamides and antibiotics. Ciprofloxacin, containing a fluorine atom at the 6-position and an amine function at position 7, belongs to the most recent generation of the quinolone series (Figure 7.13). The quinolone drugs target the protein enzyme DNA gyrase which unwinds the tightly packed double-helical structure of bacterial DNA. Such action prevents bacterial replication and transcription.

7.3   Heterocycles and Viral Infections

In our generation, significant scientific effort has been devoted to finding a cure for AIDS (acquired immunodeficiency syndrome). It is still hoped that an anti-AIDS drug will be discovered as effective against pulmonary inflammation as penicillin and sulfanilamide are against bacterial infection. However, the results to date have been rather disappointing. This is not surprising considering that the well-known viral diseases influenza, hepatitis, encephalitis and poliomyelitis are not yet effectively controlled by drugs.

Despite enormous financial input, the chemotherapy of viral infections is still in its infancy. While several antiviral drugs are currently available, their number and, in particular, their mode of action do not give us reason for great optimism at present. As an example, we consider amantadine, a derivative of the hydrocarbon adamantane (Figure 7.14). Amantadine is considered to be an antiinfluenza drug. However, it is ineffective in the treatment of patients who have already developed the illness, and therefore serves only as a prophylactic for a limited number of people. Other antiviral drugs suffer from related drawbacks.

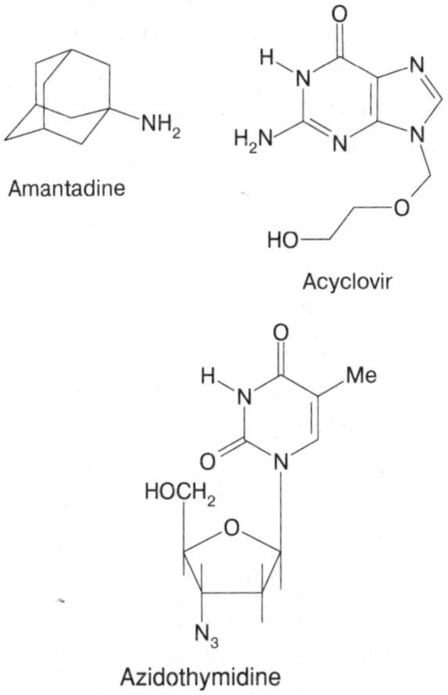

Amantadine

Acyclovir

Azidothymidine

Figure 7.14 Some antiviral drugs.

What is the main barrier to effective drug therapy of viral diseases? Surprising as it may seem, it is the relative simplicity of the structure and of the biochemistry of the replication of viruses that is paramount. Bacteria possess rather complicated organizations with significant differences between their cells and those of animals. We have already discussed the dependence of bacterial growth on p-aminobenzoic acid and some of the structural peculiarities of bacterial cell walls. However, there are further significant differences. Bacterial cells can multiply independently without penetration into a human cell. This is not the case for viruses. Viruses are living organisms that are smaller than bacteria. They are usually composed of a protein shell which envelops a DNA or RNA molecule in much the same way that a sarcophagus encases a mummy (such species are called virions).

A virus can multiply only inside a host cell. To this end, a viral carrier approaches the host cell, becomes attached to the wall and then penetrates the host cell. Once inside, the virus inserts its single DNA molecule, which contains a coded genome, into the DNA of the host cell. Now the invader commences the process of viral DNA replication and the rapid assembly of protein envelopes for the newly formed viral DNA from the building blocks within the host cell. In these processes the parasitic guests destroy the host cell. A multitude of viruses are thus formed which are then released to invade further cells of the infected organism. In RNA-containing viruses, the RNA molecule is bound to the ribosome rather than to the mRNA of the host cell. This leads to the synthesis of virus proteins.

The biological action of viruses makes it very difficult to create an effective drug as the therapeutic agent would need to be able to differentiate between the biological target, the virus, and the host cell. However, some progress has been made. For the synthesis of the highly specific antiviral drug acyclovir (Figure 7.14), the first of its kind, the Americans, Elliot and Hitchings, were awarded a Nobel prize in 1988 (note that the discoveries of the sulfa drugs and of the antibiotics were previously recognized with Nobel prizes). Acyclovir is highly effective against herpes infections. In the USA, approximately 60×10^6 people are infected with this virus. Zoster encephalitis is especially dangerous, and only 25% of patients with brain inflammation previously survived. The survival rate has increased to 75% with the aid of acyclovir. Acyclovir's activity results from its structural similarity to deoxyguanosine, a nucleoside necessary for DNA assembly (see Section 3.1). Once inside an infected cell, acyclovir is phosphorylated at its hydroxy group to produce the triphosphate (Figure 7.15). The first phosphorylation proceeds under control of the enzyme thymidine kinase formed by the virus itself, while addition of the remaining two phosphate residues is mediated by cellular kinases. During the replication of viral DNA the viral enzyme DNA polymerase mistakenly inserts acyclovir triphosphate into the growing viral DNA chain instead of deoxyguanosine triphosphate. In contrast to deoxyguanosine, the drug contains no 3'-OH group to which the following nucleotide can be linked. Thus, the chain is prevented from growing further.

The reasons why acyclovir is not incorporated into the host cell DNA are not yet completely clear. We do know that human cellular kinases are unable to catalyze effectively the first phosphorylation and, furthermore, that the DNA polymerase of herpes viruses has a much greater affinity toward acyclovir triphosphate than the host cell DNA polymerase. Thus, the possibility of creating an antiviral drug with high specificity has been realized in principle. The first and most difficult step has thus now been taken to the production of truly effective antiviral agents.

Elliot and Hitchings introduced another antiviral drug, azidothymidine or AZT (Figure 7.14) into medical practice. AZT, now officially named zidovudine, is useful in the treatment of AIDS. It impedes the human immunodeficiency virus replication process but does not cure the disease completely. The drug is an analogue of the naturally occurring nucleoside

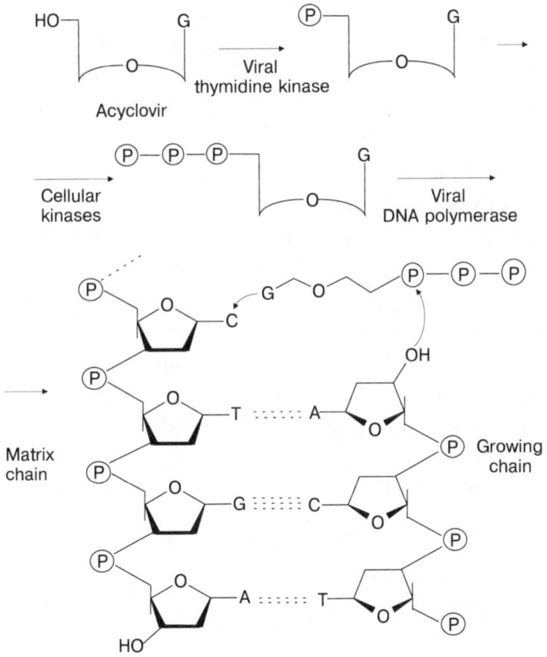

Figure 7.15 Mechanism of acyclovir action (adapted from Hirsh, M. S. and Kaplan, J. C., *Sci. Am.*, 1987, **256**, 76, with permission. © Scientific American Inc., All rights reserved).

thymidine and its mechanism of action is similar to that of acyclovir, i.e. consistent with the metabolite concept.

The successful use of acyclovir and azidothymidine suggests that potential antiviral drugs could resemble purine and pyrimidine nucleosides. However, this does not represent the only approach. Viruses may have other weaknesses that could be exploited. For example, the ability of viruses to attach to the host cell wall could, in principle, be hindered. This would prevent the virus from penetrating and therefore infecting the cell. Alternatively, the mechanism of shedding the protein envelope could perhaps be inhibited. Such activities could be shown by drugs containing nonnucleosidic structures. However, the identification and utilization of suitable structures lie in the future.

In closing this section we wish to remind readers that the first compounds to be utilized in the struggle against viral infections were heterocycles.

7.4 Heterocycles and the Diseases of Our Century

Since the turn of the twentieth century, the two major causes of death in industrially developed countries have been cardiovascular disease and cancer.

These diseases were promoted to their present dominant positions owing to the great successes in the control of infectious disease on the one hand, and because of increasingly poor dietary habits combined with stress and abnormal environmental factors on the other. Heart disease and cancer, together with nervous system disorders, are therefore often referred to as the 'diseases of the twentieth century'. Some scientists consider them to be 'specific', i.e. programmed into the genetic code to limit the lifespan of an individual organism. Nevertheless, it is generally held that the human race has not yet reached the upper limit of its potential life-span which has been suggested to be 120 or even 150 years. Improvements in social and economic conditions and achievements in medicine, biology and chemistry have already increased the average life expectancy to 80 years in some countries. A central role in this progress has been played by modern medicinal chemistry. The treatment of the majority of known diseases can today be carried out reasonably successfully. As regards the 'diseases of the twentieth century', our achievements are still modest in the case of malignant tumors, but are improving in the therapy of cardiovascular illnesses, and especially in the case of nervous system diseases.

7.4.1   HETEROCYCLES TO CURE STRESS

The mechanisms of nervous processes in humans are extremely complicated and cannot be discussed here. We merely note that the transmission of nervous impulses is always accompanied by a release of neuromediators at the nerve fiber endings. These mediators are chemical substances which affect receptors and induce a particular biochemical response, such as gastric juice secretion or elevation of arterial pressure. As a rule, each mediator interacts with several (usually two to four) specific receptors. Acetylcholine, noradrenaline, adrenaline and serotonin are the main neuromediators (Figure 7.16). Serotonin functions mainly in the central nervous system (CNS) and the other three mediators act on the peripheral nerve endings. Drugs generally act on the nervous system by the inhibition of neuromediator release or by interaction with specific receptors owing to the similarity in the structure of the drug to that of the mediator. Where a drug mimics a mediator, the drug can be either an antagonist or an agonist of the mediator. Many mediators and receptors exist. Therefore, a drug can act on a number of biotargets. Among the many known nervous system drugs, many different structures are encountered ranging from aliphatic to heterocyclic, but it is the heterocycles which predominate.

As already mentioned, neurotropic activity is an inherent property of many alkaloids, and the first drugs of this type were indeed alkaloids. The mechanism of their neurotropic action and their biological targets are well known. For example, atropine inhibits choline receptors, thus interrupting

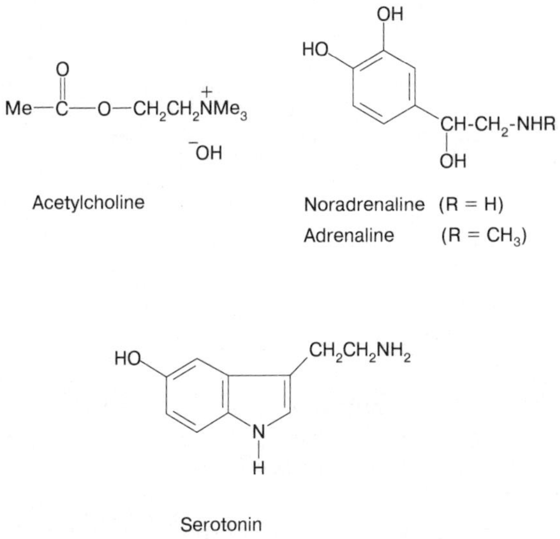

Acetylcholine

Noradrenaline (R = H)
Adrenaline (R = CH₃)

Serotonin

Figure 7.16 Neuromediators.

their interaction with acetylcholine and causing muscle relaxation. Since acetylcholine is a peripheral neuromediator, atropine exerts a powerful local action and was formerly used clinically for treating intestinal, urinary and bronchial spasms. The narcotic effect of the alkaloid morphine is accounted for by its blockage of serotonin receptors in the CNS.

The nature of the stimulatory action of caffeine and other purine alkaloids is intriguing and is related in a rather complex way to noradrenaline and adrenaline function. During emotional excitation the adrenal cortex is known to enhance sharply the release of these two mediators, resulting in an increased flow of blood to all organs. Adrenaline and noradrenaline interact with the enzyme adenylate cyclase which seems to be one of the receptors. Activated by the neuromediators, adenylate cyclase thus transforms adenosine triphosphate first into the monophosphate form and then into cycloadenylic acid (Figure 7.17). Cycloadenylic acid plays the role of a secondary mediator (hormone) by activating phosphorylase, an enzyme which stimulates physiological processes such as cardiac activity and glycogenosis in the liver. As a cyclic diester, cycloadenylic acid can be hydrolyzed to AMP. This hydrolysis is catalyzed by a widely distributed natural enzyme. Caffeine and other purine alkaloids are thought to bind the enzyme concerned, resulting in an increased cycloadenylic acid concentration and thus in the onset of a stimulating effect. There is little doubt that purine alkaloids serve as antimetabolites of adenine, which probably interacts with the enzyme in question under normal conditions.

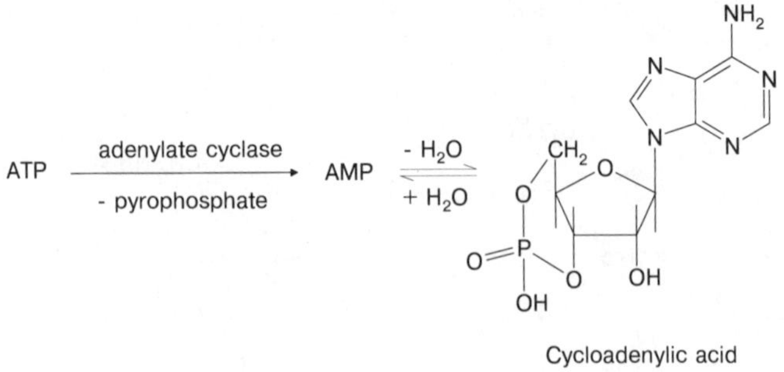

Cycloadenylic acid

Figure 7.17   Conversion of ATP into cycloadenylic acid.

Barbiturates, which are 5,5-disubstituted derivatives of barbituric acid (Figure 7.18), were the first synthetic drugs found to exert significant action on the central nervous system (CNS). Barbiturates have a hypnotic effect and suppress the CNS. Consequently, the main uses of such compounds are as tranquilizing and soporific agents. Two examples of typical barbiturates, barbital[†] and Luminal (phenobarbital), are shown in Figure 7.18. The parent of these compounds, barbituric acid, was synthesized by the famous German chemist von Baeyer in 1864, but it does not markedly affect the CNS.

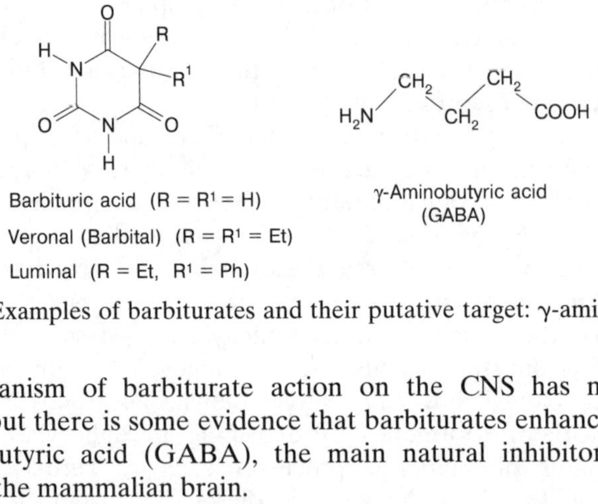

Barbituric acid (R = R^1 = H)

Veronal (Barbital) (R = R^1 = Et)

Luminal (R = Et, R^1 = Ph)

γ-Aminobutyric acid
(GABA)

Figure 7.18   Examples of barbiturates and their putative target: γ-aminobutyric acid (GABA).

The mechanism of barbiturate action on the CNS has not yet been ascertained but there is some evidence that barbiturates enhance the activity of γ-aminobutyric acid (GABA), the main natural inhibitor of nervous processes in the mammalian brain.

---

†Veronal (barbital) was the first sleep inducer to be introduced into clinical practice by Mering in 1903. Mering named the drug after the Italian town Verona. The origin of the name 'barbituric' acid is also curious. Willstaetter, who had been a student of A. von Baeyer, revealed that his then young teacher von Baeyer was attracted to a girl named Barbara, and that the first part of the acid's name was derived from her pet name Barbi. The second part of the name is much more prosaic and originates from one of the two raw materials (urea and ethyl malonate) utilized by von Baeyer to synthesize the heterocyclic ureide, malonylurea.

Although barbiturates continued to be used in clinical practice, derivatives of 1,4-benzodiazepine such as diazepam, nitrazepam, phenazepam and others (Figure 7.19) became increasingly important from the beginning of the 1960s. In a short period of time these derivatives gained worldwide acceptance, judging from their per capita consumption. 1,4-Benzodiazepine tranquilizers reduce or suppress fear, stress and anxiety and have also been used in surgical, pediatric and obstetric applications. Such drugs have been routinely prescribed by the military for the treatment of emotional stress caused by the extreme circumstances in which soldiers live and work. 1,4-Benzodiazepine derivatives, like barbiturates, probably increase the inhibitory action of GABA in the cerebral cortex via their own specific receptors.

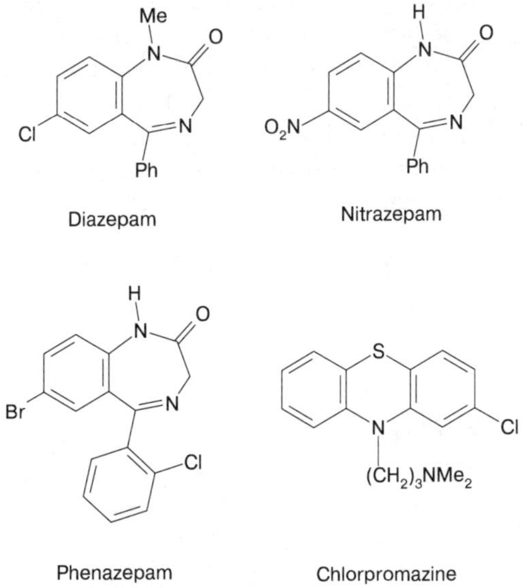

Diazepam   Nitrazepam

Phenazepam   Chlorpromazine

Figure 7.19   Tranquilizers and neuroleptics.

Our historical account skipped a few pages when we first mentioned the 1,4-benzodiazepine tranquilizers because the revolution in psychopharmacology was initiated with the derivatives of another heterocyclic system, namely phenothiazine (Figure 2.8). Phenothiazine derivatives were first introduced into clinical practice in the early 1950s. Chlorpromazine (aminazine) is the outstanding representative of the class (Figure 7.19) and has been widely used for the treatment of various mental disorders including schizophrenia. Phenothiazine-based drugs can reduce aggressiveness, phobias and reactions to external stimuli. Unlike the 1,4-benzodiazepines, phenothiazine derivatives are able to halt episodes of delirium and hallucinations, and phenothiazines are also devoid of pronounced somniferous (sleep inducing) side-effects. Compounds exhibiting such activity are called neuroleptics. They are believed to cancel the excitatory effects of adrenaline and, in particulary serotonin when the levels of these amines in the cerebral cortex are raised by a metabolic disorder.

The discoverer of ascorbic acid, Nobel prize winner Szent-Györgyi, observed that phenothiazines, like adrenaline and serotonin, are strong electron donors (see Section 2.4.3). He therefore came to the conclusion that their biological effect is due not only to their shape but also to their electronic features. The phenothiazine system apparently forms a donor–acceptor type molecular complex with a site on the biological target.

The action of many widely used anesthetics is manifested through the CNS. It has already been mentioned that morphine-type narcotic analgesics are extensively employed in the treatment of acute pain. These narcotics block the transfer of pain impulses and suppress the centers of pain perception in the cerebral cortex. The nonnarcotic analgesics amidopyrine, antipyrine, Analgin (metamizole sodium) and others (Fig 7.20) have a similar but milder effect. These pyrazolone preparations have long been used in the treatment of headaches and neurogenic, muscular and articular pain. Hence, anesthetic drugs may be considered 'molecules of mercy'.

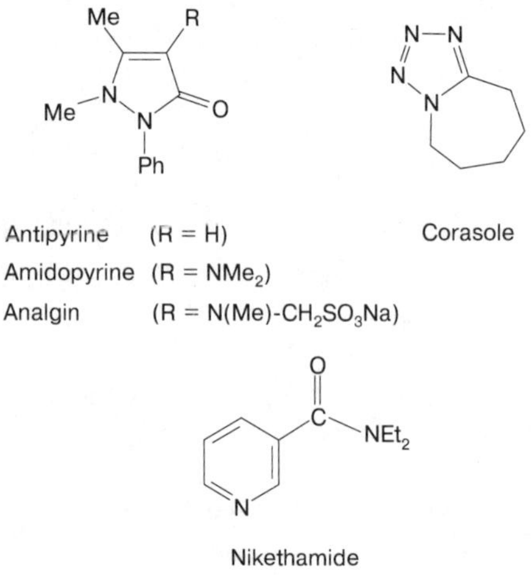

Antipyrine   (R = H)   Corasole
Amidopyrine (R = NMe$_2$)
Analgin   (R = N(Me)-CH$_2$SO$_3$Na)

Nikethamide

Figure 7.20   Synthetic analgesics and analeptics.

Drugs which have a stimulating action are used in the treatment of narcotic or hypnotic intoxications, or to increase cardiac activity when this has been slowed down during the course of an operation. Such compounds are named analeptics (Latin: *analeptica*, reanimating), and are of great importance in medicine. Analeptics influence the CNS by exciting respiratory and vasomotor centers in the cerebral cortex. Among the chief analeptics currently used, we mention specifically two heterocyclic derivatives, corasole and nikethamide, which are derivatives of tetrazole and pyridine, respectively (Figure 7.20).

7.4.2 HETEROCYCLES AND CARDIOVASCULAR DISEASES

High arterial blood pressure (hypertension), coronary spasms (stenocardia) and cardiac arrhythmia are typical manifestations of cardiovascular disorders. The drugs employed for the treatment of the above-mentioned diseases are called antihypertensive, antianginal (vasodilatory, antispasmodic) and antiarrhythmic agents, respectively. To begin, we highlight three features of cardiovascular agents. Firstly, their action is often directed toward the peripheral or central nervous system. Such nervous stimulation in turn signals the appropriate regulatory mechanism to correct the abnormal deviation. Thus, the administration of hypnotic and tranquilizing drugs, such as the alkaloid reserpine (Figure 7.4), also helps to normalize arterial blood pressure. Secondly, cardiovascular agents often produce more than one biological effect simultaneously (e.g. one preparation may halt cardiac spasms and concurrently cause a reduction in blood pressure). Thirdly, the numerous cardiovascular drugs available belong to many different classes of organic compounds.

Taking into consideration the aims of the present book, we focus primarily on the heterocyclic cardiovascular drugs. Many heterocyclic classes are represented: azines and azoles, heteroaromatic and partially hydrogenated compounds, monocyclic and polycyclic, etc.

The vasodilatory effect of alkaloids such as papaverine, theophylline, theobromine and caffeine is well known. A mild hypotensive effect[†] is observed in the case of 2-benzylbenzimidazole (bendazol), which is often used in combination with papaverine and other similar compounds. Interestingly, bendazol also has adaptogenic activity, i.e. the ability to enhance the resistance of an organism toward unfavorable influences such as catarrhal infections.

Clonidine and hydralazine, which are derivatives of imidazoline and phthalazine, respectively (Figure 7.21), exhibit marked hypotensive activity. In the last 20 years derivatives of 1,4-dihydropyridine have been intensively investigated because of their effectiveness in the therapy of hypertension and stenocardial seizures. One of the best-known preparations of the series is nifedipine (Figure 7.21). The biological action of 1,4-dihydropyridines is via inhibition of the calcium channels. Calcium ions control a multitude of intracellular processes. In particular, calcium stimulates activity of the cardiac muscle (myocardium). However, under myocardial ischemia or infarction, cardiac function needs to be facilitated and the demand of the heart muscles for oxygen must be reduced to limit metabolism and the destruction of the cell walls. These demands are met by nifedipine and its analogues which block the channels that enable calcium transport into the cell. Moreover, β-

---

[†]One should distinguish between 'hypertensive' and 'hypotensive' action. The former signifies an increase in blood pressure, whereas the latter indicates a reduction of either high or normal blood pressure.

adrenergic blocking agents block the adrenoreceptors and thus prevent
stimulation by noradrenaline and adrenaline. These blockages induce
dilation of the arteries, an increase in blood flow and a decrease in arterial

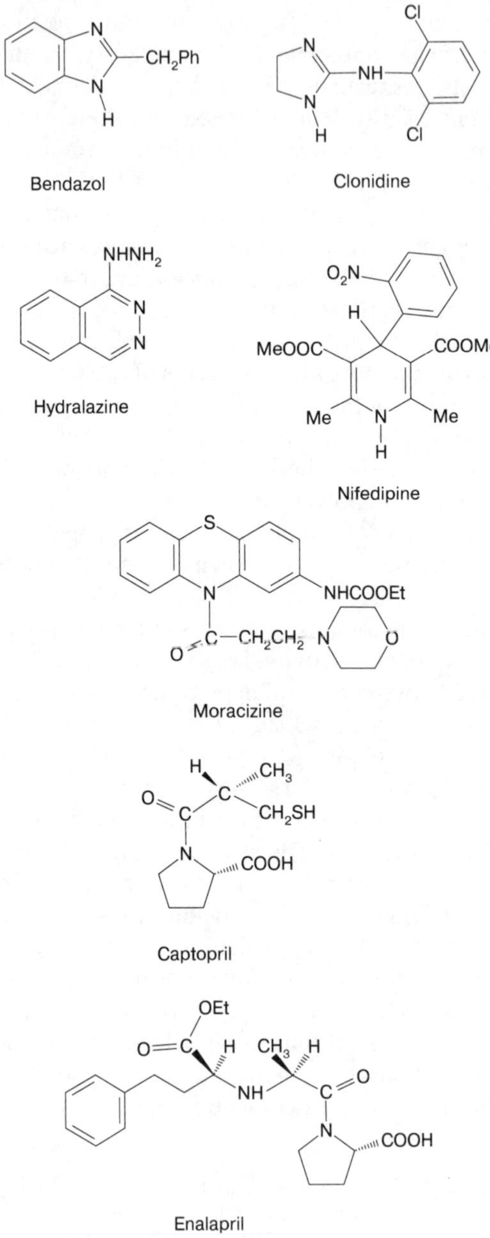

Figure 7.21   Heterocyclic cardiovascular drugs.

pressure and heart rate. These effects make up the antihypertensive and antiarrhythmic roles of the drugs. Some heterocyclic compounds are employed as strong antiarrhythmic agents. The alkaloid quinidine (see Section 7.1) and the phenothiazine derivative moracizine are two examples. Captopril and enalapril (Figure 7.21) act via inhibition of ACE (angiotensin-converting enzyme) which catalyzes the formation of angiotensin II, a powerful endogenous vasoconstrictor. These drugs are thus used as antihypertensives and as adjuncts in the therapy of heart failure.

7.4.3 HETEROCYCLES AND MALIGNANT TUMORS

The fight against cancer has been a principal focus of attention for several decades. The search for a cure for cancer is of the utmost social and economic importance. Clinically effective therapeutic procedures are being intensively sought worldwide. Regretfully, announcement of a medicinal revolution in this field of chemotherapy would be premature. Nevertheless, the chemotherapy of tumors and especially the use of chemicals in conjunction with other methods of treatment can already significantly prolong the lifespans of many patients. In some cases, chemotherapy is able to cure the patient completely, restoring them to health and to at least their natural 'three score years and ten'.

The principal challenge in the chemotherapy of cancer lies in discovering means to discriminate between cancerous and healthy (normal) cells. The main differences between these types of cell lie in the rate of DNA synthesis and replication and also in the rate of cell division: cancerous cells can divide 10^6 times faster than normal cells. These differences have been the focus of strategies aimed at developing a cure for the disease. Current approaches rely on the following considerations. Since the tumorous cells divide more rapidly, they have an increased requirement for purine and pyrimidine nucleotides (for DNA synthesis) compared with the normal cells. If we could prevent access of the nucleotides to the diseased tissues, then the malignant cells would be prevented from reproducing.

This concept has led to the search for anticancer agents among the derivatives of purine and pyrimidine and to some success. Four synthetic anticancer drugs which are currently among the most clinically effective have so far resulted from this quest: 6-mercaptopurine, methotrexate, 5-fluorouracil and ftorafurum (Figure 7.22).

6-Mercaptopurine (shown in Figure 7.22 as the dominant thione tautomer) and methotrexate are effective in the treatment of leukemia, while 5-fluorouracil and ftorafurum are used in cases of ventricular, intestinal, ovarian, pancreatic and mammary glandular tumors. The antimetabolite concept accounts for the bioactivity of all four substances. Mercaptopurine actively interferes with DNA processes by its structural similarity to adenine and

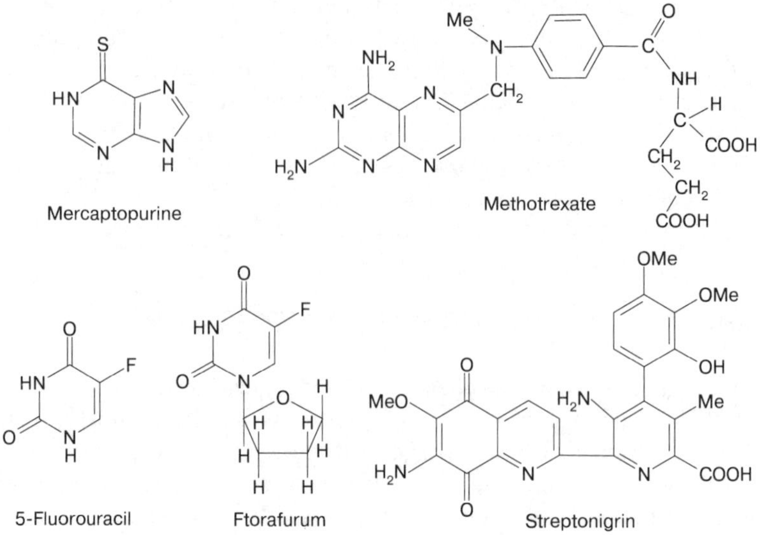

Figure 7.22 Anticancer drugs.

hypoxanthine. Methotrexate and fluorouracil inhibit the biosynthesis of thymidine and thereby reduce its content within cells. Thymidine is an essential nucleoside for the construction of DNA. The critical stage of thymidine synthesis appears to be the methylation of deoxyuridine monophosphate at the 5-position with the assistance of the enzyme thymidylate synthetase and a methylene adduct of tetrahydrofolic acid as coenzyme (see Figure 4.30). In the course of this reaction a molecule of dihydrofolic acid is formed. For the process to be reversible, dihydrofolic acid must be hydrogenated to the tetrahydrofolic form. The enzyme dihydrofolate reductase catalyzes this transformation. It is at this point that methotrexate becomes involved. Because of the similarity to dihydrofolic acid, methotrexate inhibits dihydrofolate reductase, thus inhibiting the biosynthesis of thymidine.

Fluorouracil and ftorafurum, once delivered to the malignant cells, are converted into the 5-fluoro derivative of deoxyuridine monophosphate. Because of their resemblance to deoxyuridine monophosphate (see Figure 3.3), they block the action of thymidylate synthetase and therefore also delay thymidine biosynthesis.

There are many naturally occurring heterocyclic compounds, especially among the alkaloids and antibiotics, which display anticancer activity. For instance, the antibiotic streptonigrin, containing both a pyridine and quinoline nucleus with a rather intricate substitution pattern (Figure 7.22), inhibits DNA formation and is used in the clinical therapy of lymphogranulomatosis and lymphoid leukosis.

However, a serious drawback of most existing anticancer drugs is the lack of specificity of action. As a consequence, most chemotherapeutic agents are also highly toxic toward normal cells. Thus, more effective future treatments of malignant tumors with fewer side effects necessitates a new approach to circumvent the problem of nonspecificity. One such promising method was reported around 1985. This strategy is based on the use of the biologically innocuous compound 8-methoxypsoralen (or simply psoralen). Psoralen (Figure 7.23) is injected into the bloodstream of patients suffering from such lethal forms of cancer as T-cell lymphoma. After a period of time, 500 ml of blood are taken and the malignant T-lymphocytes are separated from the other components of the blood by centrifugation. The separated T-lymphocytes are then suspended in solution under physiological conditions

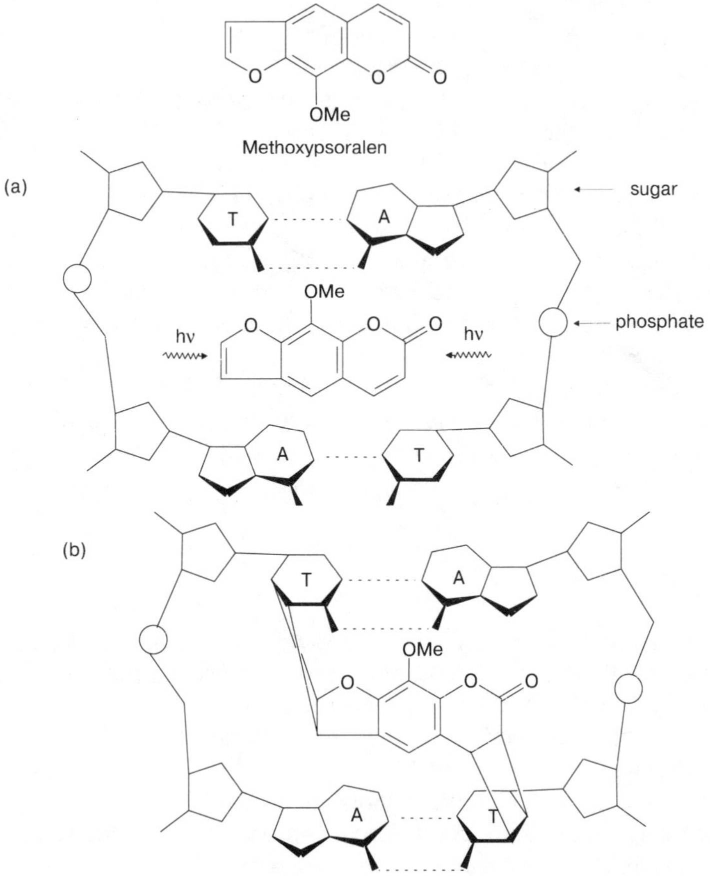

Figure 7.23 Photointeraction of the intercalated 8-methoxypsoralen molecule with thymine residues in a DNA chain: (a) before irradiation and (b) after irradiation (A = adenine, T = thymine) (adapted from Edelson, R. L., *Sci. Am.*, 1988, **259**, 68, with permission. © Scientific American Inc., George V. Kelvin).

and subjected to ultraviolet radiation. The treated lymphocytes are then resuspended in the original plasma and the mixture is injected back into the patient. The process is repeated six to seven times over a certain period and can result in a dramatic amelioration in the health of the patient.

To understand what occurs at the molecular level during such treatment, we first consider the structure of methoxypsoralen. Owing to its planar geometry, this heterocyclic molecule can intercalate between neighboring pairs of DNA bases in a parallel orientation with respect to the rings (see Figure 7.23a and Section 3.5). The ability to be activated by (UV) ultraviolet light is the second characteristic of the psoralen molecule. In the UV-activated state, the carbon–carbon double bonds of the furan and α-pyran rings are highly reactive, and induce the C-5–C-6 bonds of the thymine residues to add to the double bonds of the intercalated psoralen molecules. As a result of the photochemical reaction both DNA chains become strongly crosslinked (Figure 7.23b). These crosslinks prevent DNA replication and, consequently, malignant cell growth and division. This procedure, called photophoresis, avoids any contact of the normal cells with the photoexcited drug molecules. In the nonexcited state psoralens are perfectly harmless and are in fact found in fruits and vegetables such as figs, limes and parsnips.

The approach utilizing intercalating drugs for highly specific biological applications has been pursued by scientists with such success in the past few years that it has become an avenue for the creation of novel and versatile medicines. Polynuclear dyes of the acridinium salt type and methylene blue, which are used as antibacterial agents (Figure 7.6), are classic examples of intercalators. Note that to cause the disruption of nucleic acid synthesis, an intercalator must become covalently linked with the DNA chain (refer back to Section 3.5).

7.5 Problems

1. Acetanilide ($pK_a = 0.3$) and caffeine ($pK_a = 0.5$) are absorbed in the stomach at the rate of 30% per hour. In contrast, quinine ($pK_a = 8.4$) is absorbed slowly. Account for this difference.
2. What is the mechanism of the antibacterial effect of sulfa drugs? Of acridinium salts?
3. 6-Aminopenicillanic acid can presumably be formed from two naturally occurring amino acids. Name them and draw their structures, making the genetic linkage obvious.
4. Acute intoxication with phenobarbital ($pK_a = 7.2$) can be diminished by a factor of 15 by intravenous introduction of an aqueous bicarbonate solution, which increases blood and urine pH from the normal values of 7.4 and 6.0, respectively, to 8. Describe the chemical mechanism for the removal of this barbiturate.

5. Uric acid is usually removed from the body in urine. However, when high levels are present, it cannot be fully removed in this manner owing to its poor solubility (see Problem 10, Chapter 2). As a result, the acid begins to crystallize in joints, causing gout. Pyrazolo[3,4-*d*]pyrimidine-4-one, or allopurinol (A), is one of the most effective treatments for gout. Suggest a mechanism for the therapeutic action of allopurinol, taking into account the fact that it can also prolong the lifetime of 6-mercaptopurine (B), which is used in the treatment of leukemia (also consider the conditions described in Problem 3, Chapter 4).

A   B

6. 5-Bromouracil and 5-fluorouracil, when introduced into organisms as markers, are found to be incorporated differently: one into DNA, the other into tRNA. Which of the two uracils will bind to DNA? To tRNA? Explain your reasoning.

7. The five substances C–G promote or inhibit the bioaction of the naturally occurring mammalian neurotransmitter X.

C   D

E   F

Muscimol (C) is found in mushrooms of the genus *Amanita muscaria*. It is a functional analogue (agonist) of the neurotransmitter in question. Compound D is a competitive inhibitor of the neurotransmitter in rat brain preparations. 4(5)-Imidazolylacetic acid (E) is an agonist of X. Compound F is a drug called pyracetam which exerts a nootropic (cognitive-enhancing) effect and mimics the effects of X. The alkaloid bicuculline (G) selectively antagonizes the inhibitory action of X.

(a)   Using certain structural similarities between compounds (C)-(G) and the unknown neurotransmitter, suggest a structure for (X).

(b)   Determine, on the basis of Dreiding stereomodels, the distance between the ends of the pharmacophore group.

7.6   Suggested Reading

1. Hoeprich, P. D., Jordan, M. C. and Ronald, A. R. (eds), *Infectious Diseases*, 5th Edn, Lippincott, Philadelphia, PA, 1994.
2. Silverman, R. B., *The Organic Chemistry of Drug Design and Drug Action*, Academic Press, San Diego, CA, 1992.
3. Wermuth, C. G., Koga, N., Konig, H. and Metcalf, B. W. (eds), *Medicinal Chemistry for the 21st Century*, Blackwell, Oxford, 1992.
4. Gilman, A. G., Rall, R. W., Nies, A. S. and Taylor, P. (eds), *Goodman and Gilman's The Pharmacological Basis of Therapeutics*, 8th Edn, Pergamon Press, New York, 1990.
5. Maxwell, R. A. and Eckhardt, S. B., *Drug Discovery: A Casebook and Analysis*, Humana Press, Clifton, NJ, 1990.
6. Moberg, C. L. and Cohn, Z. A. (eds), *Launching the Antibiotic Era. Personal Accounts of the Discovery and Use of the First Antibiotics*, Rockefeller University Press, New York, 1990.
7. Korolkovas, A., *Essentials of Medicinal Chemistry*, 2nd Edn, Wiley, New York, 1988.
8. Negwer, M., *Organic-chemical Drugs and Their Synonyms (An International Survey)*, Vols 1–3, VCH, New York, 1987.
9. Albert, A., *Selective Toxicity: the Physico-chemical Basis of Therapy*, 7th Edn, Chapman & Hall, London, 1985.
10. Parnham, M. J. and Bruinvels, J. (eds), *Discoveries in Pharmacology*, Elsevier, New York, 1983.
11. Lancini, G. and Parenti, F., *Antibiotics, an Integrated View*, Springer, New York, 1982.
12. Franklin, T. J. and Snow, G. A., *Biochemistry of Antimicrobial Action*, Chapman & Hall, London, 1981.
13. Wolff, M. E. (ed), *Burger's Medicinal Chemistry*, 4th Edn, Wiley, New York, 1981.

8 HETEROCYCLES IN AGRICULTURE

> Do we need much?
> No: of bread—only one slice,
> With it a drop of milk.
> The skies provide the salt
> And clouds of white silk.
>
> V. Khlebnikov

According to his contemporaries, the Russian poet Velemir Khlebnikov was an ascetic writer who valued the spiritual aspects of existence. Modern medicine, which warns that overnutrition is harmful to our health, is thus consistent with his poem. However, malnutrition is an equally dangerous threat and today some 1.5×10^9 people worldwide suffer from diseases caused by shortages of food and parasitic infestations. The world population growth rate has reached approximately 1.6×10^6 per week with a total of 5.5×10^9 people exerting ever-increasing demands for food. Meeting these needs will require dramatic improvements in the productivity of modern agriculture.

Exacerbating these demographic problems are the losses of agricultural production due to pests such as rodents, insects, microorganisms and weeds. Almost half of all food intended for humans is consumed or spoiled by pests (30% prior to harvest and 20% during crop transport and storage). Biological methods of plant protection are of great potential importance, but chemical control is currently the main approach utilized. Chemical pest control is carried out by compounds known as pesticides, which include herbicides (substances used to kill weeds), insecticides, fungicides (agents used against fungal pathogens) and rodenticides. Stimulators and regulators of plant growth and development are also of great importance.

The driving force for continued research in this field is the need to meet severe ecological demands necessitating the creation of more effective pesticides. The mechanism of pesticide action is closely associated with pest biology and involves the disruption of vital biological functions. We have seen how heterocyclic compounds participate in many biological systems, and therefore it is not surprising that many pesticides are heterocyclic derivatives.

Worldwide, approximately 2.5 billion tonnes of pesticides are currently applied each year at an estimated cost of 15-20 billion dollars. Herbicides comprise fifty per cent of the total usage of pesticides in plant protection.

8.1   A Century of Chemical Warfare against Weeds

Though the first herbicides were applied about 100 years ago, their widespread use commenced only after World War II. Today, more than 200 different types of herbicides are utilized in agriculture. Progress in this field is evident from the decrease in the effective doses of modern herbicides (only tens of grams per hectare (ha), where 1 ha $\equiv$ 2.477 acres) compared with their earlier counterparts (tens of kilograms per hectare). Such progress would not have been possible without an understanding of the mode of herbicide action.

The modern classification of herbicides is based on the nature of the biological target of the agent and comprises three main groups: (i) compounds which prevent photosynthesis (antiphotosynthetics); (ii) chemicals which disturb the biosynthesis of, or destroy, chlorophyll and other photosensitive pigments within the cells; and (iii) materials which inhibit the biosynthesis of essential amino acids or inhibit 'dark' metabolic reactions.

Heterocyclic compounds, especially azines and azoles, are frequently employed as herbicides. Derivatives of the triazinone group, e.g. ethiozin and amethidione (Figure 8.1) are important representatives of the antiphotosynthetics. These herbicides are used to combat weeds in winter wheat crops and are characterized by rather low rates of application (0.8–1.6 kg ha^{-1}). Structurally related to the previous group are pesticides containing a triazine ring (such as prometryn, simazine and propazine) or a diazine ring (such as terbacil and chloridazone, derivatives of uracil and pyridazinone, respectively) (Figure 8.1). These herbicides are applied in slightly higher doses (1.5–1.6 kg ha^{-1}), but their action is wide ranging. Typical applications include the destruction of weeds in cotton, sugar beet, turnips, soya, peas and sunflower crops, and also in vineyards, berry plantations and orchards.

3-Amino-1,2,4-triazole, a herbicide of the azole series, has been in use for a relatively long time. Other examples from this same class include 2-trifluoromethyl-4,5-dichlorobenzimidazole (chlorflurazole) and a derivative of 1,3,4-thiadiazole named ethidimuron (Figure 8.2). Ethidimuron kills all plants, and is thus used to clear roads, aerodromes, construction sites and areas surrounding high voltage transmission lines.

All of the above-mentioned herbicides are antiphotosynthetics which interact with photosystem II (see Section 6.2). These agents bind to the membrane protein receptors of the photosynthetic complex which are in close proximity to the plastoquinone pool. As a result, electron transfer from the primary quinone acceptor to plastoquinone is interrupted. Some herbicides may disrupt the electron transport elsewhere in the electron transfer chain, but the result to the plant in both cases is the same: death.

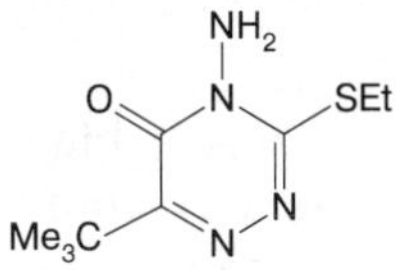

Ethiozin

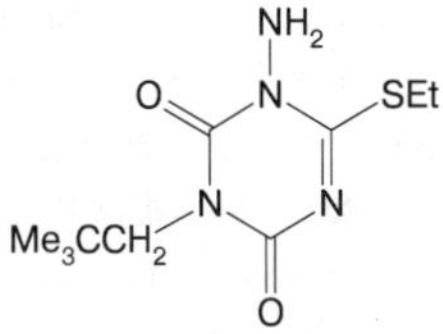

Amethidione

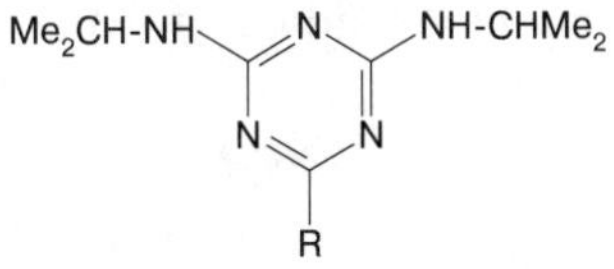

Prometryn (R = SMe)
Propazine (R = Cl)

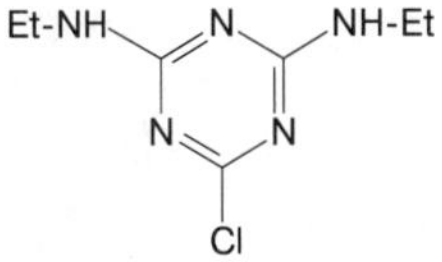

Simazine

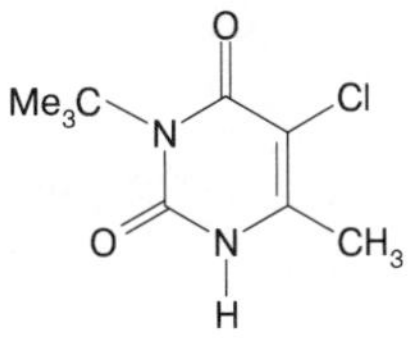

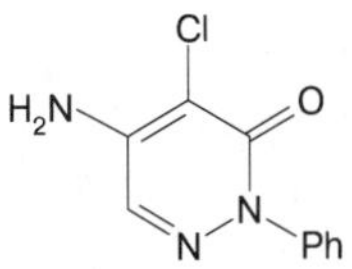

Figure 8.1 Antiphotosynthetic herbicides of the triazine and diazine series.

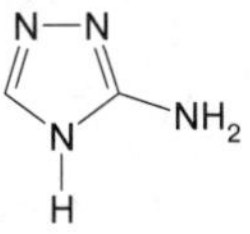

3-Amino-1,2,4-triazole

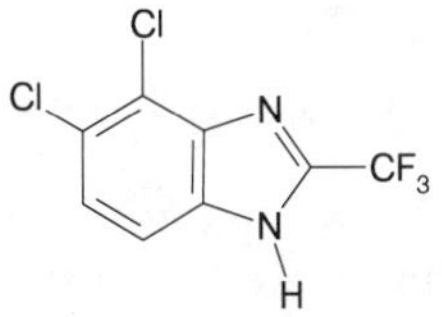

Chlorflurazole

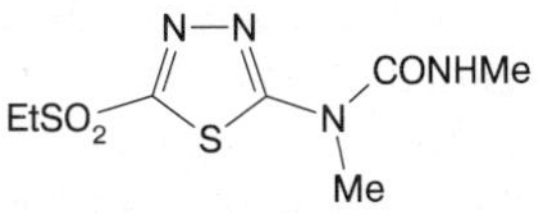

Ethidimuron

Figure 8.2 Antiphotosynthetic herbicides of the azole type.

All antiphotosynthetics contain several heteroatoms and acceptor groups. The acceptors diminish the reductive potential of the compound, and thus facilitate electron interception. Another characteristic feature of the heteroatoms and functional groups (especially C=O and NH) is their ability to bind with the amide and carbonyl functionalities of the protein receptor via hydrogen bonding and ion–dipole interactions. Herbicides which intervene with photosystem I also exist. Diquat dibromide and paraquat (methylviologen), which are quaternary salts of 2,2'-bipyridyl and 4,4'-bipyridyl, respectively, are typical representatives (Figure 8.3). They are believed to intercept electrons transmitted along photosystem I by forming stable cation-radicals. The cation-radicals reduce oxygen to the superoxide anion $O_2^{-\cdot}$ which induces the formation of hydrogen peroxide, hydroxyl radicals and singlet oxygen within the plant tissues. These species are all highly toxic and rapidly destroy plant pigments and other cellular structures. By transferring the electrons to oxygen, diquat dibromide and paraquat are regenerated. Thus, their mode of action involves electron transport.

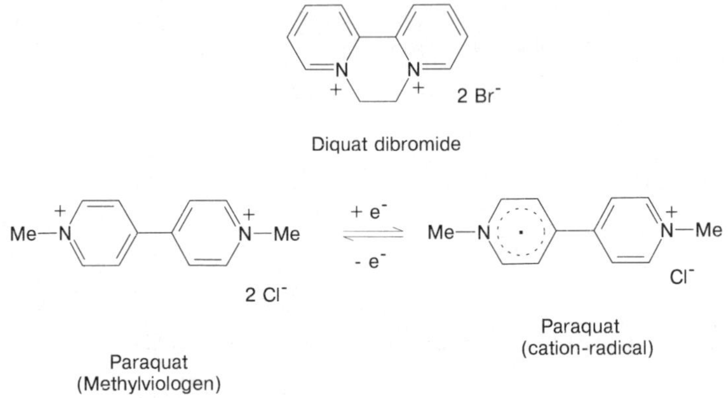

Diquat dibromide

Paraquat
(Methylviologen)

Paraquat
(cation-radical)

Figure 8.3 Herbicides which interact with photosystem I.

Extensive use of antiphotosynthetic herbicides has triggered the development of mutant weeds which are resistant to these chemical agents. The gene that codes for the receptor protein to which the herbicide attaches undergoes mutations. A mutation in which only one heterocyclic base in the gene DNA is changed, leading to the substitution of a single amino acid in the polypeptide chain (for example, serine by glycine), is sufficient to prevent adhesion of the herbicide molecule to the protein and thus to nullify its bioeffect. Thankfully, the choice of herbicides available today is sufficiently wide to allow effective control of the majority of weeds.

A serious problem remaining is that herbicides which disturb photosynthesis are still applied in comparatively high doses and can become dispersed in the environment, affecting nontargeted sectors with undesirable consequences. Therefore, since the late 1970s, interest in antiphotosynthetic herbicides has somewhat decreased. Scientists have now turned their efforts toward herbicides with a regulatory type of action. Such agents either interfere with the biosynthesis of chlorophyll, carotenoids or amino acids or change the phytohormone balance in the plant.

The biosyntheses of chlorophyll and the carotenoids in particular involve many steps. In general, disruption of any stage is detrimental to the plant. Inhibitors of photopigment biosynthesis include both heterocyclic and nonheterocyclic compounds. Difluphenican (Figure 8.4) is an example of a herbicide with such bioactivity which is used to eliminate weeds in many cereal crops. Difluphenican is valued for its low effective dose $(65–250\,g\,ha^{-1})$, high selectivity and prolonged action (a single treatment is effective for a whole season). The pyridazine herbicide norflurazone is a specific inhibitor of carotenoid biosynthesis. The pyrazole derivative Paicer (Figure 8.4) disrupts chlorophyll synthesis.

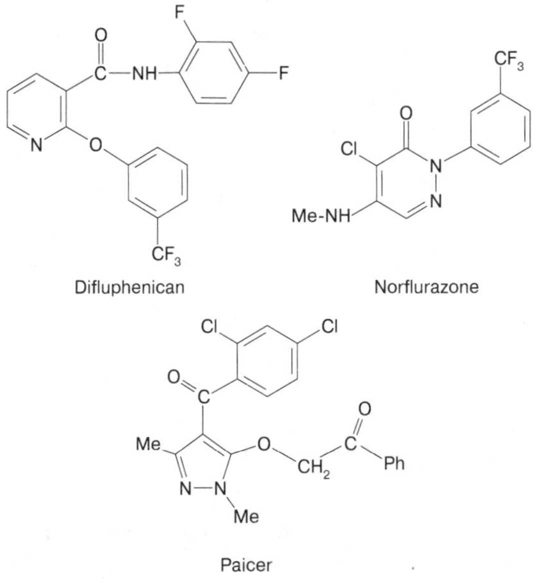

Difluphenican   Norflurazone

Paicer

Figure 8.4   Chlorophyll and carotenoid synthesis inhibitors.

High herbicidal activity is observed for some imidazolin-5-one derivatives, such as imazamethabenz (Figure 8.5). Their structures are characterized by the presence of two alkyl groups at the 4-position of the imidazoline ring, with one of these groups being branched. Herbicides of this class act as antimetabolites by deactivating acetolactate synthetase, the main enzyme involved in the biosynthesis of amino acids with branched alkyl chains (valine, leucine and isoleucine).

A recent generation of herbicides is the so-called sulfonylureas. Though their name does not reflect the presence of any heterocyclic moiety, at least one 1,3,5-triazine, pyrimidine or pyridine ring is usually present (Figure 8.5). Sulfonylureas are highly selective and are characterized by very low dose rates, which appear to be close to the theoretical limit. The optimal application rate

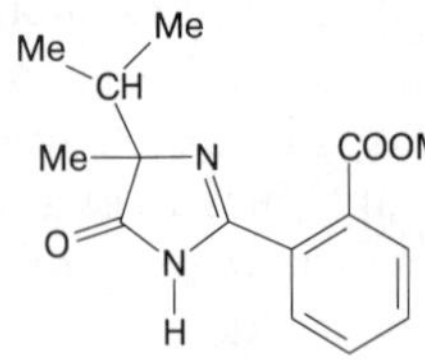

Imazamethabenz

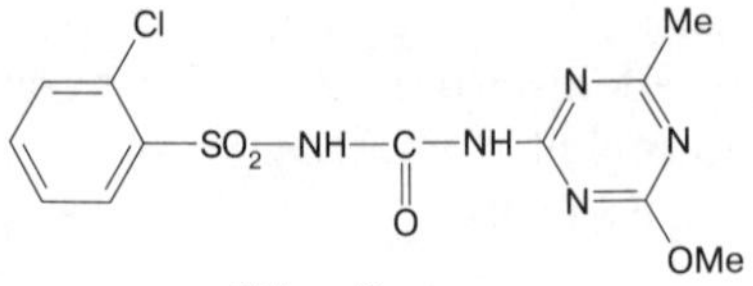

Chlorsulfuron

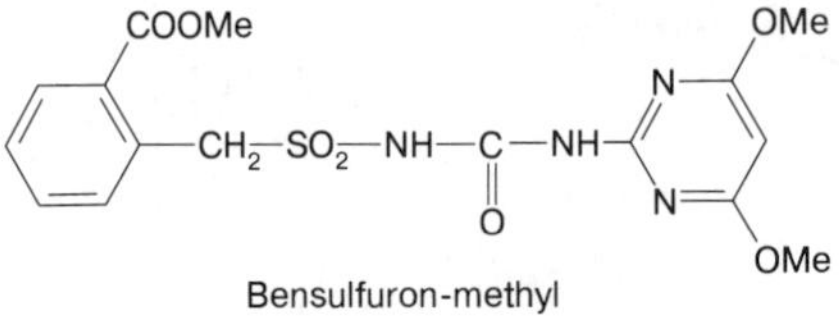

Bensulfuron-methyl

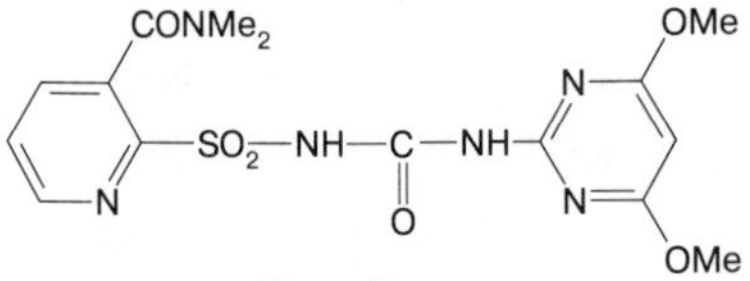

Nicosulfuron

Figure 8.5 Herbicides exerting phytohormonal effects and inhibiting amino acid synthesis.

of the most important herbicide of the series, chlorsulfuron, is $20\,\mathrm{g\,ha^{-1}}$, but a dose as low as $5\,\mathrm{g\,ha^{-1}}$ is often effective. Chlorsulfuron kills all broad-leaved weeds in grain crops. Other sulfonylureas are applied to combat weeds in soya, cotton, sunflower and corn plantations. The pyrimidine bensulfuron-methyl (Figure 8.5) is utilized for the selective destruction of weeds in rice crops.

Sulfonylureas are unique in their high activity at such low dosages, and this suggests a phytohormonal mode of action. It has been established that these herbicides affect neither photosynthesis nor DNA synthesis, but suppress the growth and division of plant cells. It is of interest that sulfonylureas, when applied in very low concentrations, stimulate seed germination in some plants, and can also retard leaf aging and enhance biomass increase. These results also testify to their effect on plant hormonal systems.

8.2 Regulators of Plant Growth

The sulfonylureas are in fact plant growth and development regulators. The agricultural application of compounds with regulatory activity began about

50 years ago, and since that time their scope has increased. Their small percentage (5% by weight) in the overall production of pesticides is largely accounted for by their effectiveness at low concentrations.

For a long time plant growth regulators were used not so much to combat weeds but to create favorable conditions for the development of cultivated plants. The range of effects of such preparations is very wide. Plant regulators can accelerate or slow the growth, flowering or ripening of plants, and make crops more drought or frost resistant. Thus, consistently high harvests can be attained. It is now well established that plants themselves produce regulators which guarantee the punctual appearance 'on the seventh day of a bright, rounded and refined leaf decorating a sprout of an old black twig', as the old saying has it. Such naturally occurring compounds, called phytohormones, cause dramatic effects when present in minute quantities. Plant hormones are subdivided by their chemical structure and mode of action into three main groups: auxins, kinins and gibberellins, together with the single compound ethylene (Figure 8.6).

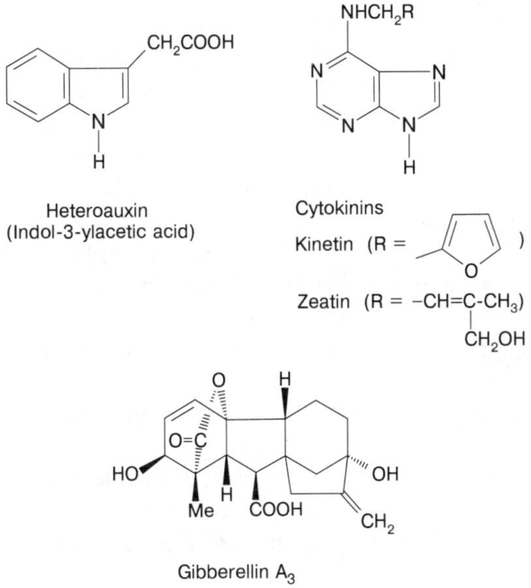

Heteroauxin
(Indol-3-ylacetic acid)

Cytokinins

Kinetin (R = [furan structure])

Zeatin (R = $-CH=C-CH_3$)
               |
               CH_2OH

Gibberellin A_3

Figure 8.6 Phytohormones.

Indol-3-ylacetic acid (heteroauxin) is one of the chief growth hormones. This auxin diffuses readily along the plant stem, and its hormonal action results in the lengthening of plant cells and stimulation of their division by increasing the rate of DNA replication. The acid induces the formation of side roots, stem sprouts and leaves, but inhibits plant growth when used in high concentrations. This acid also regulates and coordinates the growth of all plant components including the roots, stems and buds. Indol-3-ylacetic acid usually accumulates in the growing tissues of the plant: in the tips of buds and shoots, and in the young leaves and fruits. In aging tissues, synthesis of the hormone sharply decreases, and quite possibly it is this that triggers the loss of flowers, fruits and leaves. The same hormone is also found in mushrooms and some symbiotic plants. The practical uses of heteroauxin are extensive: the implantation of grafts during the vegetative propagation of fruit and berry cultures, the promotion of an increased number of side shoots in vegetables, the acceleration of ripening in fruits and the formation of seedless fruits. The main disadvantage of heteroauxin is its rapid degradation upon exposure to light. Interestingly, the synthetic homolog 4-(indol-3-yl)butyric acid is reasonably resistant toward the action of light.

Cytokinins are adenine derivatives substituted at the amino group (Figure 8.6) which enhance plant RNA synthesis and, consequently, the production of proteins in cells. Cytokinins also stimulate cell division and an increase in cell dimensions. In addition, cytokinins control the relationship between the roots and other parts of the plant, regulating hydration, adjusting to temperature changes and combating infections. Root tips and developing fruits are especially rich in cytokinins. Some plant parasites have determined that the influx of nutrients increases at the sites of cytokinin action, and have then adapted to produce these hormones and introduce them into the plant to secure themselves an adequate food supply. The mechanism by which hormones of the cytokinin series stimulate the fission and growth of cells is unknown. Chlorsulfuron and other sulfonylureas possess cytokinin-like activity. However, when these compounds are applied to weeds in optimal concentrations, plant growth is suppressed by the hypercytokinase mechanism.

The gibberellins, the third class of phytohormones, are not always categorized as heterocyclic compounds despite the γ-lactone ring in their structure. Gibberellins initiate seed germination and stem growth, and promote an increase in fruit size. These compounds are secreted by cell nuclei and promote the formation of enzymes which degrade starch and seed membranes. Gibberellins also promote the synthesis of tryptophan which is converted into indolylacetic acid in the sprout tips. Vines of seedless grapes are sprayed with gibberellins to increase the size both of the grapes and the overall bunches.

The modern agricultural arsenal contains pesticides which allow plants to be 'immunized' against almost every imaginable pest and weed. However, compounds also exist which are capable of causing disorders in gibberellin biosynthesis, consequently stunting plant growth. They do not alter the ripening schedule and can help to achieve greater harvests by reducing the height of plants and therefore their tendency to collapse, especially in heavy rain. Harvesting of the crops can be made very difficult, leading to heavy losses, if plants are blown down. Pesticides with such action are known as retardants. A typical example is uniconazole, a member of the triazole series (Figure 8.7).

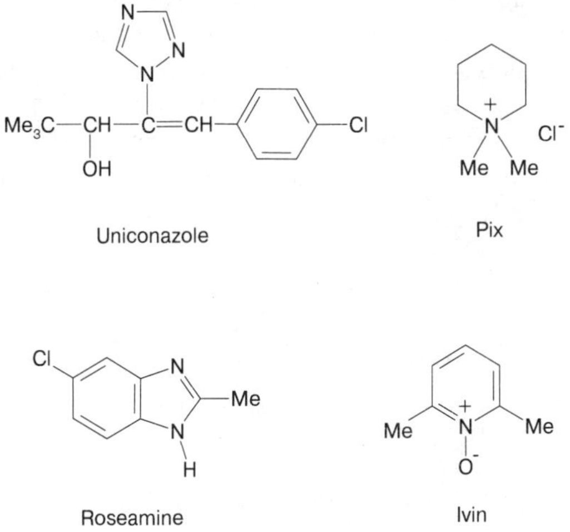

Uniconazole   Pix

Roseamine   Ivin

Figure 8.7   Synthetic plant growth regulators.

Many other synthetic regulators of plant growth and development have been prepared, each designed specifically to influence a particular biological process, such as activating or regulating photosynthesis, interfering with chlorophyll biosynthesis, stimulating plant respiration and so on. Thus, *N,N*-dimethylpiperidinium chloride (Pix, see Figure 8.7) is employed to enhance the rate of cotton boll ripening. Roseamine, or 2-methyl-5(6)-chlorobenzimidazole, is useful in preventing cotton from becoming detached from the plant. 2,6-Dimethylpyridine 1-oxide (Ivin) stimulates the growth of tomatoes and cucumbers.

8.3   The Struggle against Voracious Insects

More than 3×10^6 insect species inhabit the Earth. Of these, about 70 000 may be considered phytoparasites (or plant parasites). Such insects exact a high toll on agriculture as they devour and spoil up to 30×10^6 tonnes of grain per year. The Colorado beetle devastates potato fields, the codling moth and other parasites cause damage to orchards, and the barn weevil infests granaries. Locust swarms, sometimes occupying hundreds of square miles, are capable of destroying all crops. The arrival of these winged pests in huge numbers is, in some countries, equated with a national disaster. In ancient times these voracious creatures were referred to as 'hungerphorous' because their appearance *en masse* frequently caused human starvation.

The history of insecticides thus began long ago, and today more than 10^6 tonnes (about 25% of the total production of pesticides) are applied annually in the treatment of agricultural and other terrestrial and domestic ecosystems. The famous 'insect powder' (known in Russia as 'Persian powder') can still be found in drugstores nowadays under the name 'pyrethrum'. This preparation is recommended for combating domestic insects such as ants, bugs, and fleas. Some success against agricultural pests has also been achieved.

The use of tobacco dust as an insecticide, especially against the plant louse and other small insects, has a long history. The pyridine alkaloids anabasine and nicotine, which affect the insect CNS, are the active constituents (Figure 8.8).

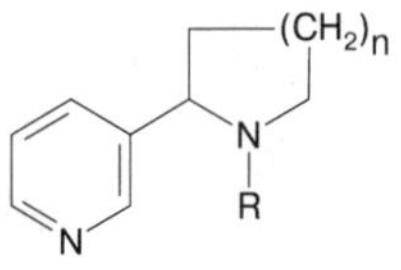

Anabasine (R = H, n = 2)
Nicotine (R = Me, n = 1)

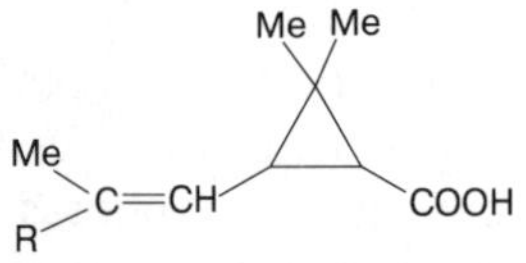

Chrysanthemic acid (R = Me)
Pyrethric acid (R = COOH)

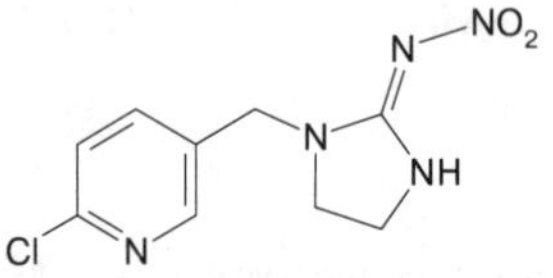

Imidacloprid

Figure 8.8 Components of naturally occurring insecticides from tobacco and pyrethrum and structure of imidacloprid.

Though nicotine and anabasine were the first heterocyclic compounds used in the chemical fight against pests, the importance of heterocycles in insecticide preparations is now not very great. Modern insecticides are largely composed of aliphatic, alicyclic and homoaromatic compounds. In particular, rapid development in the chemistry of synthetic pyrethroid insecticides was triggered after the active components of pyrethrum were determined. The esters of two cyclopropane carboxylic acids, chrysanthemic and pyrethric (Figure 8.8), were found to be responsible for the insecticidal activity.

Many insecticides belong to the chlororganic series. A notorious example of this class is DDT, which has now been banned in most countries. A number of environmentally safe replacements have since been synthesized including imidacloprid (Figure 8.8), which is a broad spectrum systemic insecticide, active as a seed dressing, for soil incorporation and foliar application for rice, cereals, corn, potatoes and sugar beet.

Contemporary chemical control of pests is largely achieved by organophosphorus compounds such as the esters of phosphoric and thiophosphoric acids. Chlorophos and fenitrothion are typical representatives (Figure 8.9). A number of compounds in this series, which is estimated to constitute about 43% of the total production of insecticides, contain heterocyclic fragments, mainly of the azine and azole types. As examples, the structures of diazinon, pirimiphos-methyl, menazon and phosalone are shown in Figure 8.9. Phosalone is widely used in Russia as an effective replacement for DDT against plant lice, ticks and other insects. Similar spectra of action are shown by diazinon and pirimiphos-methyl, which are recommended for application in fields and

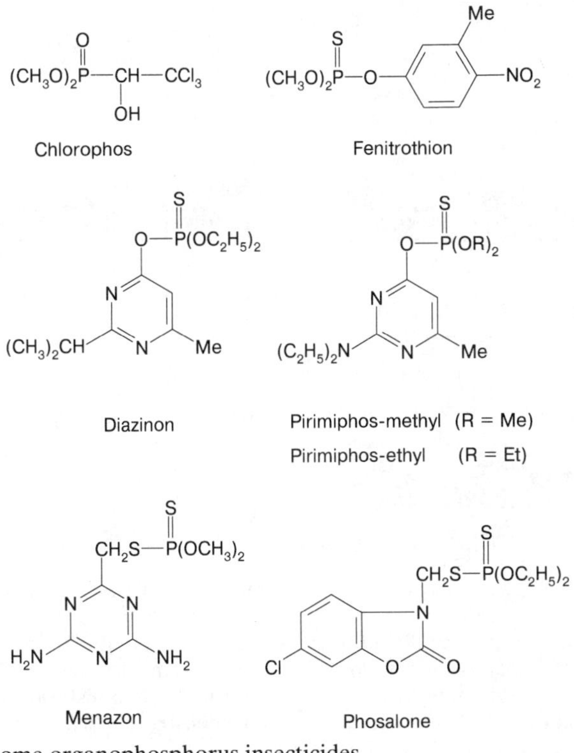

Figure 8.9 Some organophosphorus insecticides.

orchards. Pirimiphos-methyl is also used for the disinfection of storage facilities before agricultural products are introduced. A homolog (R = Et) is especially potent against soil pests, while menazon is effective against plant lice.

Organophosphorus insecticides, as antagonists of the natural neuromediator acetylcholine (Figure 7.16), paralyze the nerve system of insects. After nervous impulse transmission, the acetylcholine is degraded in the synaptic junction to make room for the next neuromediator molecule. Hydrolytic cleavage of the acetylcholine ester group is carried out under the enzymatic control of acetylcholinesterase. Acetylcholine is thought to bind electrostatically to the enzyme from two positions through its trimethylammonium group and carbonyl carbon, the former being bound to the carboxylate anion of an aspartic acid residue of the enzyme, and the latter to a serine hydroxy moiety (Figure 8.10a). In the following step the acetyl group is cleaved from the acetylcholine molecule thus transforming it into choline, a derivative of ethanolamine containing a trimethylammonium group at the β-carbon atom (Figure 8.10b). The acetyl group linked to the serine residue in the acetylcholinesterase polypeptide chain is then readily hydrolyzed and the enzyme is regenerated.

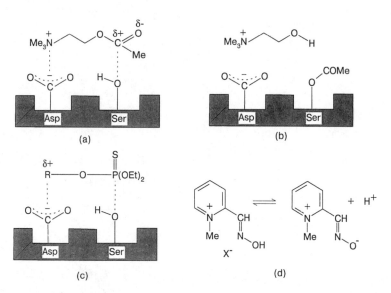

Figure 8.10 Acetylcholinesterase interaction with acetylcholine and organophosphorus insecticides: (a) two-site binding of enzyme with acetylcholine; (b) cleavage of choline portion from enzyme–substrate complex and formation of acetylated enzyme; (c) two-position binding of enzyme to insecticide; and (d) 1-methylpyridinium-2-aldoxime as a reactivator of acetylcholinesterase (adapted from Musil, J., Novakova, O., and Kunz, K., *Biochemistry in Schematic Perspectives*, Avicenum, Prague, 1977, p. 45, with permission).

All organophosphorus insecticides have in common their structural resemblance to the acetylcholine molecule which enables them to bind to acetylcholinesterase. An electrophilic phosphorus atom is attracted to the serine hydroxy group, while the R group, which extends some distance from the phosphorus atom, is attracted to the carboxylate anion (Figure 8.10c). For this association to occur, the R group must have a low electron density. Therefore, almost all insecticides contain electron-deficient aromatic (e.g. nitrophenyl) or heteroaromatic (e.g. pyrimidine, triazine) rings as the R group. By blocking acetylcholinesterase in this manner, organophosphorus insecticides prevent the acetylcholine molecule from approaching the enzyme. Therefore, nerve impulse transmission is interrupted and the insect's organs which rely upon cholinergic innervation are rendered ineffective. Unfortunately, organophosphorus preparations are also toxic to warm-blooded animals, including humans. Thus, we still need to create new, less dangerous and more highly selective insecticides.

Organophosphorus poisoning can occur, for example, among both agricultural workers and amateur gardeners. Moreover, many war gases are organophosphorus compounds with a similar mechanism of action. The availability of a suitable treatment in cases of organophosphorus poisoning is therefore important. Antidotes capable of reactivating acetylcholinesterase that has been blocked are generally heterocyclic in nature and have two common structural features, viz. an available quaternary nitrogen and an adjacent aldoxime (CH=NOH) group. 1-Methylpyridinium-2-aldoxime (Figure 8.10d) can be viewed as a typical example. The acidity of the oxime group is sufficient for ionization to the $-CH=NO^-$ anion to occur at physiological pH. The negatively charged oxygen in the anion is more strongly attracted to the phosphorus atom than is the oxygen of the serine hydroxy group, and thus the antidote displaces the organophosphorus poison from the blocked enzyme. Interestingly, if a neutral pyridine-2-aldoxime is used instead of a 1-methylpyridinium salt, the effectiveness of the reactivator decreases 1000-fold. This may imply that the reactivator, like the enzyme, acts by means of two-point contact. One hypothesis suggests that the positively charged nitrogen is necessary for the reactivator to attach to the enzyme anionic center during the expulsion of the poison. This is consistent with the fact that the distance between the positively charged nitrogen and the oxime oxygen in each antidote is close to that between the N^+ and the carbonyl carbon in acetylcholine.

Besides the pyrethroids, another class of natural insecticides has recently received much attention. These are the pheromones, which are used as a chemical means of communication between insects and which affect their behavior. Some serve as sexual attractants, others as signals of alarm or availability of food, while others are trail markers. Pheromones are produced by the exocrine glands and are then released from the organism. Once expelled, they act as chemical signals for the same biological species. Pheromone molecules, trapped by insect receptor sites usually located in the olfactory or gustatory organs, cause a specific reaction in the receiving insect. This peculiar 'pigeon post' works with amazing efficiency and accuracy, as insects are sometimes able to receive a 'postcard' from a distance of 10 or more miles from the transmitting insect. The concentration of pheromone molecules may be as low as $10^{-17}\,\mathrm{mol\,m^{-3}}$.

Pheromones have been used in agriculture for some years, and quite recently have come into household use. Pheromones are employed to entice insects into traps where they are exterminated by powerful insecticides which can thus be used in a restricted site without endangering mammals. Such a method allows pesticides to be applied in very low doses economically, with high efficiency, and without polluting the environment. However, despite all of the advantages of pheromones (high selectivity, low toxicity toward animals and man, low consumption doses) they do have one serious shortcoming, namely a high cost of production. The structures of many pheromones have been elucidated and a number have been prepared by total synthesis. The overwhelming majority are acyclic structures containing long, sometimes unsaturated, alkyl chains with a terminal functional group such as an aldehyde or ester. However, some pheromones contain heterocyclic moieties and in a number of cases the basic molecular frame is heterocyclic. The pheromone of the female bombyx (*Porthetria dispar*) contains an epoxide ring. This pheromone, disparlure (Figure 8.11), is presently manufactured on an industrial scale and is used to combat forestry pests. The pheromone of the male butterfly *Licorea ceres* and that of the 'leaf cutter' ant (*Atta texana*) are structurally simple derivatives of pyrrole (Figure 8.11). The former suppresses the motion reflexes of the female butterfly, and the latter is used by ants as a trail marker.

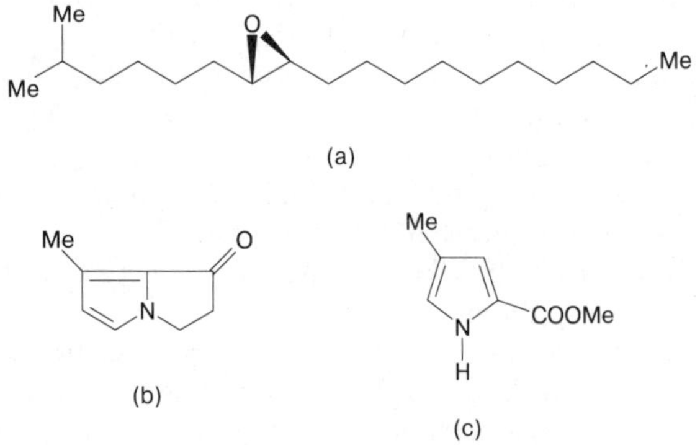

(a)

(b)

(c)

Figure 8.11 Pheromones containing heterocyclic moieties: (a) disparlure, (b) the pheromone of the *Licorea ceres* butterfly and; (c) the trail pheromone of the 'leaf-cutter' ant *Atta texana*.

8.4 Resisting the Kingdoms of Mustiness and Rot

Plant diseases caused by fungi are no less devastating than the aggressive invasion of insects, and often result in the total loss of the crop or harvest, even including the next season's seed supply. Therefore, efforts are ongoing to produce novel fungicides to control pathogenic fungal organisms successfully. Fungicides occupy third place after herbicides and insecticides in terms of volume of production (around 19% of the total). Fungicides are subdivided into contact and systemic agents based on the nature of their action. The latter are considered to be more valuable since they can move through the plant's vascular system. The majority of systemic fungicides in current use are derivatives of nitrogen heterocycles such as pyrazole, imidazole, triazole, pyrimidine, pyridine and so on. Triazole and benzimidazole preparations are the most effective of the class (Figure 8.12).

Triadimefon is the triazole-based fungicide most widely used; it has a very wide range of action and is applied during the vegetative period to combat fungal diseases of cereals, apple trees, vines, tomatoes and other plants. Triadimenol, an analogue of triadimefon containing a hydroxy instead of a carbonyl group, is utilized for seed protection and to combat mildew in

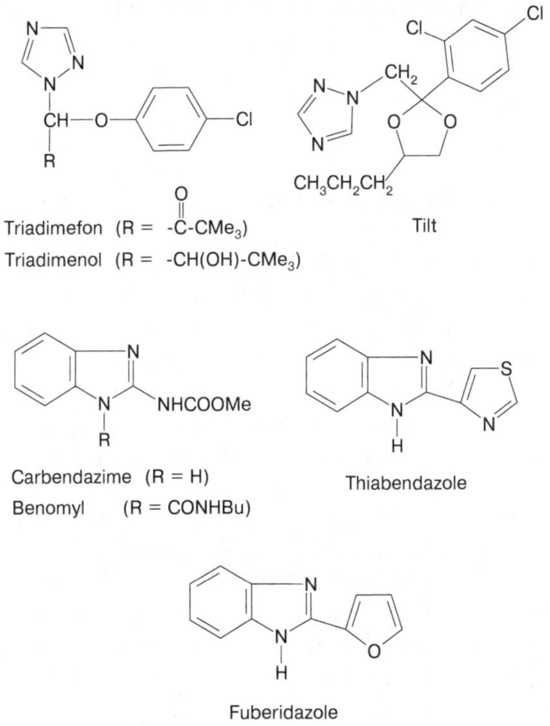

Triadimefon (R = -C(=O)-CMe₃)
Triadimenol (R = -CH(OH)-CMe₃)

Tilt

Carbendazime (R = H)
Benomyl (R = CONHBu)

Thiabendazole

Fuberidazole

Figure 8.12 Systemic fungicides.

cereals. Tilt also exhibits high activity and a broad spectrum of effects. Certain 1,2,4-triazole fungicides, such as triadimefon, triadimenol and etaconazol, show excellent activity against human pathogenic fungi. They belong to the large group of ergosterol biosynthesis inhibitors (EBIs) and their mode of action includes the inhibition of cytochrome P-450-dependent oxidative demethylation of 24-methylenedihydrolanosterol, which is a pathogen-specific precursor of ergosterol, the main sterol of pathogenic fungi. All triazole-based fungicides are of low mammalian toxicity and their application rates are very low (0.4-1.0 kg ha^{-1}). These same characteristics are inherent in numerous systemic preparations.

Benzimidazole-derived fungicides were introduced into agriculture in the early 1960s and have retained their importance despite their higher toxicities and larger dose rates per cultivated hectare. The chief bioactive compound of the series, benomyl, has a formidable range of action rarely encountered in other substances. Benomyl is used against pathogens of practically all cereals, sugar beet, vegetables, berries, fruit trees and vines, and in the treatment of cotton plantations during soil tillage. Carbendazime, thiabendazole and fuberidazole are also effective fungicides. Fuberidazole is utilized mainly in seed protection applications.

The mechanism of fungicide bioaction is highly complex and in many cases has not yet been clarified. Benzimidazole fungicides are assumed to disturb biofission of the fungal cell nuclei, whereas triazoles negatively influence the penetrability of cellular membranes.

8.5 Heterocycles in Animal Husbandry

Heterocyclic compounds have found spectacular applications in animal husbandry and veterinary medicine. Meat production from animals has been significantly increased by the use of vitamin supplements (especially thiamine) in forage and by the addition of antibiotics and tranquilizers to feed. Tranquilizers alleviate stresses in the animals, which are usually inevitable under the conditions of factory farming. The struggle against various animal parasites, especially helminthic invasions, is of tremendous importance. For many years these parasitic infections were treated with phenothiazine (Figure 2.8). Today, more efficient medications, such as the highly active thiabendazole (Figure 8.12), have superseded phenothiazine.

Surra is an extremely dangerous illness of cattle caused by single-cell microorganisms called *Trypanosoma* which invade the animal's blood and tissues. The analogous human ailment is known as trypanosomiasis. Surra is a particular problem in countries with hot climates; for a long time the breeding of cattle in central Africa was virtually impossible owing to this parasitic scourge. Compounds based on aminophenanthridinium salts were used effectively in these cases. For instance, a single dose of ethidium bromide was sometimes sufficient for the complete recovery of an animal afflicted with surra.

Ethidium bromide

Coccidiosis infections, which usually result in severe fowl losses, can make a massive negative impact on the poultry industry. To combat this parasitic disease, veterinary medicines such as nitrofuran derivatives, sulfa drugs and antibiotics are added to the chicken's drinking water. One further heterocyclic treatment that has become useful in veterinary medicine as a cure for a number of parasitic skin diseases is nicotine sulfate.

8.6   Problems

1. What physicochemical properties give rise to the herbicidal activity of paraquat and diquat dibromide? Discuss the reasons for their rather rapid inactivation in the field.
2. Picloram (A) is widely used for the control of perennial weeds and shrubs. The moderate solubility of picloram in water is dramatically increased when the herbicide is used in the potassium salt form. However, this increased solubility raises the environmental concern of potential groundwater contamination. This problem could be effectively circumvented by the formation of insoluble complexes of picloram with the metal ions present in soil and groundwater. One such nonlabile complex is readily formed by treatment of the herbicide with iron(II) ions. Indicate the structure of this complex.

A

3. Discuss the mechanism of the biological activity of the heterocyclic plant hormones. Give examples of synthetic growth regulators.

4. Nicotine is used in agriculture mainly in the monosulfate form. Give the structural formula of this salt.
5. Discuss the reasoning which led to the use of 1-methylpyridinium-2-aldoxime as an antidote for organophosphorus reagent poisoning. Why is the neutral pyridine-2-aldoxime 1000 times less active?
6. What are systemic fungicides? Give examples from the benzimidazole series.
7. Diniconazole (B) is a broad action systemic fungicide which was shown to bind stoichiometrically to cytochrome P-450 enzymes via a lone pair of electrons on one of the nitrogen atoms. Diniconazole thus disrupts ergosterol biosynthesis in fungi by mimicking the conformation of one of the intermediates, lanosterol (the cytochrome P-450 enzymes are responsible for oxidation of the lanosterol methyl group at the C-14 position). Structure B is the most active of the four possible isomers. (a) Draw the structures of the other three isomers. (b) Is there any intramolecular stabilizing interaction in these isomers? (c) Which of the three nitrogen atoms of the fungicide binds to the cytochrome enzymes?

B

8.7   Suggested Reading

1. Baker, D. R., Fenyes, J. G. and Basarab, G. S. (eds), *Synthesis and Chemistry of Agrochemicals IV*, 4th Edn, American Chemical Society, Washington, DC, 1995.
2. Tomlin, C. (ed), *The Pesticide Manual: a World Compendium: Incorporating the Agrochemicals Handbook*, 10th Edn, Royal Society of Chemistry, Cambridge, 1994.
3. Bunce, N. J., *Environmental Chemistry*, Wuerz, Winnipeg, 1990.
4. Hutson, D. H. and Roberts, T. R. (eds), *Herbicides*, Wiley, New York, 1987.
5. Ware, G. W., *Pesticides, Theory and Application*, Freeman, San Francisco, CA, 1983.
6. Sitting, M. (ed), *Pesticide Manufacturing and Toxic Material Control Encyclopedia*, Noyes Data Corporation, Park Ridge, NJ, 1980.
7. Tedder, J. M., Nechvatal, A. and Jubb, A. H., *Basic Organic Chemistry. Part 5: Industrial Products*, Wiley, London, 1975.

9 HETEROCYCLES IN INDUSTRY AND TECHNOLOGY

> I throw the garland of roses
> To the world of mysterious themes,
> And by it I plunge into chaos
> of unknown creative day-dreams.
>
> V. Bryusov

Heterocycles have been indispensable in the recent far-reaching developments in science and technology for numerous applications, including electronics, communications and aerospace technology. At the same time they remain enormously important in traditional branches of industry such as in the dye industry. The manufacture of synthetic dyes began in the second half of the nineteenth century and heterocycles immediately achieved preeminence. However, from time immemorial man had learned how to use natural pigments, of which a considerable number are heterocyclic. We begin this chapter by discussing the contribution of heterocycles to the polychromism of our world.

9.1 Heterocycles and Natural Colors

The human eye perceives the surrounding world as a multicolored picture. The existing natural colors from green grass to the palette of a butterfly wing give us immense aesthetic pleasure. We do not know whether animals share these feelings with us, but we do know that they require different colors to fulfill certain biological functions and in some cases for survival itself. Animals use their own colorings to disguise themselves, to threaten enemies, to search for mates and so on. Animals also rely on color to judge whether fruit is suitable to eat.

Colored substances usually contain an extended carbon chain composed of alternating single and double bonds. Such a chain is called a chromophore. β-Carotene (Figure 6.4), the orange pigment found in carrots and apricots, is a typical example. Carotenoids are the most widespread natural pigments. We have already seen in Chapter 6 that carotenoids are the main constituents of

183

chloroplasts found in every green plant. Another, less widely distributed, class of pigments is the quinones. 5-Hydroxy-1,4-naphthoquinone (juglone), the yellow-orange component of unripe walnut shells (Figure 9.1), is a typical representative.[†] In contrast to the carotenoids, the functional group electrons participate in the conjugation in this pigment along with electrons of the C=C bonds. When a molecule is excited by light, electrons are transported along the conjugated chain from the donor (hydroxy) to the acceptor (carbonyl) group. This process causes significant stabilization of the excited state and results in deeper coloration. Such functional groups are called auxochromes. Pyrrole-like and pyridine-like heteroatoms are efficient auxochromes in many recent far-reaching developments in science and technology. Consequently, many heterocyclic compounds are encountered among naturally occurring pigments. Green chlorophyll and red hemoglobin are classic examples.

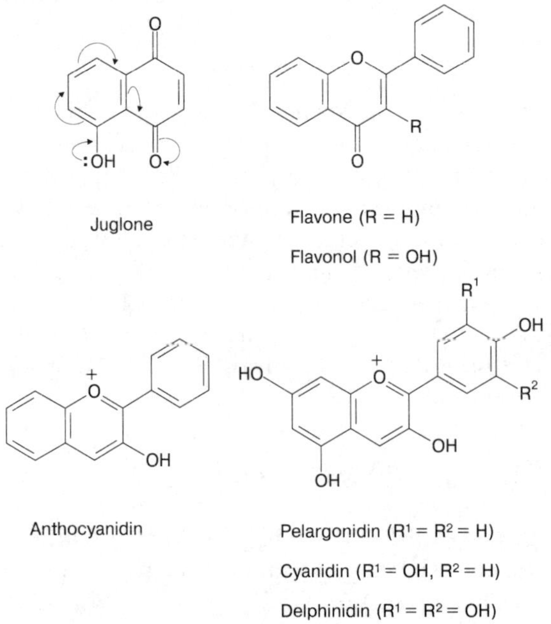

Juglone

Flavone (R = H)

Flavonol (R = OH)

Anthocyanidin

Pelargonidin (R^1 = R^2 = H)

Cyanidin (R^1 = OH, R^2 = H)

Delphinidin (R^1 = R^2 = OH)

Figure 9.1   Juglone and other flavonoid pigments.

Another class of heterocyclic pigments includes substances with the flavonoid structure, comprising derivatives of flavone, flavonol and anthocyanidin (Figure 9.1). A benzopyran skeleton forms the basis of the flavonoid structure. Substituents typically include one or more phenolic hydroxy groups. Flavonoids usually exist in a methylated or glycosidated

---

†Juglone exists in a reduced, uncolored form in the walnut shell. When the shell is broken, the reduced compound is released with the juice and is oxidized by air to form the pigment.

form in the plant. The most important flavonoids are the anthocyanidin glycosides, such as orange pelargonidin, red cyanidin and purple delphinidin. These pigments are responsible for the orange, red and purple/blue coloration of flowers. The colors of ripe cherries, strawberries, raspberries, plums and red apples are also the result of anthocyanidins.

Flavones and flavonols differ from anthocyanidins by an almost complete absence of light absorption in the visible region. Nevertheless, they produce vivid white and cream colors in a number of flowers such as tea rose, cherry, apple, plum, apricot and others. 'Just look! So many daisies are here and there,' wrote the Russian poet Severyanin, 'and many are in blossom; even too many; in full bloom. Their petals are triple-edged as wings, and white as silk.'

Flavonoid pigments are very seldom encountered in animals. The animal kingdom is rich in another family of heterocyclic pigments, namely the pterins (Figure 9.2). Pterins are frequently found in butterflies and other insects. Thus, the white color of the cabbage butterfly's wings arises from the presence of leucopterin. Chrysopterin imparts a yellow hue to the lemon butterfly, erythropterin produces the bright red color of the *Zegris f. Chr.* butterfly and xanthopterin is partly responsible for the yellow color of wasps.

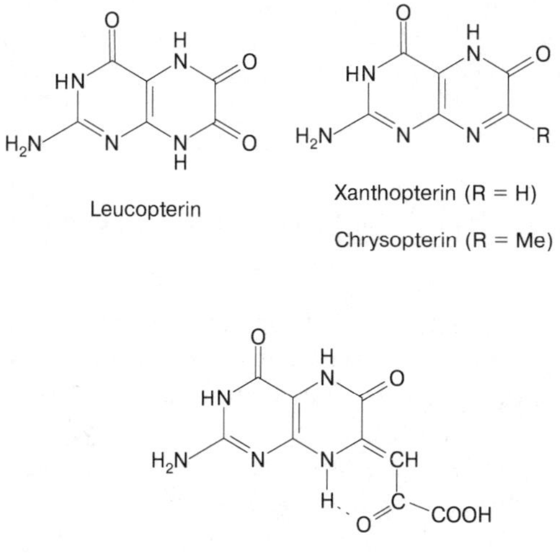

Leucopterin

Xanthopterin (R = H)

Chrysopterin (R = Me)

Erythropterin

Figure 9.2   Pterin pigments.

9.2   Dyes

9.2.1   FROM IMPERIAL CLOAKS TO JEANS

Since ancient times people have used dyes, obtained from various natural sources, to adorn their dwellings and clothes and to prepare cosmetics. In the middle ages, the yellow flavonoid pigment luteolin (Figure 9.3a) was extremely popular. Luteolin was prepared from the stems, leaves and seeds of a plant called *Reseda luteola*. The blue dye indigo was separated from the leaves of another plant, *Indigofera tinctoria*. The well-known purple dye 6,6'-dibromoindigo, known as Tyrian Purple or Royal Purple, was obtained from the Mediterranean sea mollusk *Murex brabdaris*. In ancient times, Tyrian Purple was manufactured in the Phoenician town of Tyre and was originally used to color the cloaks of pharaohs, emperors and high priests.

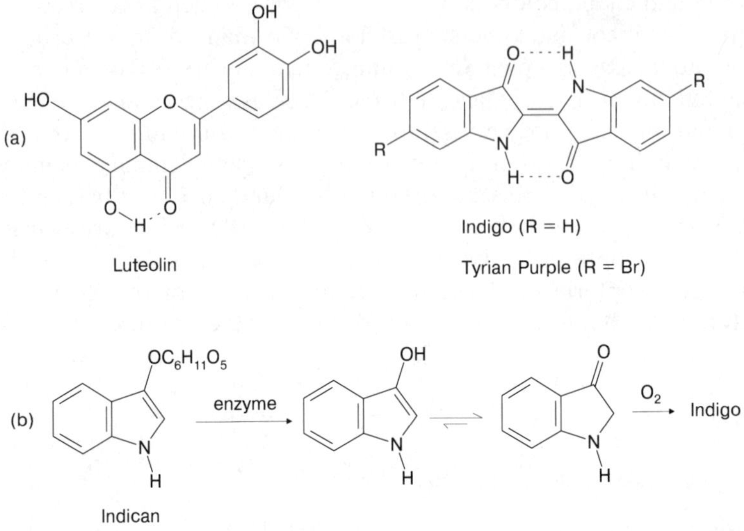

(a)

Luteolin

Indigo (R = H)

Tyrian Purple (R = Br)

(b)

Indican

Figure 9.3   (a) Heterocyclic dyes from natural sources. (b) Formation of indigo from indican.

It should be noted that indigo and dibromoindigo do not occur in nature, but are formed during the processing from natural precursors. Thus, the leaves of the indigo plant contain indican (3-hydroxyindole *O*-glycoside). When the leaves are destroyed indican is first transformed into 3-hydroxyindole (indoxyl, which exists predominantly in the tautomeric oxoform), and then into indigo (Figure 9.3b) by the action of enzymes and oxygen.

Unfortunately, almost all natural organic dyes have shortcomings. Flavonoids, for example, suffer from poor stability toward light and chemicals. Other dyes are rather expensive to produce owing to the difficulties of their separation and purification, and their low content in the raw materials; for example, approximately 10 000 mollusks had to be processed to isolate just 1 g of Tyrian Purple! It is clear why this dye was used in ancient and medieval times only by the rich.

A new era in the dye industry began in the middle of the nineteenth century when rapid advancements in organic chemistry allowed the creation of synthetic dyes. Aromatic and heterocyclic compounds 'stole the limelight'. The first entirely synthetic dye was mauveine, an ionic derivative of phenazine (Figure 9.4) produced by the English chemist Perkin via oxidation of a mixture of aniline and toluidines with potassium dichromate in sulfuric acid. Mauveine is red in color and is characterized by high stability to light, washing and mechanical agitation. In the past, the dye was widely used to color silk and wool.

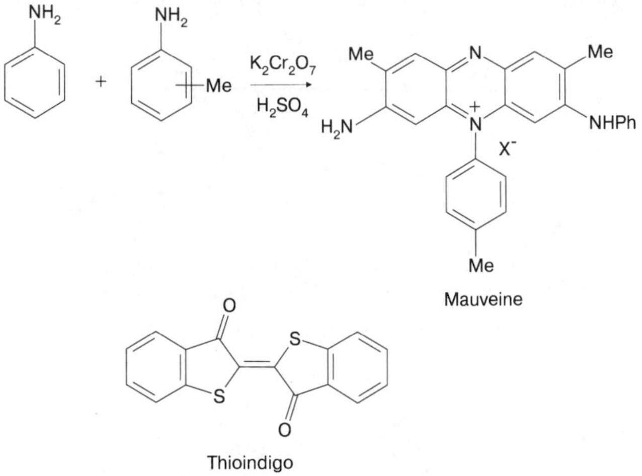

Mauveine

Thioindigo

Figure 9.4 The synthetic dyes mauveine and thioindigo.

Another significant event in the history of chemical dyes occurred during 1869–1883 when von Baeyer succeeded in elucidating the structure of indigo, thereby enabling an industrial synthesis of indoxyl and subsequently indigo; for this he received the Nobel prize in 1905.[†] Since this time inexpensive synthetic indigo has become a common dye with continuing widespread use in the textile industry, mainly for coloring jeans. The success of the indigo production (more than 10 methods of synthesis were developed) gave rise to the production of numerous derivatives and analogues known as indigoid dyes. The replacement of nitrogen atoms by other heteroatoms was often used in the modification of the indigo structure. The red dye thioindigo (Figure 9.4) serves as a typical example.

9.2.2 'CYANINE' MEANS AZURE

An important feature of heterocyclic dyes is the ready modification of their precise hue by structural changes. Other parameters (solubility, affinity to fabrics) are altered by conversion of the heteroatom to the cationic form.

[†]Incidentally, von Baeyer was incorrect in ascribing the *cis* orientation to the two C=O groups in the indigo molecule. It was not until 1926 that the error was corrected.

Discovery of this fact spurred the synthesis of a large family of cationic dyes. The first representative, a blue dye named cyanine (Greek: *kyanos*, azure), was synthesized in 1856 by G. Williams by heating a mixture of quinoline and 4-methylquinoline (lepidine) with isopentyl iodide in alkali. Although cyanine (Figure 9.5) did not find practical applications, it served as a prototype for the large group of cationic 'cyanine' dyes.

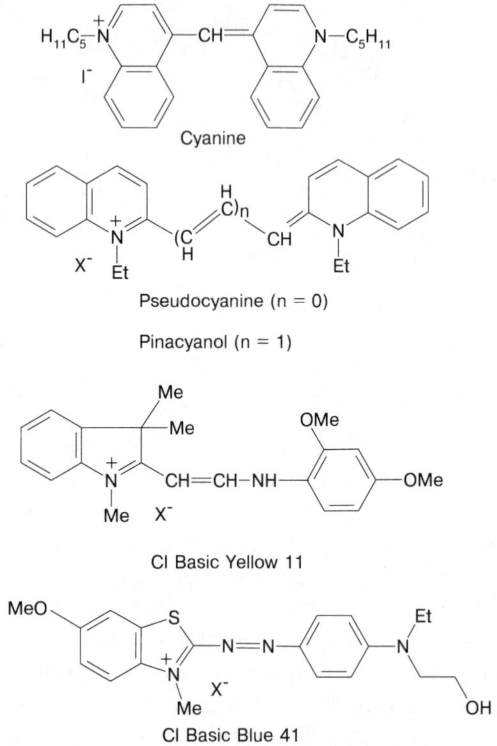

Cyanine

Pseudocyanine (n = 0)

Pinacyanol (n = 1)

CI Basic Yellow 11

CI Basic Blue 41

Figure 9.5   Examples of cyanine dyes.

Cyanine dyes generally consist of two heterocyclic ring systems connected by a bridge of conjugated carbon bonds which can vary in length. At one end, a (usually heteroaromatic) heteroatom is in a cationic state and serves as an electron acceptor; in the other, partially saturated nucleus, the heteroatom is formally electronically neutral (pyrrole-like) and functions as an electron donor. Figure 9.5 shows three such dyes: cyanine, pseudocyanine and pinacyanol.

It is to be stressed that the difference between the heteroatoms mentioned above is grossly oversimplified as it takes into consideration only one possible canonical structure (resonance structure). In reality, the positive charge is distributed more or less equally between the nuclei because of the symmetry and conjugation of the molecule. It is this delocalization of the electrons and spreading of the charge that imparts the characteristic deep color to the cyanine dyes.

Nonsymmetrical cyanine dyes are also known. In some, familiar functional groups such as NH_2, OH, OR and so on play the role of electron donor. A typical example, CI Basic Yellow 11, is widely manufactured. The cationic azo dyes, e.g. CI Basic Blue 41 (Figure 9.5), are considered to be variants of the cyanine dyes. Here, the azo group functions as the electron-transferring bridge in place of the polymethine structure previously discussed. Figure 9.6 details the synthesis of a typical example, CI Basic Red 22. In the first step, a heteroaromatic amine is diazotized and coupled with dimethylaniline or another coupling component. The azo product thus formed is transformed into a quaternary salt by heating with an alkylating agent. Owing to the nonequivalence of the three triazole ring nitrogens, two isomeric quaternary salts, in a molar ratio of 1:6, are formed. In one isomer, the two methyl groups are located at the 1-position and 4-position; in the other, they occupy the 2-position and 4-position. The commercially available dyestuff is such a mixture.

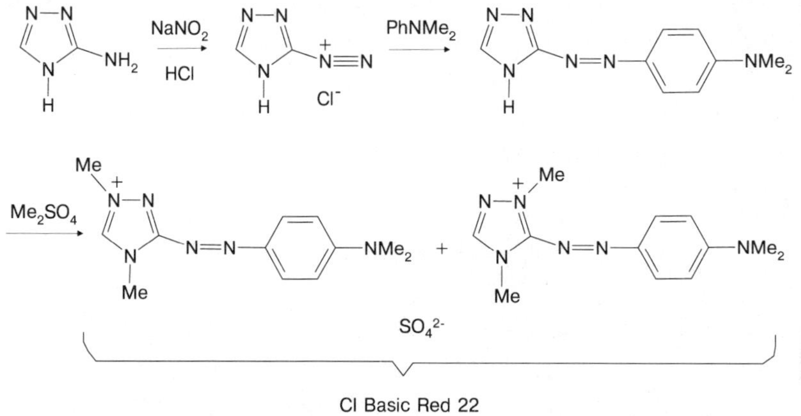

Figure 9.6   Synthesis of the dye CI Basic Red 22.

The value of cationic dyes lies in their rather high light resistance, their intense color and the wide palette available, ranging from red to violet. Moreover, they are one of the very few classes of dyes suitable for the coloration of polyacrylics, which are synthetic fibers in widespread use.

9.2.3   DYES OF THE TWENTIETH CENTURY

Despite their unique role in nature, porphyrin pigments for a long time had no industrial applications. Porphyrins and metalloporphyrins are rather weak absorbers of visible light, and their colors are insufficiently intense for these compounds to be useful as dyes. The stability of porphyrins toward the action of light and chemicals also leaves much to be desired. However, in the late 1920s and early 1930s a new class of dye was discovered with great practical importance. This porphyrin-based substance was named phthalocyanine because of its blue color and the fact that it was derived from phthalic acid. The parent compound, unsubstituted phthalocyanine (Figure 9.7), is prepared by heating the nitrile of phthalic acid with the alcoholate of a higher alcohol. The disodium salt of phthalocyanine is formed by tetramerization of phthalodinitrile. Acidification liberates free phthalocyanine. As Figure 9.7 shows, phthalocyanine has a tetrabenzotetraazaporphyrin structure. The introduction of four additional nitrogen atoms to the meso positions of the

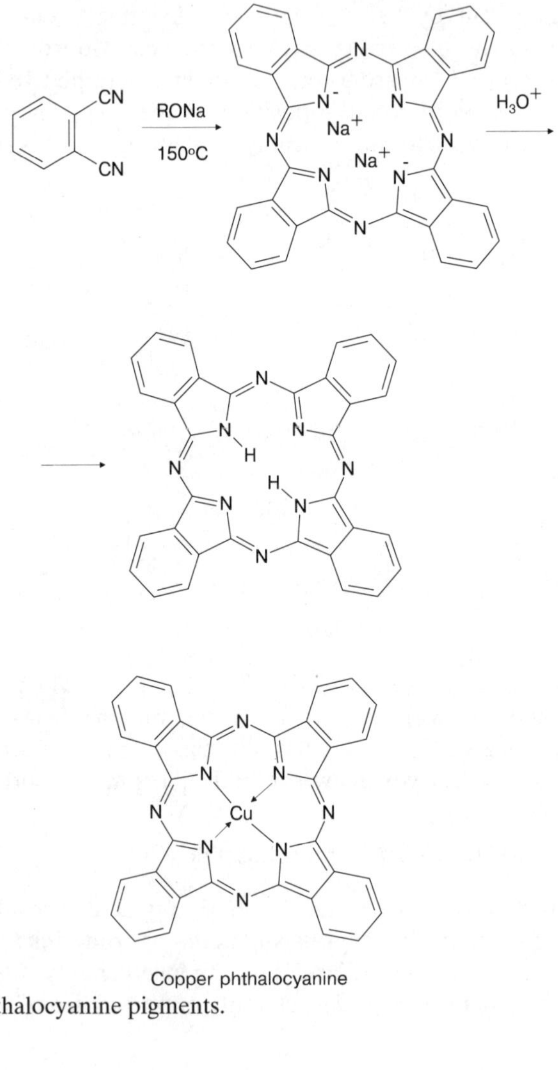

Copper phthalocyanine

Figure 9.7   Phthalocyanine pigments.

porphyrin system, together with the annelation of each of the pyrrole rings with a benzene ring, dramatically enhances light absorption in the visible region. The stability toward light, chemicals and temperature is also enormously increased. The phthalocyanines are all very strongly colored pigments which produce bright colors that are highly resistant to acids, alkalis and high temperatures. Phthalocyanines are used to color many varied materials including fibers, paper, polymers, artificial leather and so on, and to produce varnishes, inks and dyes in solution form. Phthalocyanine itself, its complex with the copper(II) ion (copper phthalocyanine) and a derivative containing fully chlorinated benzene rings are now the most widely used members of the class. They generate green-blue, bright blue and green colors, respectively, and are chiefly utilized as pigments, i.e. dyes which are insoluble in the medium in which the dyeing process is carried out. The worldwide production of phthalocyanines amounts to many thousands of tonnes. The discovery of phthalocyanines was a truly outstanding achievement in the dye chemistry of the twentieth century.

A number of other classes of heterocyclic dyes including the phthaloperinone, quinacridone and triphenodioxazine systems (Figure 9.8) are also used for the preparation of pigments.

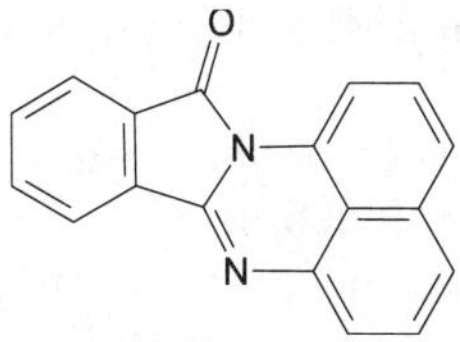

Phthaloperinone

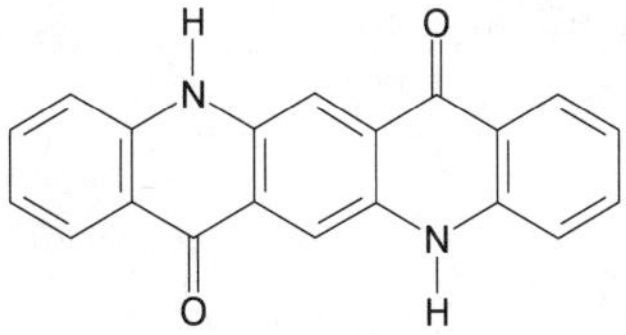

Quinacridone

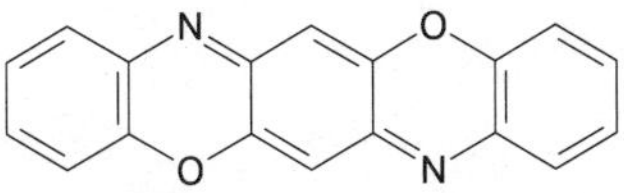

Triphenodioxazine

Figure 9.8   Polynuclear chromophores used in pigment manufacture.

9.2.4   THE ANCHORING OF DYES

As chromophores, heterocycles can, in addition to providing color for a dye, help to fasten this color to the fiber. This was clearly demonstrated by the so-called 'reactive dyes' first synthesized in the mid-1950s. These dyes bind to a fiber by forming covalent bonds rather than simply via absorptive forces and nonbonding interactions. The effect of the binding may be compared with the effect of a strong anchor that secures a ship during a violent storm.

The structure of a reactive dye may be represented in a general way as Chr–M–X, where Chr is the molecular chromophore, X is the reactive functional group which interacts with the fiber and M is an intermediate link whose main function is to attach the X substituent. Reactive dyes are effective for coloring wool and synthetic polyamide fibers, but are most often used on cellulose fibers such as cotton. The mode of action of reactive dyes is based on the following sequence in which they combine with the hydroxy groups of the cellulose (Cel–OH) or the amido groups of the amide fibers (Am–NH)

$$\text{Chr–M–X} + \text{Cel–OH} \longrightarrow \text{Chr–M–O–Cel} + \text{HX}$$
$$\text{Chr–M–X} + \text{Am–NH} \longrightarrow \text{Chr–M–N–Am} + \text{HX}$$

Any known chromophore, e.g. a phthalocyanine, anthraquinone or, most often, an azo dye, may function as the chromophore group, X is generally Cl or F, and the M fragment is usually a heterocyclic system. Since the interaction of a reactive dye with a fiber essentially involves nucleophilic displacement of the halide, the function of the M group is to activate this displacement, i.e. M should be an electron acceptor. Highly π-deficient azines such as 1,3,5-triazine, pyrimidine, quinoxaline and so on meet this requirement. Many reactive dyes include a triazine nucleus which is typically introduced into the dye molecule via 2,4,6-trichlorotriazine, also known as cyanuric chloride (Figure 9.9). Treatment of dyes containing an amino group

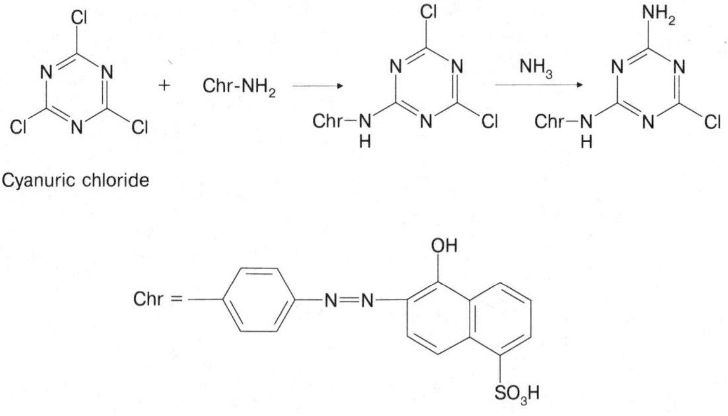

Figure 9.9   Synthesis of reactive dyes.

with cyanuric chloride generates a reactive dye. However, replacement of one further chlorine by an amino group is sometimes carried out. Anchoring of the dye to the fabric occurs via the remaining chlorine atom. Figure 9.9 depicts the synthesis of a deep red reactive dye. An alternative method of anchoring dyes relies on vinyl sulfone intermediates like Chr–M–$SO_2CH=CH_2$.

9.3 Fluorescent Agents

Many organic compounds, including some heterocycles, possess a property which causes them to glow under the action of ultraviolet or short wavelength visible light. Such compounds are named luminophores, and the process by which light is radiated is called photoluminescence, or simply luminescence.[†] Luminophores, or fluorescent agents, have found diverse applications in industry, technology and science. To understand better the principles of their use, we first examine a number of the specific features of luminophores.

9.3.1 WHY THEY SHINE

Almost all organic luminophores, including the aromatic hydrocarbons stilbene, terphenyl and anthracene (Figure 9.10), contain an extensive conjugated system. The greater the number of π-electrons participating in the conjugation, the narrower the energy gap between the highest occupied molecular orbital (HOMO) and the lowest unoccupied molecular orbital (LUMO). As a result, in a compound containing extended conjugation the

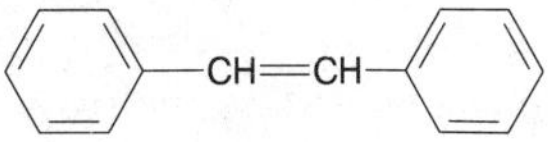

Stilbene

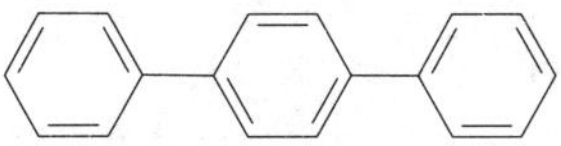

para-Terphenyl

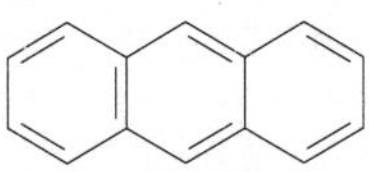

Anthracene

Figure 9.10 Examples of aromatic luminophores.

[†]In certain cases luminescence can be caused by sources of excitation other than light, such as radioactivity (radioluminescence), electric fields (electroluminescence), mechanical forces (triboluminescence) and so on.

energy of near-ultraviolet or even visible light becomes sufficient for an electron to migrate from the HOMO to the LUMO. We have already discussed the dissipation of energy by an excited molecule in Section 6.1 (see Figure 6.5) and are aware that some of the absorbed energy may be lost in various nonradiative processes, e.g. vibration and rotation. Therefore, when such molecules emit a photon (by spontaneous emission; see Figure 6.5d), the energy of the photon emitted is lower than the energy of the photon which caused the excitation (this phenomenon is known as the Stokes shift). Consequently, the light emitted by the molecule will have a longer wavelength than the light absorbed. Thus, anthracene absorbs light of wavelength 380 nm and emits at 434 nm; that is, upon irradiation in the near-ultraviolet, the otherwise colorless anthracene emits blue light.

There are two kinds of luminescence: fluorescence and phosphorescence. The spin of the excited electron does not change during fluorescence (Figure 6.5d). In other words, the molecule is in the singlet excited state S_1 (the ground states S_0 of molecules other than radicals are also singlet states). In the case of phosphorescence, the spin of the excited electron changes and therefore two electrons with unpaired spins exist in the excited state of the molecule. Such a condition is called a triplet state and is designated T_1. A triplet state has a much longer lifetime than a singlet state because $T_1 \longrightarrow S_0$ is a forbidden transition by the so-called selection rules, in contrast to the $S_1 \longrightarrow S_0$ allowed transition. This leads to an increase in the duration of phosphorescence (up to tens of seconds), whereas fluorescence usually lasts less than 10^{-6} s. The majority of organic luminophores irradiate mainly via fluorescence, although there may be some concurrent phosphorescence.

Structural rigidity in a molecule, coupled with an extensive delocalized π-electron system, favors luminescence. The inclusion of electron-donating and electron-withdrawing groups in the conjugated system also has a favorable effect. Structural rigidity is necessary to minimize loss of excitation energy into vibrational modes of the various molecular fragments. The role of the donor–acceptor groups involved in the conjugation is evident: the energy difference between the frontier orbitals is lowered and the intensity of the light absorption and emission increases. Heterocyclic rings contribute to the rigidity of the molecular framework, while the heteroatoms actively participate in the conjugation. This explains the fact that among heterocycles there are many luminophores with practical applications. We now consider a number of these in detail.

9.3.2   SAFETY AND AESTHETICS

Bright dyes that stand out on illumination (e.g. by headlights) at night as well as in daylight are extremely useful. Such compounds are in demand for modern advertisements, decorative art, printing, textiles, road markings, aerodrome signs and navigation. Ordinary dyes are unsuitable because their brightness is due to reflected light. By contrast, luminescent dyes not only reflect light but also transform a portion of the absorbed light into luminescent radiation. Fluorescence, in addition to reflected light, greatly improves the brightness and intensity of the radiation.

Cyanine dyes based on 3,3-dimethylindolinium (e.g. CI Cationic Rose 2C), rhodamines B and 6G (nitrogen analogues of fluorescein), and 1-alkylaminoanthrapyridones (Figure 9.11) are among the compounds frequently utilized as luminophores for the preparation of fluorescent dyes. Such luminophores are used in conjunction with a polymeric substrate, special adhesives and very often in combination with other dyes and luminophores. Thus, any dye color desired can be achieved with increased brightness.

Luminescent dyes are extensively utilized to color plastic materials and synthetic fibers. Such compounds are applied to clothes or insignia worn by road workers, mine-workers, air force pilots and so on as a safety measure. Among the dyes used for this purpose, many derivatives of 1- and 6-aminoanthrapyridone and 1,8-naphthoylene-1,2-benzimidazole (Figure 9.11) can be found.

9.3.3 HOW TO CONVERT WHITE INTO SNOW-WHITE

Perfect whiteness can rarely be achieved in the manufacture of linen, paper and plastic coatings as the starting materials often have a yellowish tint which frequently intensifies during use. This tint is caused by the absorption of some long wavelength blue light because blue is the complementary color to yellow. To increase the whiteness of a material, one must manipulate it to reflect or radiate blue rays.[†] The procedure based on blue light reflection has a long history. In earlier times, ultramarine ('blue') or a small quantity of indigo carmine was added to the water during the washing of linen. The material thus treated ('blued') then reflected a small excess of blue radiation and appeared to be whiter. However, the whiteness thus achieved was far from ideal as the textile took on a grayish hue.

In modern times the application of so-called 'optical bleachers' has achieved the desired goal. Their effectiveness is a result of the emission of blue light by luminescence. Optical bleachers are in fact colorless fluorescent dyes. They absorb light in the near-ultraviolet range and re-emit it by fluorescence in the blue region of the visible spectrum. The application of such compounds creates the appearance of intense whiteness. It is familiar to us that white clothing appears luminescent in restaurants and clubs where soft blue light is used; this is a result of optical bleachers.

The majority of optical bleachers are derivatives of heterocycles, although

---

†A chemical method of eliminating yellowish tints in textiles involves treatment with an oxidizing agent. Unfortunately, the oxidation is not restricted to the yellow color bodies and thus some fiber damage is unavoidable.

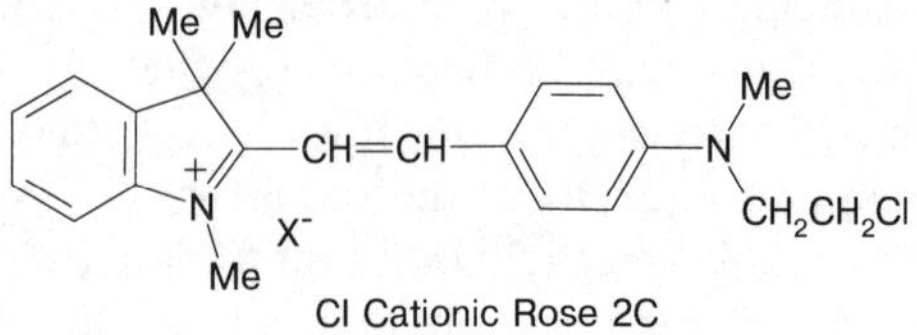

CI Cationic Rose 2C

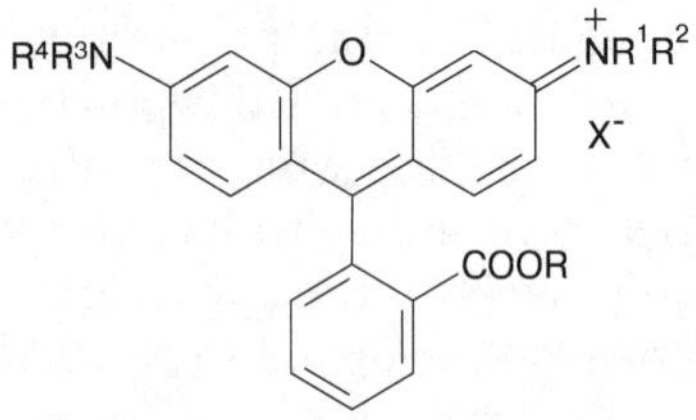

Rhodamine B ($R = H$, $R^1 = R^2 = R^3 = R^4 = Et$)

Rhodamine 6G ($R^1 = R^3 = H$, $R = R^2 = R^4 = Et$)

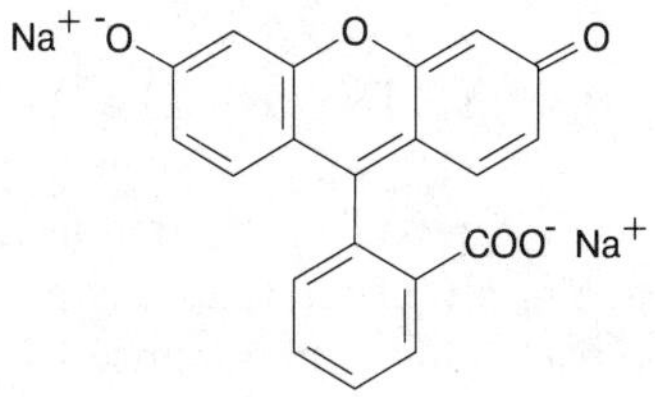

Fluorescein
(disodium salt)

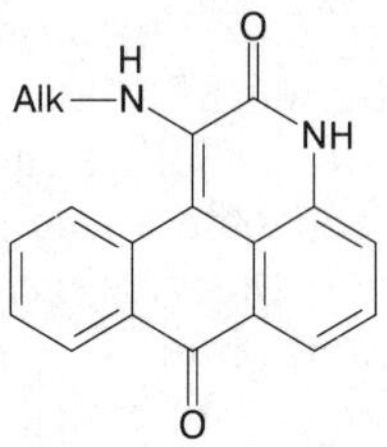

1-Alkylaminoanthrapyridone

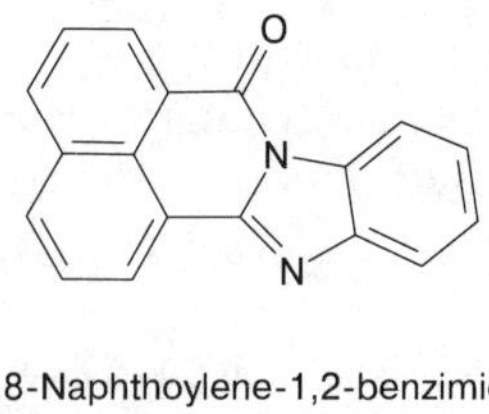

1,8-Naphthoylene-1,2-benzimidazole

Figure 9.11 Luminophores used for the preparation of fluorescent paints.

their heterocyclic nuclei are not always responsible for the luminescence. Whiteners of the 4,4'-diaminostilbene-2,2'-disulfonic acid series are employed widely. The most useful of these contain triazine substituents at the amino groups (Figure 9.12, structure (1)). If chlorine atoms are retained in the triazine rings, the bleacher becomes tightly fastened to the fiber in a manner similar to that of the reactive dyes.

Optical bleachers are added to plastic materials or synthetic fibers at the temperature of the melt. Therefore, bleachers must be thermostable, as are, for example, whiteners constructed from the heterocyclic derivatives of stilbene (Figure 9.12, structures (1)–(3)). Coumarin and 1,3-diarylpyrazoline analogues (Figure 9.12, structures (4) and (5)) are also useful whitening agents.

A familiar application of optical bleachers is their use in washing powders and other cleaning preparations.

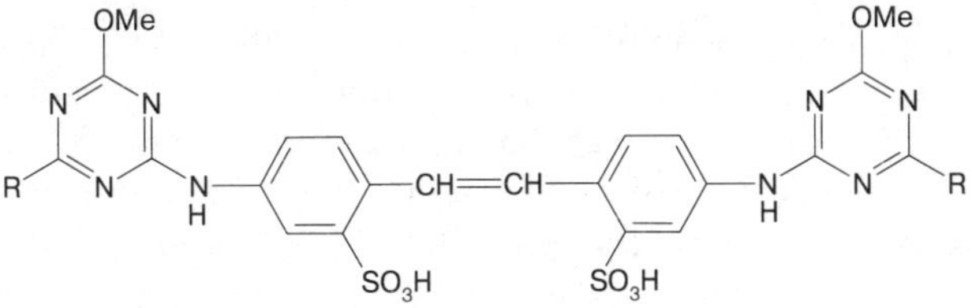

1 (R = Cl, C_6H_5NH)

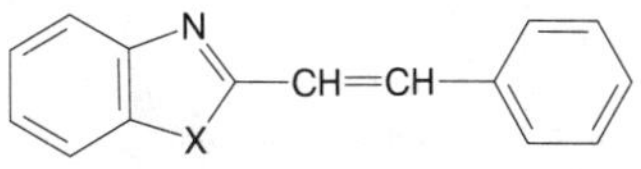

2 (X = NH, O, S)

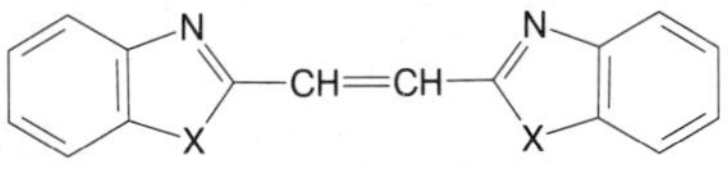

3 (X = NH, O, S)

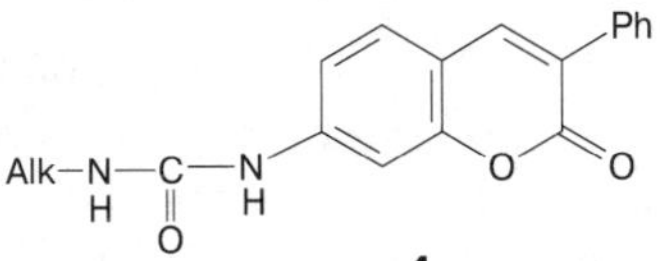

4

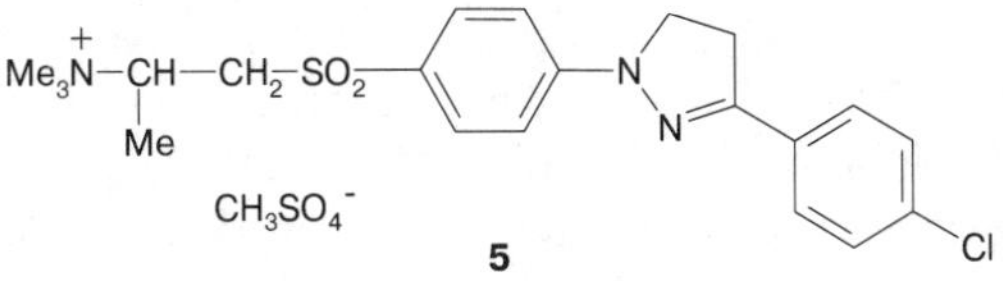

5

Figure 9.12   Examples of optical bleachers.

9.3.4   MARKERS, INDICATORS AND DIAGNOSTIC AGENTS

Luminophores have proven to be irreplaceable as components for a multitude of applications, including some involving living cells. For instance, fluorescein is used in geological and hydrological studies to determine the directions of underground water flows and their connections to points of emergence above ground. Fluorescein, rhodamine and 1,8-naphthoylene-1,2-benzimidazole are used in luminescent flaw detectors to locate microscopic fissures and other superficial damage in industrially manufactured metallic, ceramic and concrete materials. The article being tested is immersed in a luminophore solution for a period, then washed and dried. While the surface of the article may appear to be free of luminophore, a sensitive fluorescence detector allows ready detection of the compound in microcracks.

A similar approach is employed in medical applications. Thus, a minute quantity of fluorescein is injected into the blood to check the permeability of the blood vessels. Luminophores have provided a powerful impetus for advances in medical diagnostics. Numerous fluorescent tracers capable of selective attachment to nucleic acids, lipids, polysaccharides, antibodies, cellular membranes, damaged cells and so on have been synthesized. For instance, the luminescent dye acridine orange 'marks' healthy and cancerous cells differently and therefore has been used in the diagnosis of malignant tumors.

Acridine orange

Organic luminophores have also been used in devices that register ionizing particle flux: α-, β-, and γ-rays, neutrons and even neutrinos can be detected. In such cases luminophores serve as scintillators, i.e. substances which produce short-lived flashes (scintillations) when struck by the ionizing particles. The number of scintillations produced is recorded by a photomultiplier. Scintillators are used in diverse branches of science and technology including nuclear and space research. Among heterocyclic scintillation agents, 2,5-diphenyloxazole, 2-phenyl-5-(4-biphenylyl)-1,3,4-oxadiazole and 1,3,5-triphenyl-Δ^2-pyrazoline (Figure 9.13) dissolved in organic solvents are the most often used as liquid scintillators. As such solutions can be prepared in virtually unlimited volume, luminophores are especially useful

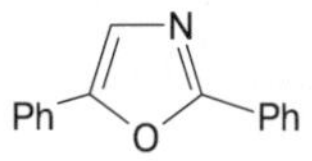

2,5-Diphenyloxazole

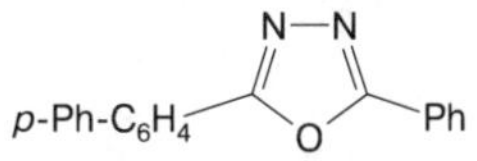

2-Phenyl-5-(4-biphenylyl)-
1,3,4-oxadiazole

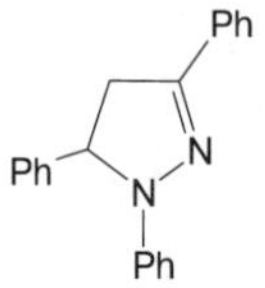

p-Ph-C$_6$H$_4$

Ph

1,3,5-Triphenyl-Δ^2-pyrazoline

Figure 9.13   Heterocyclic luminophores used as scintillators.

in detecting the presence of particles at very low flow densities.[†]

Fluorescent materials have also been applied as thermo-indicators in industry, as fluorescent indicators in analytical chemistry, as luminescent biological markers and so on. The use of luminophores as active principles in lasers deserves separate mention and is covered in the next section.

9.3.5   LASERS CONTAINING HETEROCYCLIC LUMINOPHORES

When we speak of 'lasers', our imagination conjures up pictures of bright, highly focused beams of light capable of cutting metal, initiating thermonuclear synthesis, rapidly reading stored information (compact disks, magnetic cards, etc.), transferring information along glass filaments, determining distances between distant objects (e.g. the Earth and moon), surgically incising live tissue and so on. These and other uses of lasers have revolutionized modern industry, electronics and information handling, as well as medical and scientific research.

A laser is a source of light, or indeed of any form of electromagnetic radiation. Candles and electric lamps are also sources of electromagnetic radiation, but the difference is that they produce noncoherent light, i.e. radiation composed of photons of different frequencies and direction or phase. Laser beams, on the contrary, are characterized by perfect coherence (Figure 9.14).

---

†Solid scintillators are also made in the form of single crystals of certain organic luminophores such as anthracene, stilbene, diphenylacetylene and others (their solid solutions in polymers can also be used). These types of scintillators, in addition to their inorganic counterparts, are usually used to measure flows of high energy radiation.

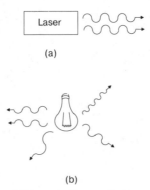

(b)

Figure 9.14 (a) Coherent and (b) noncoherent light sources (Kovalenko, L. J., and Leone, S. R., *J. Chem. Educ.*, 1988, **65**, 681).

In a laser, some other form of energy is converted into coherent light. The substance which transforms the energy is called the laser active medium. Lasers are classified, according to the aggregation state of their active medium, as gas, solid or liquid lasers. In gas lasers the active medium consists of atoms, ions or molecules of various compounds. In solid lasers, rare-earth metals or chromium(III) (ruby laser) in the crystalline or glassy form are used, and the excited ions serve as the source of radiation. The active medium in liquid lasers is composed of a solution of an organic compound. Complex compounds of rare-earth ions with various organic ligands have long been used as active media. However, in the late 1960s it was found that laser effects could be achieved by organic luminophores alone, especially heterocyclic derivatives. We first examine the mechanism of laser action and then discuss structures.

For a laser to function, the active medium first has to be excited. This process, called laser pumping, is carried out by an additional energy source, such as an impulse lamp. There are three types of interactions between matter and light. We have already discussed two of these: photon absorption

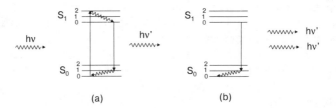

Figure 9.15 Stimulated emission of radiation: (a) molecular absorption of a photon with frequency $h\nu$ and spontaneous emission of a photon with frequency $h\nu'$ by fluorescence; and (b) emission of second photon with frequency $h\nu'$ stimulated by the initial photon's effect on a second excited molecule.

and spontaneous emission (see Sections 6.1 and 9.3.1). The third, called stimulated emission, is the type of interaction on which laser action is based. When a photon, emitted during fluorescence with a frequency hv', acts upon another excited molecule, the emission of a second photon of the same frequency, direction and phase occurs (Figure 9.15). The process is propagated in a manner similar to a chain reaction. Two emitted photons now stimulate the emission of four photons and so on; thus, the emission of coherent radiation increases many-fold. This phenomenon is summarized by the acronym 'laser': light amplification by stimulated emission of radiation.

Under stimulated emission, the excited molecules radiate photons and are converted to the nonexcited state S_0. Of course, the process terminates when no excited molecules remain and the radiation ceases. To prevent this, the population of the excited state $S_1(0)$ has to be maintained at a sufficiently high and stable level. Such a state, an inverted population (see Figure 9.16), is achieved by (i) systematic pumping of the active medium together with (ii) the construction of a system of mirrors which directs the majority of the laser radiation to reconvert molecules to the excited state.[†]

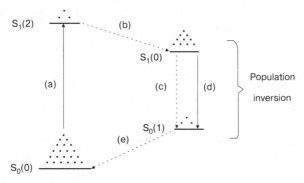

Figure 9.16 A four-level laser (the marking of energy levels is conventional and correlates to that of Figure 9.15; each point in the figure represents one molecule): (a) excitation, (b; e) rapid relaxation, (c) slow relaxation and (d) laser radiation (Kovalenko, L. J. and Leone, S. R., *J. Chem. Educ.*, 1988, **65**, 681).

Rather stringent demands are made on the compound used in the laser. The luminophore must achieve a high quantum yield of fluorescence and must be photostable. Moreover, the absorption band should not overlap with the fluorescence band appreciably. In rhodamine 6G (Figure 9.17) only minor overlap occurs, and rhodamine 6G is therefore suitable as a laser dye.

Figure 9.16 shows that a substance capable of generating laser radiation should possess at least four appropriate energy levels between which the

---

†The chamber in which the laser radiation is generated contains an outlet for a portion of the radiation for the required purpose.

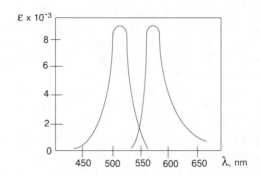

Figure 9.17   Spectra of absorption (left) and fluorescence (right) for rhodamine 6G (in ethanol, $\lambda_{exc} = 365\,nm$) (adapted from Krasovitskii, B. M., and Bolotin, B. M., *Organic Luminescent Materials*, VCH, Weinheim, 1988, p. 287, Figure B22, with permission).

necessary transitions can take place. Additionally, relaxation of the molecules from the S_1 to S_0 state should be relatively slow, whereas the passages $S_1(2)\longrightarrow S_1(0)$ and $S_0(1)\longrightarrow S_0(0)$ should occur rapidly.

These requirements are met by a number of heterocyclic luminophores which are usually classified by the wavelengths at which they generate laser radiation (Table 9.1). The greatest generation of energy (radiation in the near-ultraviolet and blue ranges of the spectrum, i.e. 300–400 nm) is achieved by 2,5-diaryloxazoles, 2,5-diaryl-1,3,4-oxadiazoles and 2-arylbenzoxazoles. Compounds of the 1,4-bis(5-phenyloxazol-2-yl)benzene type generate radiation at somewhat greater wavelengths (400-440 nm), still within the blue region. Coumarins, especially their 7-hydroxy and 7-dialkylamino derivatives, are extensively used as luminophores in the blue and green regions of the electromagnetic spectrum.

Xanthene dyes such as rhodamines 6G and 3V and 7-hydroxy-3*H*-phenoxazin-3-one are almost ideal for generating laser radiation in the red region of the spectrum (520–650 nm). Organic compounds which lase in the infrared region (800–1200 nm) are also known. Such compounds are usually cyanine dyes with long conjugated polymethine chains. The opposing termini of the chains contain strong electron-donating and electron-withdrawing groups, usually heteroaromatic cations and their partially reduced π-excessive counterparts. Indoline, benzoxazoline, benzthiazoline, pyrylium and thiapyrylium nuclei are the heterocycles most frequently incorporated into the polymethine chain. The dye (6), based on the 1-thiabenzopyrylium cation, holds a record for this series of compounds as its solution in dimethyl sulfoxide generates laser radiation in the range 1150–1240 nm.

Table 9.1   Some heterocyclic compounds used as laser active media.

Structural formula	Name	Wavelength of radiation generated (nm) (region)
 R^1 —〈 〉— C(X–N)=C(O) —〈 〉— R X = CH, N; R = R^1 = Ph R = Ph, R^1 = H	2,5-Diaryloxazoles 2,5-Diaryl-1,3,4-oxadiazoles	300-400 (near-ultraviolet and blue)
(2-Arylbenzoxazole structure) R	2-Arylbenzoxazoles	300–400 (near-ultraviolet and blue)
Ph (oxazolyl–benzene–oxazolyl) Ph	1,4-Bis(5-phenyloxazol-2-yl) benzene	400–440 (blue)
Me R (coumarin structure) O=C–O R = OH, NEt₂, etc.	Substituted coumarins	440–500 (blue and green)
R\ /R Et–N (xanthene) N⁺–Et Cl⁻ COOEt R = H (Rhodamine 6G); R = Et (Rhodamine 3V)	Xanthene dyes	520–650 (red)
HO (phenoxazinone structure) O	7-Hydroxy-3*H*-phenoxazin-3-one and other oxazine dyes	500–800 (red)
(benzazolium)–(CH=CH)ₙ–CH=(benzazolium) R⁺ N N R X = NPh, O, S, Se, CMe₂	Polymethine dyes	800–1200 (infrared)

6

Lasers based on organic dyes cannot be very powerful in terms of energy. However, they do offer one significant advantage over solid and gas lasers. The width of the luminescence band in conjugated organic compounds, which may approach 200 nm (Figure 9.17) allows the frequency of the laser radiation generated to be changed smoothly over the operating range, while other types of lasers lack this ability.

9.4 Fire Retardancy

We cannot guarantee the authenticity of the story which follows, but it seems to be an appropriate introduction to this section. At a conference, the general manager of a chemical fibers factory, after lighting a cigarette, casually allowed the burning lighter to ignite the necktie of an old friend. When the horrified friend recoiled, the manager said with feigned surprise, 'Mine doesn't burn. It's nonflammable,' and immediately proved his claim. In this way he succeeded in advertising his company's new product. The case was particularly impressive as the 400 °C temperature of the cigarette lighter flame instantly scorched silk, wool and cotton, and melted nylon and many other fibers. The manager's tie fabric was woven from a special type of chemical fiber distinguished by its exceptional thermostability. Obviously, the manufacture of neckties was not the major application intended for such fibers; the necktie was selected only to distinguish the manufacturer's fiber from others. Heat resistant fibers have become essential in the modern world for the production of fireproof clothing for firemen, welders, foundrymen, pilots and astronauts. These fibers are also utilized in the manufacture of parachutes, conveyer belts, heat-insulating material, and asbestos replacements. The development of heat resistant fibers has significantly influenced progress in aerospace and aeronautics technology.

We now discuss the composition of such thermostable compounds. Chemically, these substances are polymers composed of aromatic and heteroaromatic residues. Polybenzimidazoles, and to some extent polyquinoxalines, have become especially useful. Benzimidazole and quinoxaline are highly stable molecules which do not decompose at temperatures of up to 600 °C (of course, at this temperature they exist as gases). As constituents of polymer chains these two nuclei render the macromolecule inherently highly heat resistant and stable. The presence of N–H bonds and amide (NHC=O) groups in such polymers imparts additional durability.

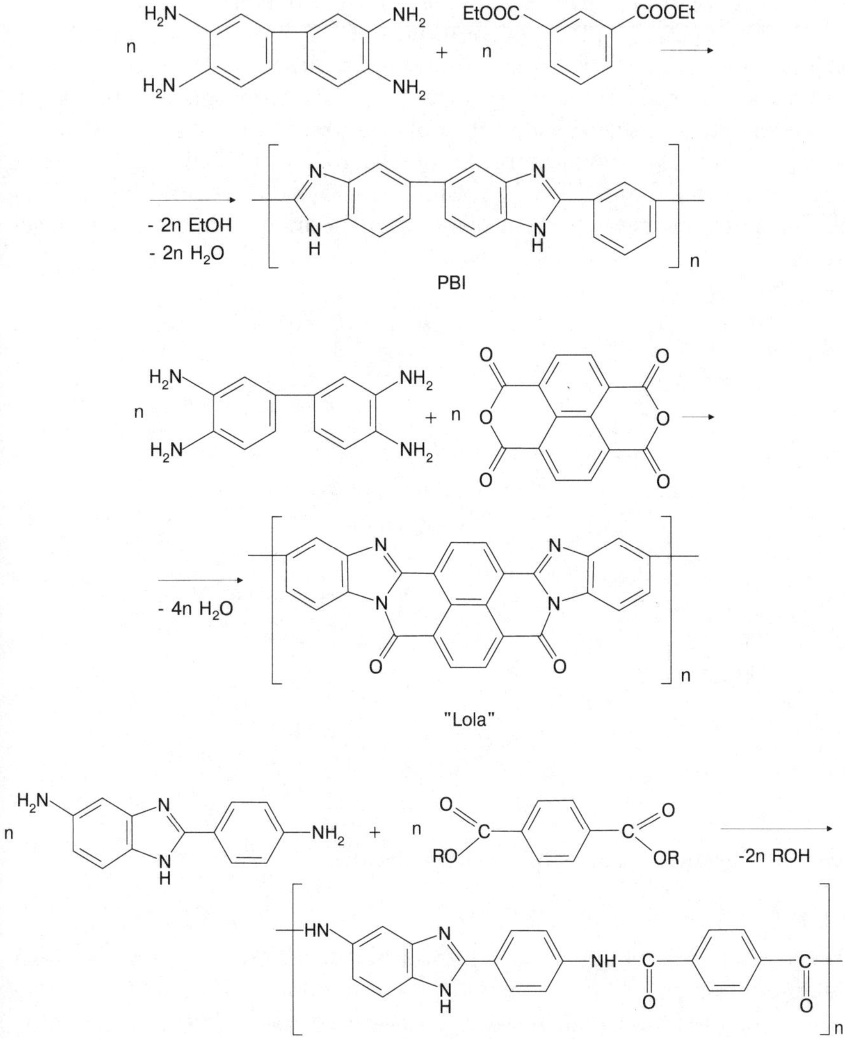

Figure 9.18 Thermostable polybenzimidazoles.

In the USA, the Celanese company produces several thousand tonnes each year of the polybenzimidazole known as PBI by the condensation of 3,3′,4,4′-tetraaminobiphenyl with diethyl isophthalate. Another heat resistant polybenzimidazole, used in Russia to manufacture a fiber named 'Lola', is synthesized by the condensation of tetraaminobiphenyl with the dianhydride of naphthalene-1,4,5,8-tetracarboxylic acid. The preparation of polymers from readily available benzimidazole-based monomers is also possible. In particular, very strong fibers can be obtained from polymer (**7**), which is itself formed from the polycondensation of 5-amino-2-(*p*-aminophenyl) benzimidazole with derivatives of aromatic dicarboxylic acids, such as terephthalic esters (Figure 9.18). Articles made from polybenzimidazole fibers retain their properties during prolonged periods at 200–300 °C in the presence of air. At temperatures from 300 to 350 °C their heat resistant properties are maintained for 24 h, while at 400–450 °C several hours is the limit. Such fibers exceed the heat resistance of sodium and potassium silicate glasses.

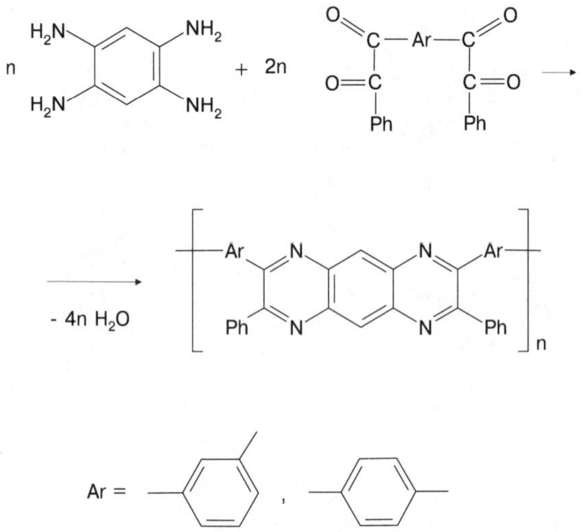

Figure 9.19   Synthesis of thermostable polyquinoxalines.

Thermostable polyquinoxalines are formed from the reaction of 1,2,4,5-tetraaminobenzene with aromatic bis-α-diketones (Figure 9.19). The resulting fibers exhibit high resistance toward corrosive agents and heating at temperatures up to 300 °C. Further thermostable fibers have been synthesized which contain other heterocyclic systems including benzoxazole units.

9.5   Photographic Materials and Recorders of Information

The versatile array of chemicals available for use in photographic processes is largely based on heterocyclic compounds. Silver bromide crystals, which are usually used as the photosensitive material, frequently need to have their sensitivity to light enhanced by the addition of silver sulfide. As a result of such activation the unexposed sites of the film can become slightly exposed causing haziness in the photograph. To avoid this, one must add antifogging substances to the standard photographic regimen of chemicals employed nowadays. Salts and complexes are formed with the silver particles which appear in the unexposed regions of the film, thus eliminating fogging and restoring clarity and purity to the picture. Most contemporary antifogging substances and photoemulsion stabilizers are heterocyclic in nature. Examples include benzotriazole, 5-nitrobenzimidazole, 1-phenyl-5-mercaptotetrazole and 5-hydroxy-7-methyl-1,2,4-triazolo[2,3-a]pyrimidine (Figure 9.20).

A further disadvantage of silver bromide is that it is sensitive only toward the blue and violet region, i.e. the higher energy radiation of visible light. To widen the range of sensitivity to include the entire visible region of the spectrum, one must add so-called optical sensitizers photoemulsions. The

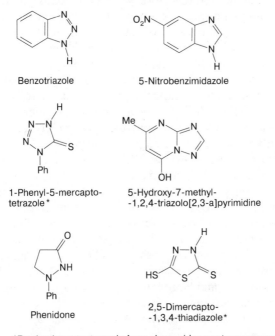

Benzotriazole   5-Nitrobenzimidazole

1-Phenyl-5-mercapto-   5-Hydroxy-7-methyl-
tetrazole*   -1,2,4-triazolo[2,3-a]pyrimidine

Phenidone   2,5-Dimercapto-
                            -1,3,4-thiadiazole*

*Predominant tautomeric form shown (does not correspond
to common name)

Figure 9.20   Heterocyclic compounds used in photography.

photosensitizers are usually cyanine dyes, such as pseudocyanine or pinacyanol discussed earlier (Figure 9.5). The former enhances the sensitivity of the photoemulsion towards blue and green light, while the latter has a similar effect with respect to red light.

Heterocycles are also utilized as auxiliary compounds during other steps in the photographic process. Thus, phenidone, a derivative of pyrazole, is used in combination with hydroquinone as a developer, and 2,5-dimercapto-1,3,4-thiadiazole (Figure 9.20) is an effective toner. Toners form colored compounds with the silver particles; therefore, by careful selection from a range of compounds, black, brown and other tints can be emphasized.

New silver-free reprographic materials are eagerly sought, in part owing to fears of a shortage of, and increased price for, silver. A classic silver-free reprographic printing process is based on the application of light sensitive diazo compounds. When exposed to light, the diazo species decompose with the liberation of nitrogen. If the exposed print is further processed with a phenol, the unexposed sites undergo azo coupling of the phenol and the unaffected diazonium salt. The final result of this conversion is the formation of an azo dye which provides color to the picture. Correspondingly, the exposed sites remain colorless. The light sensitive diazo compounds are generally homoaromatic, and therefore we will not discuss them further.

Heterocyclic compounds play a leading role in the creation of another type of silver-free reprographic material which has attracted the close attention of scientists over the last 20 years. This material is based on photochromic substances which reversibly change color by the action of light. Indolylspiropyrans are distinguished by their quality and effectiveness among the various classes of organic photochromes. An example is compound (**8**) shown in Figure 9.21, which is produced from the condensation of 1,2,3,3-tetramethylindolium iodide with 5-nitrosalicylic aldehyde. On irradiation with light of the appropriate wavelength, spiropyran (**8**) is transformed into the intensely purple valence isomer (**9**)[†] which structurally resembles the cyanine dyes. Compound (**9**) is sufficiently stable for the color of a picture to be preserved over long periods. The reverse reaction, i.e. closure of the pyran cycle to give the colorless form (**8**), is possible only upon repeated irradiation of isomer (**9**) by a powerful light source of a different wavelength.

Despite their comparatively low light sensitivity, photochromes possess a few remarkable advantages over silver halide based materials. Firstly, photochromic transformations have a molecular nature that provides exceptionally high clarity and large information storage capacity to the image. Thus, the use of micro imaging enables the information contained in a large library to be stored in compact format. Another advantage is that as a clear

---

†Valence isomers are substances which interconvert as a result of electron and bond shifts (in contrast to other types of isomerism and tautomerism in which atoms or groups migrate).

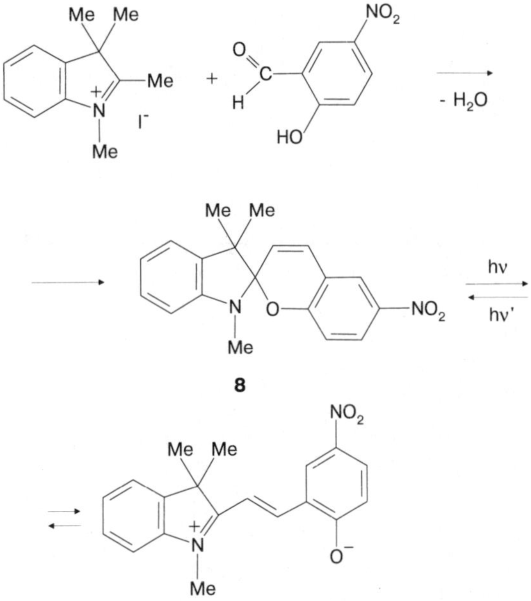

Figure 9.21 Indolinospiropyran (**8**) and photoisomerization into the colored form (**9**).

color image of the subject is received immediately following exposure, the need for the traditional operations of developing and fixing is eliminated. Last but not least, photochromic materials can be highly economical as in some cases they can be utilized repeatedly after deletion of the original image.

9.6 Other Applications

Various heterocyclic compounds are now used in the food industry as dyes, aromas and flavorings. For example, of the seven dyes cleared for use in the USA in drugs, food products and cosmetics, the following are heterocycles: red erythrosine, yellow tartrazine and blue indigo carmine (Figure 9.22a).

The pleasant odors of many foods are generally not attributable to a single compound. The aroma is a bouquet—a mixture embracing up to 100 volatile components as in coffee, wine and smoked foods. However, the aroma 'profile' of such a mixture is determined by a relatively small group of substances, and the components have been established in many cases. Thus, it has been determined that 8-methylpyrrolo[1,2-a]pyrazine is the major constituent of the odor of roasted meat; 2-methoxy-3-methylpyrazine imparts the fragrance of roasted ground nuts, and of coffee and cacao beans;

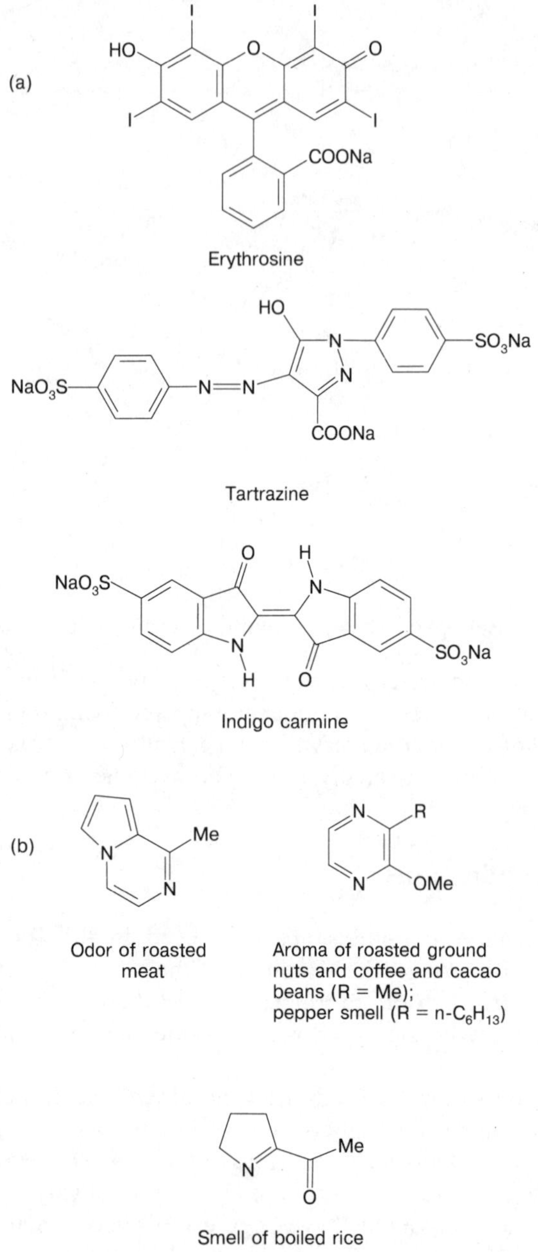

(a)

Erythrosine

Tartrazine

Indigo carmine

(b)

Odor of roasted
meat

Aroma of roasted ground
nuts and coffee and cacao
beans (R = Me);
pepper smell (R = n-C_6H_{13})

Smell of boiled rice

Figure 9.22   Heterocyclic compounds used as (a) food dyes and (b) aromas.

2-methoxy-3-*n*-hexylpyrazine simulates the aroma of pepper; and 2-acetylpyrroline is responsible for the smell of boiled rice (Figure 9.22b).

Another industrial use of heterocycles is for preservation. Isopsoralene has been added to liqueurs and cosmetics, and 2-(5-nitrofuryl-2)-acrylic acid was suggested as an effective preservative for wine. Poly(*N*-vinylpyrrolidone) (Figure 9.23) is produced industrially as a blood substituent, an extender of drug action and a hairspray ingredient.

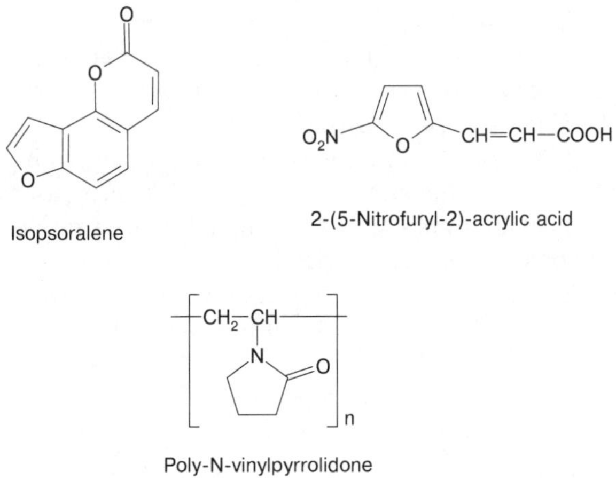

Isopsoralene

2-(5-Nitrofuryl-2)-acrylic acid

O_2N —CH=CH—COOH

Poly-N-vinylpyrrolidone

Figure 9.23   Examples of heterocyclic compounds used in the food industry and medicine.

Heterocycles occupy an important place in analytical chemistry. They are used in the determination of numerous metal ions and of inorganic and organic compounds, as well as in the extraction of metal ores. Analytical reagents for the determination of metal ions usually contain a chelating agent composed of several heteroatoms or a heteroatom and a functional group. Three classical reagents of this type are 2,2'-bipyridyl (Figure 1.6), *o*-phenanthroline and 8-hydroxyquinoline (Figure 9.24). The last of these is widely used for the estimation of cobalt(II), chromium(III), iron(II), vanadium(V) and other ions. The procedure is based on the formation of chelated complexes with the participation of the pyridine nitrogen and hydroxy oxygen. These complexes are colored. Therefore, the concentration of the extracted metal ion can be determined quantitatively by measuring the intensity of the colored solution spectrophotometrically following extraction with a suitable solvent. 2,2'-Bipyridyl and *o*-phenanthroline are especially useful in the analysis of the iron(II) ion. In its complexes with these compounds, iron(II) is coordinated to three ligands, causing its external electronic shell to resemble that of krypton. The complexes have an

octahedral structure as depicted by

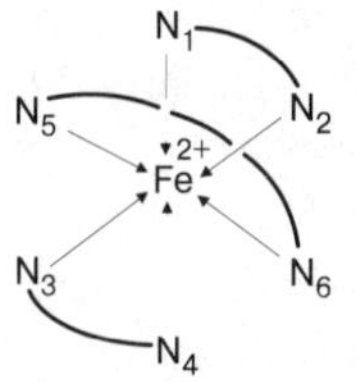

Many heterocyclic analytical reagents contain potential thiol (SH) groups, e.g. 2-mercaptobenzimidazole, 2-mercaptobenzoxazole and 2-mercapto-benzothiazole (Figure 9.24).[†] Owing to ionization of the S–H (or N–H) bonds, sparingly soluble salt-like complexes are formed (with participation of the sulfur atoms) with the ions of heavy metals such as cadmium, lead, copper and gold. Determination of the metal concentration is carried out gravimetrically or spectrophotometrically.

2-Aminoperimidine is an interesting example of a relatively new analytical reagent (Figure 9.24). Treatment with sulfuric acid yields the sulfate which is

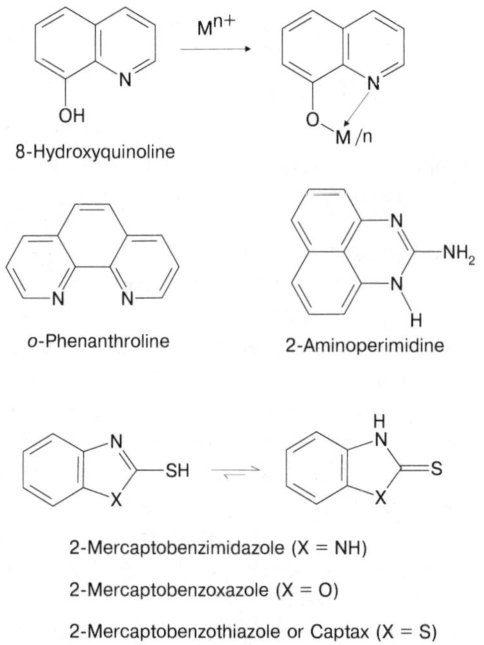

8-Hydroxyquinoline

o-Phenanthroline

2-Aminoperimidine

2-Mercaptobenzimidazole (X = NH)

2-Mercaptobenzoxazole (X = O)

2-Mercaptobenzothiazole or Captax (X = S)

Figure 9.24   Several heterocyclic analytical reagents.

[†] All of these derivatives, although commonly named as thiols, in fact exist preferentially in the thione tautomeric form containing C=S and NH groups (see Figure 9.24).

distinguished by an incredibly low solubility (it is less soluble than the familiar barium sulfate). However, the main point to note is that perimidinium sulfate, in contrast to $BaSO_4$, forms highly stable colloidal suspensions, thus allowing rapid and precise determination by the nephelometric method. 2-Aminoperimidine is utilized to estimate quantitatively the air sulfur dioxide and sulfuric acid content, which is of great importance in environmental control and protection. Additional information concerning analytical reagents is given in Chapter 10.

Heterocycles are utilized in many other spheres: liquid crystals, polymeric materials, rubber stabilizers, vulcanization accelerators and so on. For example, copolymers of butadiene and 2-methyl-5-vinylpyridine are used in the production of rubbers resistant to the action of heat, oils, lubricants and gasoline. Captax (Figure 9.24) is not only an analytical reagent but also an effective accelerator of rubber vulcanization, for which its current industrial manufacture is impressive. A number of probable future applications of heterocycles are discussed in the next chapter.

9.7 Problems

1. What are the main industrial and technological applications of heterocyclic compounds?
2. *trans*-Thioindigo and *trans*-N,N'-dimethylindigo are easily converted into the corresponding *cis* isomers. In contrast, indigo itself undergoes the analogous conversion with difficulty. Explain.
3. Dilute acids and alkalis have little influence on the color of indigoid dyes, in contrast to concentrated solutions of acids and alkalis. Thus, in concentrated sulfuric acid thioindigo changes from red (λ_{max}546 nm) to blue (λ_{max}641 nm). Indigo becomes green in sodium butoxide solution. Account for these observations.
4. Hydrogenation of one, two or even three of the outer double bonds of the pyrrole nuclei of porphyrins does not cause a change in color. Suggest an explanation for this fact.
5. Interaction of a reactive dye with water leads to a decreased ability to adhere to the fiber being dyed. What is the mechanism of this undesirable reaction? To illustrate your answer, you should use the formulas shown in Figure 9.9.
6. What is the basic principle of optical bleacher use? Give some examples of heterocyclic optical bleachers.
7. What properties are required by laser dyes?
8. What is a photochrome? What is the principle behind the use of photochromic substances for the recording of optical information? Include a heterocyclic photochrome in your answer.

9. Dinitrobenzylpyridines A and B are photochromic substances, whereas C is not. Explain this observation, giving consideration to the structures of the photoexcited species.

A

B   C

9.8   Suggested Reading

1. Leznoff, C. C. and Lever, A. B. P., (eds), *Phthalocyanines: Properties and Applications*, VCH, New York, 1993.
2. Dürr, H. and Bouas-Laurent, H. (eds), *Photochromism: Molecules and Systems*, Elsevier, Amsterdam, 1990.
3. El'tsov, A. V., *Organic Photochromes*, Consultants Bureau, New York, 1990.
4. Zollinger, H., *Color Chemistry*, VCH, Weinheim, 1987.
5. Gordon, P. F. and Gregory, P., *Organic Chemistry in Colour*, Springer, Berlin, 1983.
6. Tedder, J. M., Nechvatal, A. and Jubb, A. H., *Basic Organic Chemistry. Part 5: Industrial Products*, Wiley, London, 1975.

10 MODERN TRENDS AND PROSPECTS OF DEVELOPMENT

> At the bottom of a pitcher
> Dwarfy's met a giant teacher.
> 'How's a dwarf so small as thee
> Got into this, just say to me?'
>
> G. Sapghir

In the preceding chapters our attention was focused mainly on the central role of heterocycles in nature and on their practical applications. This chapter discusses the varied directions of contemporary development in heterocyclic chemistry, beginning with a number of fundamental investigations. The boundary between pure and applied research is now less defined, and many important theoretical works are influenced by the challenges encountered in applied chemistry. The chemistry of macrocyclic compounds is undoubtedly now of great significance based on the volume of literature produced. This is not surprising as earlier research into the porphyrins, phthalocyanines and other macrocycles demonstrated their overwhelming importance.

10.1 Macrocycles as Molecular Containers

In 1967 Pedersen published a series of articles in the *Journal of the American Chemical Society*. The then 63 year old chemist from Du Pont described the synthesis of a new type of heterocyclic compound which could be classified as a cyclic ether. The new system was based on alternating oxygen atoms and CH_2CH_2 bridges (Figure 10.1a). In total, Pedersen synthesized more than 60 cyclic polyethers which contained from four to 20 oxygen atoms. The rings ranged in size from 12 to 60 members.

The most intriguing feature of these macrocyclic polyethers was their ability to form unusually stable crystalline complexes with alkali metal ions. The practical importance of this discovery was immediately obvious as scientists did not previously have at their disposal efficient reagents for the extraction and separation of these ions.

215

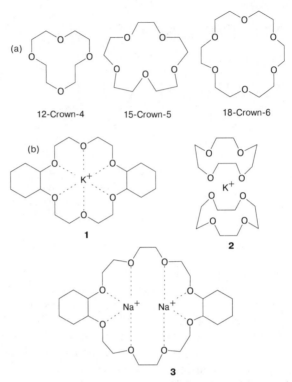

Figure 10.1 (a) Crown ethers and (b) their complexes: (**1**) is a 1:1 complex of dicyclohexano-18-crown-6 with KI; (**2**) is a 2:1 complex of 12-crown-4 with KI; and (**3**) is a 1:2 complex of dicyclohexano-24-crown-8 with NaI.

The strength of the complex, with a preferred 1:1 polyether:ion ratio, is determined by the electrostatic attractions between the metal ion contained within the macrocyclic cavity and the ether oxygen atoms (Figure 10.1b, structure (**1**). Pedersen noticed the similarity between a royal crown and the macrocyclic polyether 'crowning' an ion. To simplify the rather complicated nomenclature of these macrocycles, Pedersen proposed the new name crown ethers, which has now become universally accepted. The names of crown ethers include two numerals: the first designates the size of the ring, and the second the number of oxygen atoms it contains.

It was further found that the stability of each complex largely depended on the correlation between the internal cavity size of the crown ether and the ionic radius of the cation. The data in Table 10.1 indicate that 12-crown-4 forms a stable complex with the Li$^+$ ion, 15-crown-5 is the most suitable ether for complexation with Na$^+$, and 18-crown-6 is suitable for coordination with K$^+$, NH$_4^+$, and Rb$^+$.

Table 10.1 Cation diameters and cavity sizes of optimum crown ethers (adapted from Vögtle, F. and Weber, E. (eds), *Host Guest Complex Chemistry: Macrocycles: Synthesis, Structures, Applications*, Springer, Berlin, 1985, Chap. 1, p. 18, Table 1, with permission).

Cation	Cation diameter (Å)	Crown ether	Cavity diameter (Å)
Li^+	1.36	12-Crown-4	1.2–1.5
Na^+	1.90	15-Crown-5	1.7–2.2
K^+	2.66	18-Crown-6	2.6–3.2
NH_4^+	2.86	18-Crown-6	2.6–3.2
Rb^+	2.94	18-Crown-6	2.6–3.2
Cs^+	3.38	21-crown-7	3.4–4.3

A mismatch in cation diameter and cavity size, however, does not preclude complex formation. If the cavity is too small for a cation (e.g. K^+ for 12-crown-4), complexation may still occur. In some such complexes a 2:1 ratio of constituents is observed in which one cation simultaneously coordinates with two molecules of crown ether (Figure 10.1b, 'sandwich'-like structure (**2**)). If the macrocycle is too large, the cavity may be occupied by two cations at the same time, forming a 1:2 molar ratio complex (Figure 10.1b, structure (**3**). The cation may be enveloped by the macrocycle like a pearl in a half-open oyster shell. In each of these three cases just mentioned, the stability of the complex is significantly lower than when there is optimal correlation between the cation and crown ether dimensions.

The stability constant, K_s, is a measure of complex stability. K_s is obtained by application of the law of mass action to the corresponding equilibrium

$$K_s = \frac{[L \cdot M^+]}{[L][M^+]}$$

In this equation, $[L \cdot M^+]$, $[L]$ and $[M^+]$ are the concentrations of the complex, free ligand and cation, respectively. For instance, the $\log K_s$ values for the sodium and potassium complexes of dicyclohexano-18-crown-6 are 6.4 and 8.3, respectively. Their ratio, approximately 2, is a logarithmic measure of the selectivity of 18-crown-6 toward Na^+ and K^+ ions. In other words, a solution of the crown ether to which equal concentrations of Na^+ and K^+ ions are added will contain only one bound sodium ion for every 100 ions of potassium involved in complexation.

Crown ethers have found applications in many fields of science and technology. They are effective in the separation of alkali and rare-earth metal ions which is essential for their analysis, extraction and purification. Owing to the solubility of crown ethers in nonaqueous media, alkali metal salts can be solubilized, i.e. transferred from an aqueous to an organic phase. For example, potassium permanganate is itself moderately soluble in water but practically insoluble in organic solvents. However, in the presence of 18-

crown-6, $KMnO_4$ becomes readily soluble in benzene forming so-called 'purple benzene', which has exceptionally high oxidizing properties greatly surpassing those of permanganate in aqueous solution. This may be explained by the fact that in water the MnO_4^- ions are surrounded by a dense solvation shell, whereas in the nonpolar solvent, benzene, they are unsolvated or 'naked'. Purple benzene, in contrast to aqueous $KMnO_4$, readily oxidizes alkenes, alcohols, aldehydes and even alkyl groups in alkylarenes at room temperature to form carboxylic acids in almost quantitative yield. The oxidation of cyclohexene under such conditions provides adipic acid in 100% yield

$$\underset{\text{18-crown-6, benzene}}{\overset{KMnO_4}{\longrightarrow}} \quad HOOC\text{-}(CH_2)_4\text{-}COOH$$

The development of crown ethers has been closely monitored by the medical profession. Substances capable of selectively binding cesium ions in the presence of Na^+ and K^+ would be potential treatments for human exposure to the very dangerous $^{137}Cs^+$ radioactive ion, a widespread radionuclide. Similar dangerous effects of $^{90}Sr^{2+}$ and other ions have been observed. Reports from many countries regarding successful research in radionuclide entrapment have recently become available.

The crown ethers proved to be the first synthetic analogues of naturally occurring substances which could transport alkali metal ions (Na^+ and K^+) through cellular membranes. Natural ion transporters, ionophores, act according to the same principle as the crown ethers, but the former have substantially more complicated structures. Thus, one of the best-known ionophores, the antibiotic valinomycin, has a macroheterocyclic structure composed of six α-amino acid and six α-hydroxy acid residues connected by alternating amide and ester functions (the structure can be found in the suggested further reading material on biochemistry). Valinomycin is a specific potassium ion carrier having a K^+/Na^+ selectivity of about 10^4. In contrast to the crown ether complexes, the K^+ ion coordinates with valinomycin through the carbonyl oxygen atoms of the ester groups to form an octahedral structure of the type[†]

[†]One further mode of ion penetration through the membranes involves tubular channels. The alkaline-earth ions, particularly Ca^{2+}, are preferentially transported in this manner.

Valinomycin, like other natural cation carriers, is a reversible ionophore. Such ionophores, having penetrated a cell, liberate the cation under the effect of certain interactions and are then rapidly returned to the outside of the cell to bind another cation. The rate of these trans-membrane migrations can reach several thousand per second, and they may even operate against a concentration gradient.

The forces driving the movement of the ionophore include changes in the pH of the medium, redox potential, irradiation and other factors. Many types of reversible crown ethers which mimic natural ionophores have been prepared. Thus, the 18-crown-6 derivative containing a long side chain with a terminal NH_2 group (Figure 10.2, structure (**4**)), responds well to pH changes. This crown ether forms complex (**5**) with K^+ in neutral or slightly alkaline media. However, upon acidification the amino group is protonated and the ammonium ion formed expels the K^+ cation from the ether cavity using its 'arm'-like chain with an ammonium 'hand' (see Figure 10.2, intramolecular complex (**6**)). Figuratively speaking, the crown ether 'bites its tail'. Since K^+ and NH_4^+ ions are similar in size (Table 10.1), displacement of K^+ by NH_4^+ in acidic media must be the result of entropy factors which favor intramolecular complexation of the alkylamino group. If the ammonium complex is returned to a neutral medium containing excess K^+, the potassium ions enter the cavity owing to deprotonation of the ammonium group. Thus, crown ether (**4**) can be compared to a shuttle moving back and forth between the external and internal walls of the membrane. By this mechanism K^+ ions are transported into the cell and H^+ ions out of the cell (Figure 10.2, structure (**7**)).

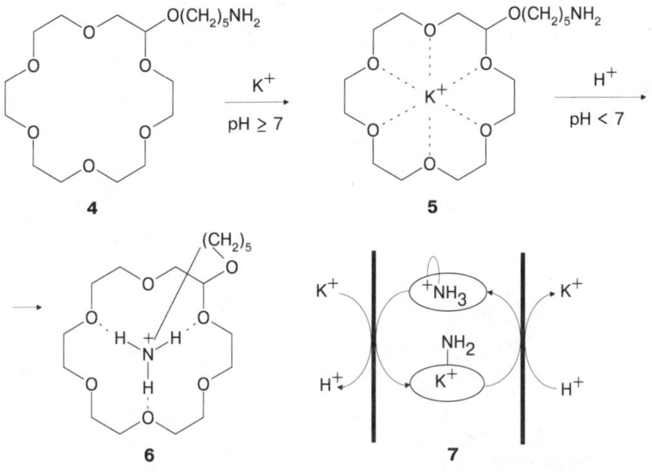

Figure 10.2 pH-dependent crown ether properties (Nakatsuji, Y., Kobayashi, H. and Okahara, M., *J. Chem. Soc., Chem. Commun.*, 1983, 800).

The functioning of natural ionophores is, of course, much more complex and efficient than that of the best synthetic reversible-type crown ether presently known. Moreover, the action of synthetic ionophores has been tested only on artificial polymer membranes. Nevertheless, progress has been substantial.

The discovery of crown ethers stimulated an army of chemists, in the literal sense of the word, to attempt the synthesis of new, more effective and more selective complexation reagents for the alkali metals. Much attention to detail was required, in particular to create internal cavities that were better organized and more capacious. While an ether macromolecule can be likened to a crown or a hat, it seemed reasonable then that other three-dimensional molecular containers for metal ions such as molecular cups, jugs, pots, saucepans, barrels and so on could be prepared. Intuitively, the contents would be easier to capture and maintain inside such 'vessels'. Gradually, this line of reasoning gained momentum and finally came to be known as 'container chemistry'. Major contributions to its development were made by the French chemist Lehn and the American researcher Cram who, together with Pedersen, were awarded the 1987 Nobel prize for chemistry.

Lehn began his investigations in 1968 by synthesizing three-dimensional aliphatic amino ethers (**8**) (Figure 10.3) which he named cryptands (Greek: *krypte*, cave, cavern). The association with caves resulted from the presence in such compounds of an internal cavity limited by three ether chains which met at two bridged nitrogen atoms. Structure (**9**), [2.2.2]cryptate, is formed by complexation of a metal ion to cryptand (**8**) ($m = n = 1$). Complex (**9**) is the most intensively studied cryptate because its internal cavity size is

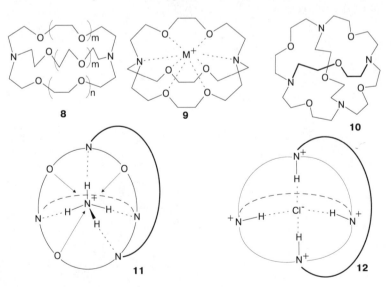

Figure 10.3   Cryptands and their complexes.

suitable for many cations including Na^+ and K^+. The cations are retained within the cavity not only by the walls of this 'cavern' but also by electrostatic attraction between the cation and the electron pairs of the six oxygen and two nitrogen atoms. It is not surprising that the stabilities of [2.2.2]cryptand complexes with Na^+ and K^+ ions (picrate counter ion: $\log K_s = 10.6$ and 13.2, respectively) are four or five orders of magnitude greater than those of the analogous 18-crown-6 complexes.

The formation of complexes with rather unusual compositions and structures is not rare. Metallic sodium is known to dissolve in liquid ammonia to form a dark blue equilibrium solution of sodium atoms, ions and solvated electrons. However, sodium is much less soluble in alkylamines, e.g. ethylamine. It has been shown that if [2.2.2]cryptand is added to ethylamine, sodium dissolution occurs much more readily, and over time precipitation of golden crystals of [2.2.2]cryptand·2Na occurs. X-Ray analysis has indicated that one sodium atom (as a cation) is situated inside the cryptand cavity, whereas the second is located outside the cavity as an Na^- anion. Thus, [2.2.2]cryptand stabilizes the separation of charge in the Na^+Na^- ionic pair, which is of considerable significance.

Compound (10) is another type of cryptand which resembles the shape of a football. The size of the cavity best fits the cesium or ammonium ion. Indeed, its complex with Cs^+ is the most stable of all known complexes of this cation. The stability of the complex of compound (10) with an ammonium ion (see Figure 10.3, schematic structure 11) is largely as a result of the fact that the four hydrogens of the tetrahedral NH_4^+ ion are directed toward the four cryptand nitrogens, thus allowing the formation of hydrogen bonds. Interestingly, the hydrogen bonding, together with steric shielding and electrostatic effects, results in a decrease in acidity of the NH_4^+ ion in complex 11 by six orders of magnitude compared with uncomplexed aqueous NH_4^+.

Cryptands may be adapted for anion complexation if the cavity is surrounded by positively charged centers. For instance, the chlorine anion is well suited to the internal cavity size of protonated cryptand 10 resulting in a complex represented schematically by structure (12), in Figure 10.3. Since a bromide anion could hardly occupy the same cavity because of its increased size, compound 10 is a good reagent for the separation of Cl^- and Br^- ions.

Cram commenced his study of molecular cavities in the mid-1970s. His attention was attracted by one seemingly negligible deficiency of the crown ethers and cryptands. X-Ray analysis demonstrated that both groups of compounds were not organized well enough to accept the desired guest ions because their structures were somewhat 'crumpled' (Figure 10.4, structures 13 and 14). Therefore, entry of a cation into the cavity would necessitate additional energy expenditure to reorganize and smooth out the structure, which would be reflected in the stability of the complex. Cram and his coworkers were successful in their engineering of molecular containers void of this shortcoming. In a complicated series of steps, aromatic ethers (15 and 16) with structures preorganized for complexation were prepared. The new compounds were christened spherands and cavitands. To illustrate more fully this chemical class, we include non heterocyclic spherands (in particular 15).

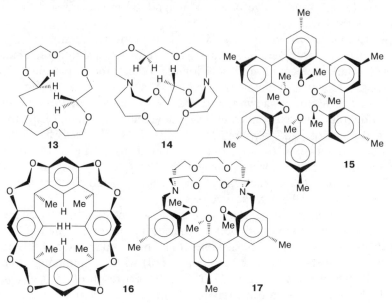

Figure 10.4 Unorganized structures of crown ether (13) and cryptand (14). Preorganized structures of spherand (15), cavitand (16) and cryptaspherand (17).

Spherands and cavitands are a type of molecular container in which the walls are lined with aromatic nuclei, the complexation being achieved by oxygen atoms. Ionic molecular or atomic guests are attracted and bound by these atoms. The containers may even be thought of as having short 'legs' if one attributes this role to the external methyl groups (structures (15) and (16)). The synthetic strategy employed by Cram proved to be highly successful. Spherand (15), for example, forms a complex with the Na^+ cation with $\log K_s = 14.1$ (see Table 10.2), which indicates that the spherand complex is much more stable than the analogous complexes with 18-crown-6 and [2.2.2]cryptand. However, the most remarkable property of compound (15) is its unprecedented Na^+/K^+ selectivity (10^{10}), which even surpasses all natural ionophores. Once inside spherands and cavitands, alkali metal ions are shielded to a great extent by the methyl or methylene groups linked to the oxygen atoms. These groups prevent solvation of the cation. Cavitand (16) can trap and hold prisoner small neutral molecules such as CH_2Cl_2, MeCN, SO_2 and others.

Table 10.2 Stability constants ($\log K_s$) of macrocyclic complexes with Na^+ and K^+ (picrate counterion) (Bell, T. W., Firestone, A. and Ludwig, R., *J. Chem. Soc., Chem. Commun.*, 1989, 1902).

Ion	Structures[a]				
	(1)	(9)	(15)	(17)	(19)
Na^+	6.4	10.6	14.1	9.9	14.7
K^+	8.3	13.2	4.4	13.9	14.3

[a]See Figures 10.1, 10.3, 10.4 and 10.6.

Many other molecular containers have been synthesized. Of these, cryptaspherand (17), a cryptand–spherand hybrid, is also characterized by a high K^+/Na^+ selectivity (Table 10.2) which is similar to that of the natural antibiotic valinomycin and exceeds that of many other synthetic ionophores.

Cram *et al.* more recently synthesized molecules with structures resembling two closely associated hemispheres. These compounds were obtained from two different cavitands, one containing CH_2SH groups on opposite peripheries, the other containing CH_2Cl groups. As expected, nucleophilic displacement of chloride occurred when equimolar quantities of these cavitands were mixed in dimethylformamide in the presence of cesium carbonate (the latter being required for mercapto group ionization). This reaction joined the two hemispheric molecules together on both sides (Figure 10.5). Of course, these two hemispheres or 'cups' do not form a tight seal and a small clearance between them still remains. However, the gap is too narrow for even the smallest molecules trapped in the cups during the reaction to escape. The analogy with a prison or cage is rather appropriate and is reflected in the name given to such macromolecules—carcerands (Latin *carcer*, a prison cell). Carcerands were the first organic compounds synthesized with capacious cavities entirely isolated from the exterior. It has been shown that if small molecules are 'accidentally' trapped inside the carcerand at the moment of closure, they will remain inside as 'jailbirds' and neither crystallization, chromatography nor any other conventional method can release them. Thus, during carcerand synthesis Cram and his coworkers found that diverse types of small particles present in the reaction mixture (dimethylformamide, cesium and chloride ions) were captured inside the carcerand.[†]

In the mid-1980s interest began to shift toward aza analogues of crown ethers and cryptands. The aza group (–N=), with a specifically oriented unshared pair of electrons, has a more rigid configuration than an amine

†The only species capable of exiting the carcerand upon heating are small linear molecules such as CO_2 or O_2.

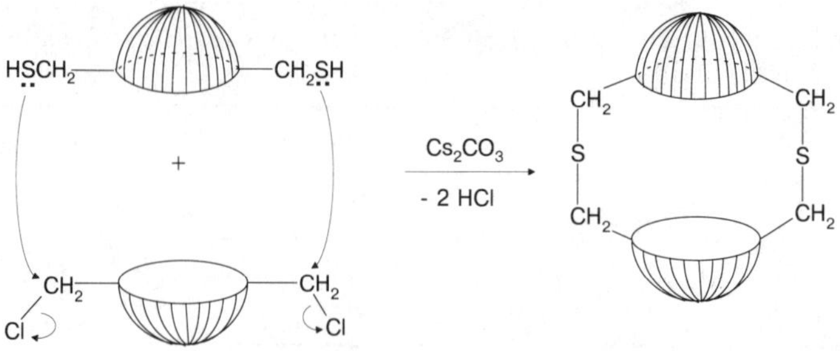

Figure 10.5   The principle of joining two cavitand molecules to form a carcerand.

nitrogen or ether oxygen. Azamacrocycles should therefore be capable of more precise design for the reception of guest ions. Indeed, azacrown (**18**) shown in Figure 10.6 reacts with the K$^+$ cation in a 1:1 molar ratio to give a complex which is measurably more stable than the similar complex with 18-crown-6. However, record-breaking stability was attained by the Na$^+$ and K$^+$ complexes of azacrown (**19**) (also called torand), which have stability constants that are even higher than for the cavitands (Table 10.2). Unfortunately, the flat and somewhat simple structure of the azacrowns is also a disadvantage because it results in low Na$^+$/K$^+$ selectivity.

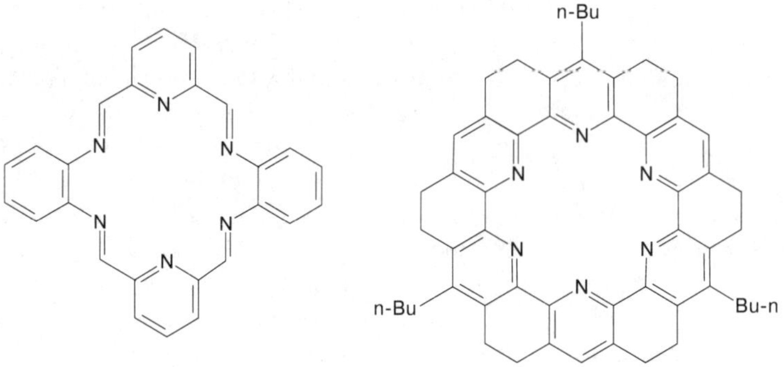

18

19

Figure 10.6   Azacrown ethers.

Azacrown (**20**) (Figure 10.7) possesses some interesting properties which are useful in the determination of Li$^+$. In nonpolar solvents such as methylene chloride (**20**) exists in the red tautomeric form (**20b**). The compound loses its color in polar media (e.g. methanol) owing to conversion to the fully aromatic tautomer (**20a**). When lithium salts are added to a solution of azacrown (**20**) in CH$_2$Cl$_2$, Li$^+$ cations begin to displace protons from the internal cavity to form complex (**21**). Since the latter is also colorless, the process can be monitored spectrophotometrically.

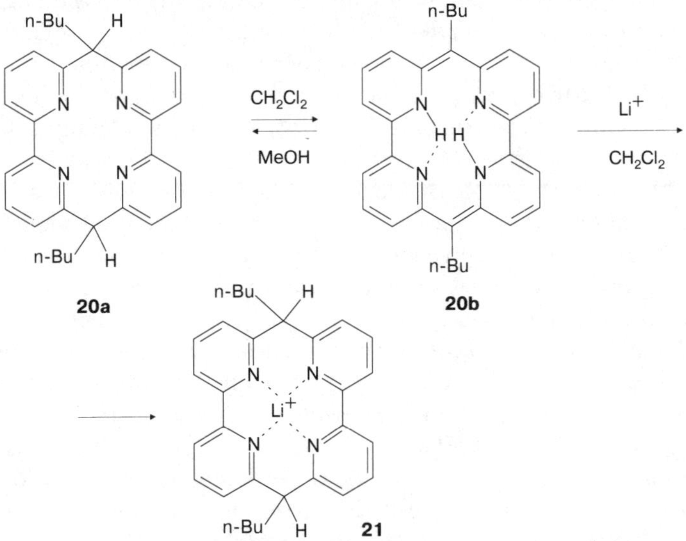

Figure 10.7 Azacrown ethers used for Li$^+$ ion detection.

Research involving the construction of cascade or coreceptive macrocyclic ligands is being vigorously pursued. Such compounds are intended to possess two different complexation sites: one soft (polarizable), the other rigid (almost nonpolarizable), which would act predominantly electrostatically. A typical example is compound (**22**) shown in Figure 10.8. Macrocycle (**22**) has

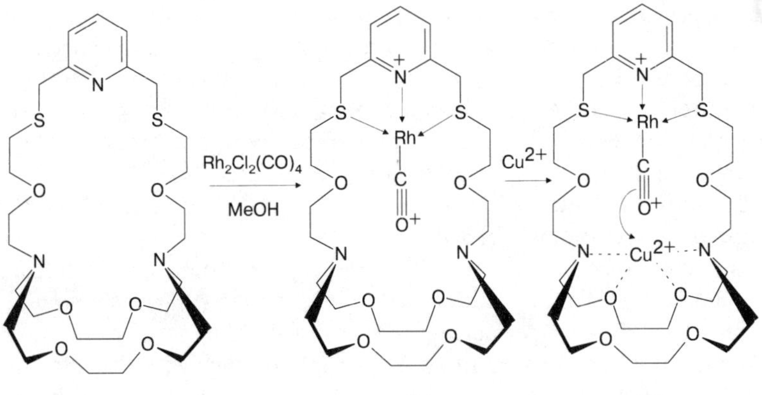

Figure 10.8 Formation of a macrocyclic cascade ligand.

a soft coordinative center composed of two sulfur atoms and an adjacent pyridine nitrogen atom. The rigid center is formed by the cryptand portion of the molecule. On treatment with a mild Lewis acid such as rhodium carbonyl, compound (22) forms complex (23) in which the metal ion is coordinated to the soft center with the carbonyl oriented toward the inside of the macrocycle. Further addition of a copper(II) salt effects coordination of the copper(II) ion with the cryptand fragment, with the ion being closely located to the carbonyl oxygen. The copper(II) ion also coordinates with the C≡O⁺ group thus activating it toward nucleophilic addition. Such chemical activation imitates the function of some metal-containing enzymes.

The binding of two identical ions (e.g. two copper ions) to both centers of a cascade ligand is also possible. In this case the two cations vary substantially in the final complex owing to the different chemical environments. For example, the reductive potentials of the two ions are now significantly different.

The synthesis of macrocycles capable of forming complexes of the 'guest–host' type with organic molecules has become the focus of intense investigation in the last decade. Such macrocycles are necessary for the separation and activation of organic compounds, for the creation of a new generation of drugs and for resolving many other chemical and biochemical problems. Thus, the search for ligands capable of selectively binding urea was initiated by the acute need for efficient methods of blood purification for kidney patients. Work is being carried out on the synthesis of macrocyclic agents capable of complexation with sulfa drugs, catecholamines, amino acids, peptides and so on. The protonated bicyclic system (25) depicted in Figure 10.9 forms rather stable complexes with α-amino acids and other betaines in neutral aqueous solution. The crown ether portion becomes linked to the ammonium group of the acid (see Figure 10.2, structure 6), while the positively charged polyamine ring coordinates with the carboxylate anion. In the resulting complex, ligand 25 has the conformation of a half-open book. Organic molecules have a much more complicated structure than metal cations. Therefore, figuratively speaking, on complexation with a crown ether, the latter are satisfied by a molecular 'sack for a sphere', whereas the former are more demanding in this regard and require the crown to be an exact fit, as, for instance, 'a violin to its case'.

Metal ions are retained inside a macrocycle predominantly by strong ion–dipole interactions, whereas the complexation of neutral organic molecules relies on much weaker forces including hydrogen bonding, charge transfer and dipole–dipole interactions. Under these circumstances, steric shielding by the host molecule is of special importance in holding the guest molecule. Azacrown ether (26) forms a rather stable 1:1 complex with the nitromethane molecule in which the guest is oriented almost perpendicular to the average plane of the cyclic polyether. The nitromethane methyl group is oriented toward the macrocycle and forms hydrogen bonds with the pyridine nitrogen

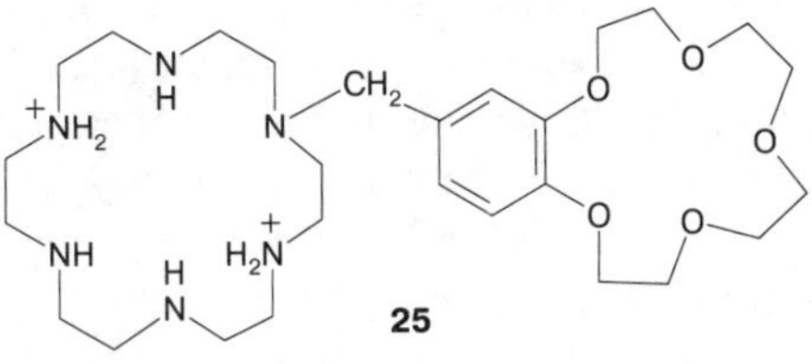

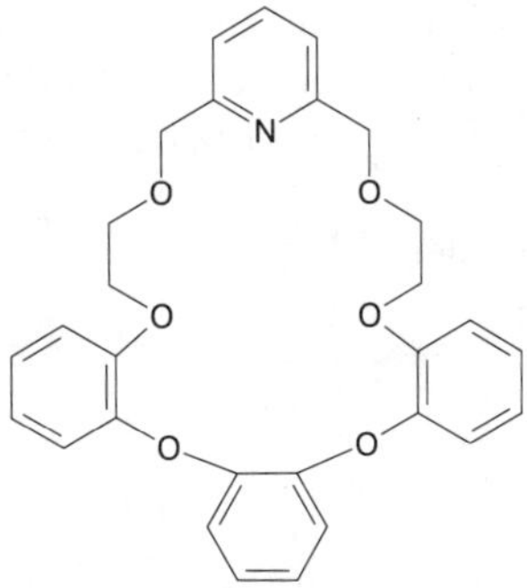

26

Figure 10.9 Macrocycles used to bind betaines (**25**) and nitromethane (**26**).

and ether oxygen atoms. The nitro group is encircled by the three phenyl rings. As a result, the $MeNO_2$ molecule is held captive in a semispherical cavity with a radius of about 350 nm. Nitroethane, dimethylformamide, dimethyl sulfoxide and many other solvents do not form complexes with this crown ether, the selectivity toward nitromethane is very high.

10.2 Self-assembling Molecular Systems

The formation of complexes between macrocyclic compounds and small organic molecules resembles the relationship between enzymes and receptors whose structures are also adapted for recognizing and binding strictly determined substrates. Container chemistry is therefore often called 'receptor chemistry' or supramolecular chemistry. Lehn defined supramolecular chemistry as the chemistry of intermolecular bonding. Complexes of two or more chemical entities and the structures of such complexes are encompassed by this field. This field extends beyond the boundaries of molecular chemistry, which is limited to the structures, properties and transformations of discrete molecules.

In addition to container chemistry, there now exists another rapidly growing main division of supramolecular chemistry which is even more exciting. This is the creation of molecules and molecular assemblies capable of self-organization. Thus, another link has been established between chemistry and the astonishing phenomena of living matter, which include the self-assembly of nucleic acids, matrix syntheses of proteins and the formation of antibodies.

As in the case of 'host–guest' chemistry, heterocyclic compounds have occupied leading positions in the design of artificial self-assembling molecular systems. Ligands of type (27), consisting of a number of 2,2'-bipyridyl units which are linked by relatively flexible ether groups, were an early example of such compounds (Figure 10.10). In the presence of copper(I) or silver(I) salts, two threads of such a ligand wind around the metal ions and around each other to form a double-stranded helicate (28). The driving force for this process is the well-known tendency of copper(I) and silver(I) ions to realize tetrahedral coordination geometry. As a result, each ion binds to a bidentate fragment of each of two molecules (27).

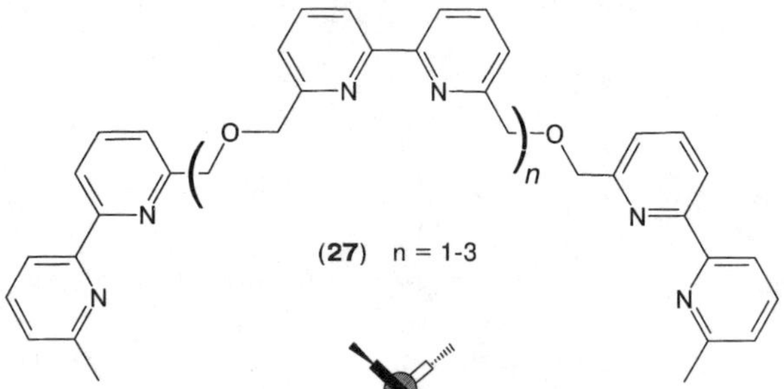

(27)   n = 1-3

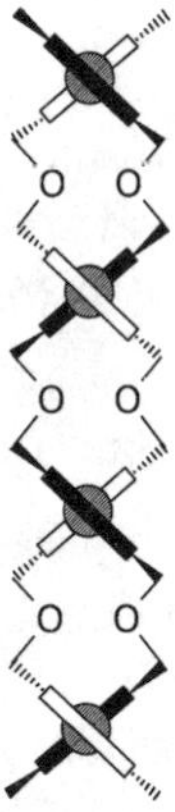

(28)

Figure 10.10   Formation of a double-stranded helicate (shaded circles are Ag$^+$ or Cu$^+$ ions; white and black rectangles are 2,2'-bipyridyl fragments) (Lehn, J.-M. and Rigault, A., *Angew. Chem., Int. Ed. Engl.*, 1988, **27**, 1095).

Besides coordination bonds, self-assembly can also be induced by hydrogen bonding or electrostatic interactions (Figure 10.11). Thus, macrocyclic tetracationic salt (29) and polyether (30) cocrystallize as a very stable self-organized complex ($K_a = 11\,150\,M^{-1}$) of the pseudorotaxane type, existing in two conformations, **31a** and **31b**. During the association the linear molecule (**30**) passes through the macrocycle in a manner which recalls thread passing through the eye of a needle: the linear molecule is held inside the macrocycle by ion–dipole interactions between the positively charged 4,4′-bipyridyl residues and the electron-rich 1,5-dioxynaphthalene fragments.

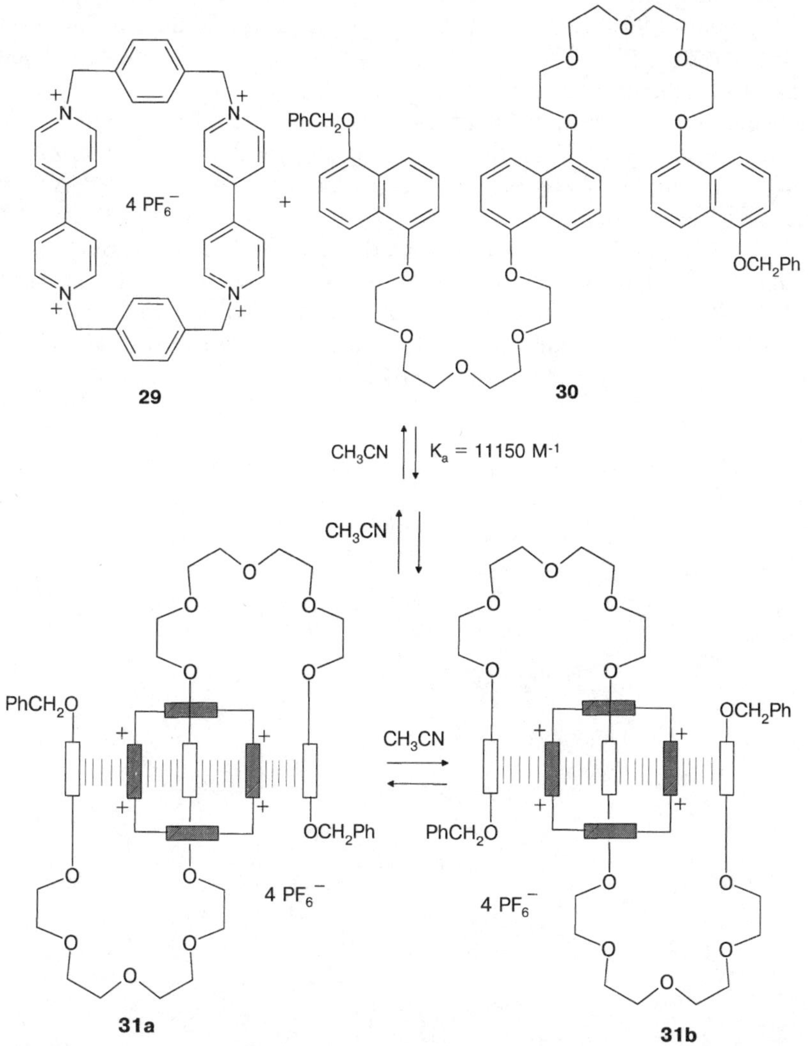

Figure. 10.11 Self-assembly induced by electrostatic interactions (shaded and white rectangles are 4,4′-bipyridyl and 1,5-dioxynaphthalene fragments, respectively) (Ashton, P. R., Philp, D., Spencer, N., Stoddart, J. F. and Williams, D. J., *J. Chem. Soc., Chem. Commun.*, 1994, 181).

Many other examples of self-assembling molecular systems are now known, and most of them are heterocyclic.

10.3   Enzyme Models

The creation of simple, efficient and cheap enzyme models has been an ongoing goal. Such research has been a powerful incentive for the further development of biotechnology and medical chemistry, and there is hope that it will also engender new energy-saving and ecologically friendly technologies. Since the majority of coenzymes are derivatives of nitrogen heterocycles, research in this field involves the synthesis of complex, carefully designed heterocyclic compounds. For example, the thiazolium salt (**32**) shown in Figure 10.12 has recently been synthesized as a potential thiamine pyrophosphate model. Salt (**32**), in the presence of K^+ or Na^+ (but not Li^+) cations, enhances the rate of pyruvic acid decarboxylation (see Section 4.2.2) by one order of magnitude compared with the 3-ethyl-4-methylthiazolium cation (**33**). The role of the crown ether seems to involve complexation of the Na^+ and K^+ ions and stabilization of an intermediate of proposed structure (**34**), resulting in a reduction of the activation energy of the reaction.

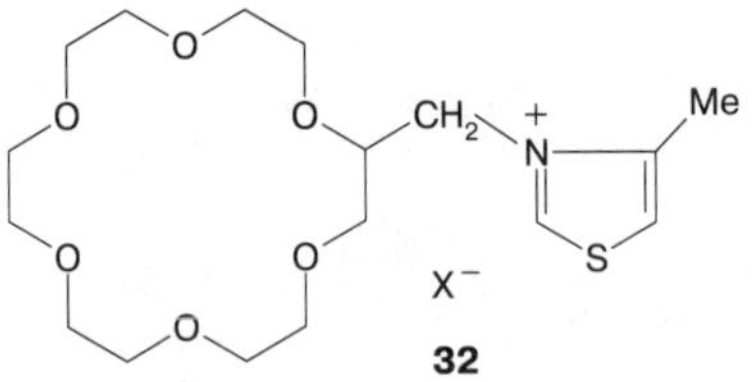

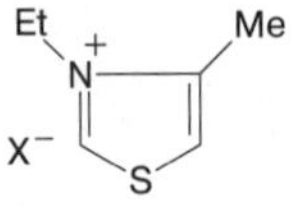

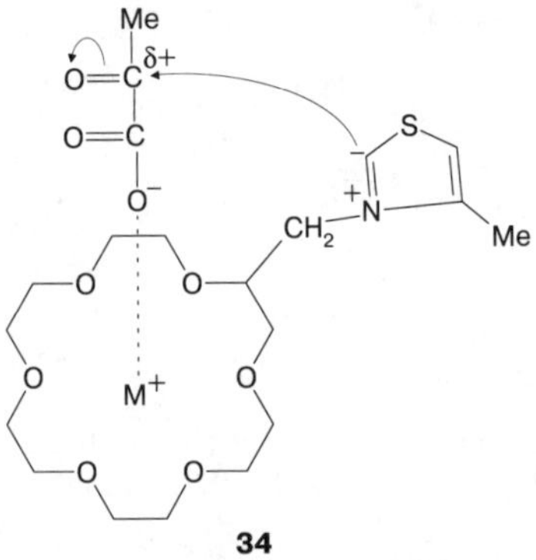

Figure 10.12   Thiazolium salts which model thiamine pyrophosphate functions.

Numerous investigations have been designed to model the NAD-H coenzyme (see Section 4.2.1). The characteristic structure of NAD-H arises because the 1,4-dihydropyridine ring has a boat conformation, and different substrates recognize the nonequivalent faces of the heterocycle (designated as A and B). Most substrates are reduced from the A face which is inside the boat, but some react only on the opposite B face. Nonbonding interactions which stabilize either the first or second intermediate complex account for this difference. In both cases, the reductions proceed with very high stereospecificity.

A number of NAD-H models have been developed which provide high stereospecificity during the reaction of carbonyl-containing substrates. One such model is represented by compounds of type (35) shown in Figure 10.13. When R is H or CO_2Et, methyl benzoylformate is reduced to the (+)-enantiomer of methyl mandelate, but when R is $CONMe_2$, the (−)-enantiomer is formed (Figure 10.13). The optical purity of the reduction product can be as high as 97%. It is believed that during reduction to the (+)-enantiomer the transition complex (36) is formed. As a result, nonbonding interactions (electrostatic attraction and hydrogen bonding) between the 3-hydroxymethyl group of the 1,4-dihydropyridine and the methoxycarbonyl group of the substrate dominate. In the reduction to the (−)-enantiomer, the dominant role is played by electrostatic interactions between the dimethylamino group and the methoxycarbonyl group, as shown in Figure 10.13 (structure (37)). Reductions by compounds of type (35) are catalyzed by magnesium ions, as in the case of the NAD-H coenzyme. These ions appear to take part in the formation of a triple complex (the magnesium ion is represented by a shaded circle in structures (36) and (37)).

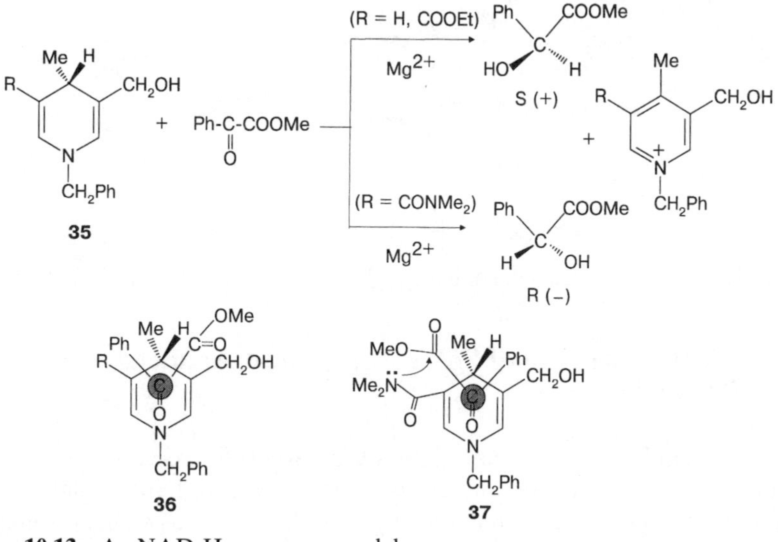

Figure 10.13 An NAD-H coenzyme model.

A great deal of research has targeted heme and oxygenase models that would be capable of storing or activating molecular oxygen. The cores of these models usually consist of synthetic metalloporphyrins or other complexes of transition metals with other ligands, e.g. cryptands. As a rule, the structures are rather complicated and need to fulfill the following requirements: (i) to bind an oxygen molecule strongly but reversibly; (ii) to possess a metal ion to which the oxygen molecule can be attached, coordinated on the opposite face to an axial ligand; and (iii) to be itself stable toward significant oxidation of the metal or ligand dimerization by the oxygen molecule.

Biochemical molecules fulfill these requirements by 'fencing' a porphyrin ring to create a spherical cavity which holds the oxygen molecule. Similar cavities exist in all hemoproteins, including hemoglobin. In constructing synthetic models, it is more important to imitate the apoenzyme rather than the coenzyme. The iron(II) porphyrin depicted in Figure 10.14, structure **38**, is an interesting example of a stable complex in which the oxygen molecule is bound to the iron cation by only one atom. The structure of such a ligand is sometimes compared with a 'picnic-basket' in porphyrins of the generalized structure **39** in Figure 10.14.

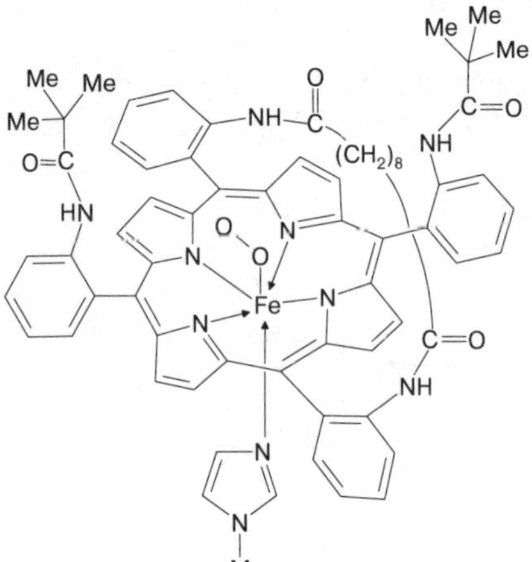

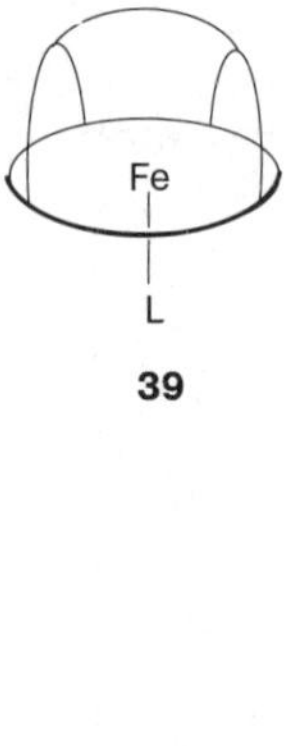

Figure 10.14   A synthetic model of hemoglobin.

10.4   The Conversion of Solar Energy and Artificial Photosynthesis

Another technically important field of heterocyclic research is directed toward reducing worldwide energy shortages by creating molecular systems capable of photochemically reducing water to hydrogen as a cheap source of fuel. The tris(bipyridyl)ruthenium(II) complex has been demonstrated to behave as such a photocatalyst and has currently become the focus of considerable attention (Figure 10.15, structure **40**). As a consequence, the study of other diazines and their complexes has been actively pursued. A complex of the ruthenium(II) ion with 2,2'-bipyrazine shows promising photocatalytic and electrophysical properties.

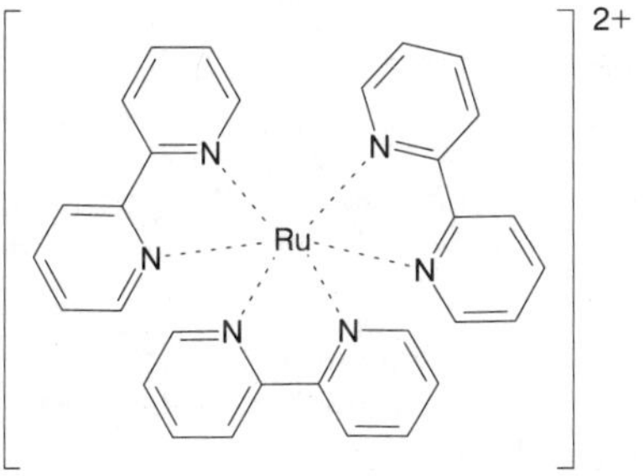

40

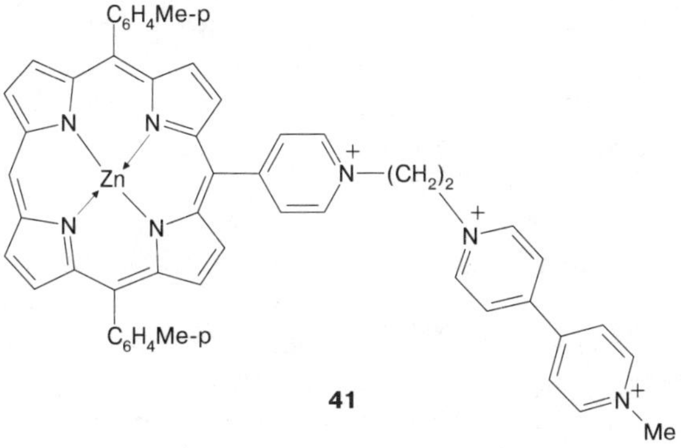

41

Figure 10.15 Synthetic catalysts used for the photochemical production of hydrogen from water.

To obtain hydrogen from water, we require two dark-reaction catalysts in addition to a photocatalyst. The first is an electron transporter, while the second is a reduction catalyst. Additionally, an electron donor which can regenerate the oxidized photocatalyst is needed. Methylviologen (MV^{2+}) (Figure 8.3), is widely used as the electron carrier, and colloidal platinum as the reduction catalyst. The role of the electron donor is usually played by triethanolamine or ethylenediaminetetraacetic acid (EDTA). Generally, the chemical reactions represented in Figure 10.16 form the basis of the catalytic system.

The reaction sequence (Figure 10.16) involves the following stages. (i) The tris(bipyridyl)ruthenium(II) complex absorbs a quantum of light and is promoted to the excited state in which its electron-donating properties are

1) $Ru(bpy)_3^{2+}$ $\xrightarrow{\text{hv}}$ $\overset{*}{Ru}(bpy)_3^{2+}$

2) $\overset{*}{Ru}(bpy)_3^{2+}$ + MV^{2+} $\longrightarrow$ $Ru(bpy)_3^{3+}$ + $MV^{+\bullet}$

3) $MV^{+\bullet}$ + H_2O $\xrightarrow{\text{coll. Pt}}$ MV^{2+} + $1/2\ H_2$ + OH^-

4) $Ru(bpy)_3^{3+}$ + $EDTA$ $\longrightarrow$ $Ru(bpy)_3^{2+}$ + $EDTA^{+\bullet}$

Figure 10.16   Fundamental reactions involved in the reduction of water to molecular hydrogen by tris(bipyridyl)ruthenium(II) (Prasad, D. R. and Hoffman, M. Z., *J. Am. Chem. Soc.,* 1986, **108**, 2568).

dramatically enhanced. (ii) The complex consequently liberates an electron which is trapped by methylviologen, converting it to the cation-radical $MV^{+\bullet}$. (iii) The cation-radical subsequently reduces water to produce molecular hydrogen and hydroxide ion. In the process, methylviologen is regenerated. (iv) In the final stage EDTA assists the one-electron reduction of the ruthenium(III) complex, thus regenerating the photocatalyst. Note that during the course of this sequence, EDTA is oxidized to a cation-radical and is further decomposed. EDTA cannot be regenerated—it is irreversibly lost during the production of hydrogen.

In the search for efficient catalysts, chemists have constructed complex molecules which contain more than one catalytic center; for example, compound (**41**) (Figure 10.15) incorporates both a photocatalyst (zinc porphyrin fragment) and an electron transporter (methylviologen analogue). The photodecomposition of water performed in the presence of (**41**) also requires the participation of colloidal platinum and a 1,4-dihydronicotinamide derivative in place of EDTA as the electron donor.

Intense study has also been directed toward the production of photovoltaic cells using heterocyclic compounds such as metalloporphyrins, quinacridones and cyanine dyes (see Sections 9.2.2 and 9.2.3). The operating principles of photovoltaic cells are schematically illustrated in Figure 10.17. A 'sandwich', consisting of a photoabsorbent pigment layer positioned between two multilayered electrodes is deposited on a transparent glass plate. A polymer filler is usually added to ensure pigment layer uniformity. One of the electrodes receives electrons, while the other produces electrons, thus generating a current. Therefore, a photovoltaic cell made from an aluminophosphate complex of phthalocyanine (**42**) contains aluminum and silver electrodes (Figure 10.17). When the pigment absorbs visible light and becomes excited, an electron is liberated which is trapped by the aluminum electrode, thus generating a current. The corresponding positive charge on the

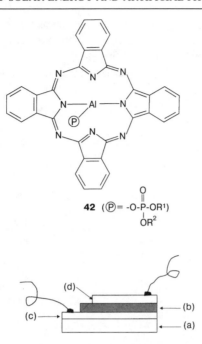

42 ($\textcircled{P}$= -O-$\overset{\overset{O}{\|}}{\underset{\underset{OR^2}{|}}{P}}$-OR¹)

Figure 10.17 Photoabsorbent phthalocyanine pigment (**42**) and (below) a photovoltaic cell: (a) glass support, (b) pigment, (c) silver electrode and (d) aluminum electrode (Tomida, M., Kusabayashi, S. and Yokoyama, M., *Chem. Lett.,* 1984, 1305).

pigment is compensated for by an electron released from the silver electrode.

Investigations concerning solar energy conservation are also of great importance. Certain organic substances, including heterocycles, are able to absorb sunlight and to store the solar energy by transforming it into an energy-rich product. In practice, all such transformations are represented by valence or *cis–trans* isomerizations. For example, 1-methyl-5-phenyl-Δ^2-pyrazoline undergoes valence isomerization to form energy-rich 2-phenylcyclopropylazomethane (Figure 10.18a). *N,N*-Diacyl derivatives of indigo similarly undergo *trans–cis* photoisomerization (Figure 10.18b). The stored energy resulting from these conversions amounts to around 10 kcal mol^{-1}. A nonheterocyclic norbornadiene containing a quadricyclane system (Figure 10.18c) holds the record for the largest energy conservation capacity (26 kcal mol^{-1}). The isomerized energy-rich products have to be reasonably stable under ordinary conditions, and for the compounds to be of practical use an additional requirement is that the stored energy should be liberated on demand, preferably in the form of heat. The energy-rich product must usually be stimulated, e.g. by catalysts or mild heating, to effect energy release. Once the energy is discharged in the form of heat, the molecule reverts to the original low energy state. Such transitions can be of practical importance provided that the substance can be cycled repeatedly. However, experience has demonstrated that in due course, most substances are gradually destroyed. Thus, the search for materials capable of withstanding many energy conversion cycles has begun in earnest.

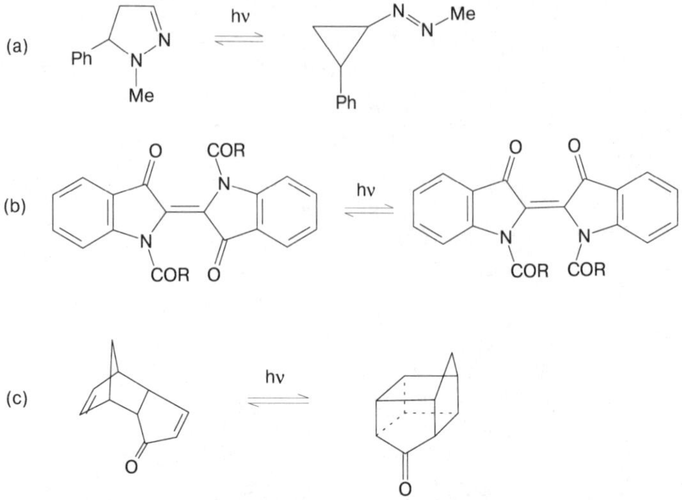

Figure 10.18 Examples of photoisomerizations utilized for solar energy conservation (Scharf, H.-D., Fleischhauer, J., Leismann, H., Ressler, I., Schleker, W. and Weitz, R., *Angew. Chem., Int. Ed. Engl.*, 1979, **18**, 652).

Reversible phototransformations could also potentially be employed in small heating systems such as domestic heaters. On bright sunny days a substance placed in a specialized tank on the roof would accumulate solar energy which could then be utilized to heat the dwelling at night or in cool weather. Figure 10.19 is an energy diagram of the processes which occur during the storage of solar energy. An initial product R, upon absorption of a quantum of light, is transferred to an excited triplet state (R*). Further stabilization is achieved by conversion to the high energy product P which stores part of the absorbed energy (ΔE) as chemical bond energy.

Figure 10.19 Energy diagram for the conservation of light energy.

10.5 Organic Conductors

The synthesis of versatile conductive materials including 'organic metals' and superconductors has been a major achievement of heterocyclic chemistry in the last two decades. These materials are all constructed from ion-radical salts or charge transfer complexes. The preparation of such materials requires a strong electron donor which can be readily polarized and a powerful electron-withdrawing agent. The well-known tetrathiafulvalene (TTF) (Figure 2.8) was the first electron-donating component widely used for such materials. Its ion-radical salt with tetracyanoquinodimethane (TCQDM) (see Figures 2.9 and 2.11a), first obtained in 1972, was one of the earliest organic compounds to display conductivity of the metallic type.[†] This observation initiated the search for more effective electron donors and acceptors.

The search for electron acceptors was initially focused on π-deficient heterocycles which contain strong electron-withdrawing groups. 2,4,6-Tricyano-1,3,5-triazine (**43**), 3,6-dicyano-1,2,4,5-tetrazine (**44**) and the bisthiadiazole system (**45**) based on TCQDM serve as examples (Figure 10.20). However, donor systems have proved to be more important: their structural features usually involve substitution of the sulfur atoms in TTF by selenium and extension of the conjugated system of TTF by incorporating multiple bonds and additional electron donor groups. The investigation of tetramethyltetraselenofulvalene (TMTSF, structure **46**) and bis-(ethylenedithiolene)tetrathiafulvalene (ET, structure **47**, Figure 10.20) provided particularly valuable results. More than 10 superconductors were discovered among the ion-radical salts of compounds **46** and **47**. The counteranion was not necessarily an organic anion-radical of the TCQDM$^{-\cdot}$ type; indeed, enhanced properties were shown by inorganic anions of the NO_3^- or PF_6^- type for the TMTSF-based salts and by the linear anions I_3^-, IBr_2^-, AuI_2^- and so on for the ET salts. Thus, the ion-radical salt TMTSF$^{+\cdot}$PF$_6^-$ becomes a superconductor at 2 K under a pressure of 4 kbar. The salt (ET$^{+\cdot}$)$_2$[Cu(SCN$^-$)$_2$]Cl holds the record for superconductivity among organic compounds.[‡] The critical temperature (T_c) for transition into the superconductive state is 12.8 K (at 0.3 kbar). Such salts are usually prepared by controlled electrochemical oxidation of the organic donors in the presence of the corresponding inorganic salts.

---

†Most organic compounds are electroinsulators, e.g. the conductivity of anthracene is less than $10^{-22}\ \Omega^{-1}\,cm^{-1}$. Some organic compounds (dyes and charge-transfer complexes) are semi-conductors (conductivity in the range 10^{-8}–$10^{-5}\ \Omega^{-1}\,cm^{-1}$). The conductivity of organic metals is between 10^1 and $10^4\ \Omega^{-1}\,cm^{-1}$. Although this is markedly lower than the electroconductivity of copper, organic metals can become superconductors at low temperatures (see text).
‡Fullerene, discovered in 1985, consists of 60 carbon atoms arranged in the form of a soccer-ball. It is considered as the third form of pure carbon and contains unique intermediate hybridization of the s- and p-orbitals. When doped with alkali metals in 1991, it became for a time the superconductor with the highest recorded T_c values: 18 K for K_3C_{60} and 28.6 K for Rb_xC_{60}. Heterocyclic fullerenes incorporating as many as four boron atoms (called dopeyballs: $C_{56}B_4$) have also been prepared.

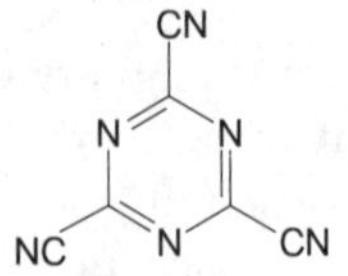

43 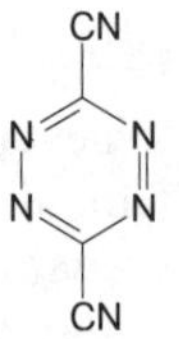   **44** 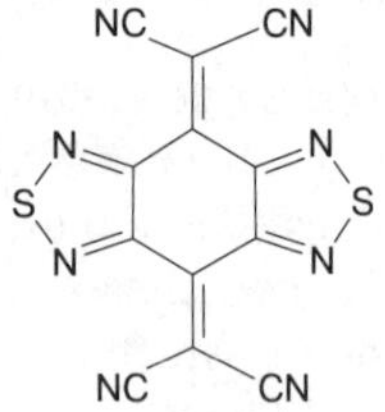   **45**

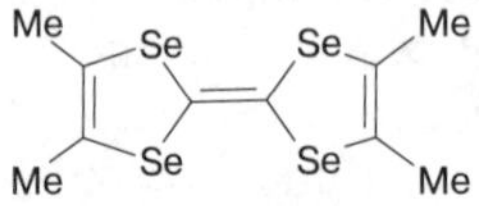

46 (TMTSF)

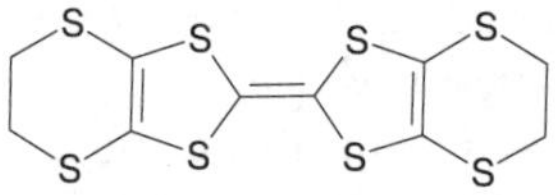

47 (ET)

Figure 10.20 Electron acceptors (**43–45**) and donors (**46** and **47**) used for the preparation of organic superconductors.

At the beginning of the 1970s, metallic or, more correctly, semimetallic conductivity of organic substances was known only for graphite. Conductivity occurs only along the plane of the condensed benzenoid rings, as expected for a large two-dimensional molecule. Theoretically, the probability of intermolecular metallic conductivity induction in organic materials was assumed to be very low. We now discuss how such conductivity arises.

X-Ray analysis has provided much insight into this phenomenon, demonstrating that in organic metals the ionic moieties are assembled in a parallel stacking arrangement with the cations and anions being separated into different piles (Figure 10.21a). Conductivity occurs through each stack which has quasimonomeric character, reflecting the nature of crystalline 'organic metals'. A possible mechanism for the conductivity of cation-radical piles is depicted in Figure 10.22. Electron transfer within such stacks results in the simultaneous formation of neutral donor molecules and dications (Figure 10.22a). The appearance of dications is supported by the fact that all compounds known to donate electrons to 'organic metals' possess very small differences between their first and second ionization potentials. Under chemical or controlled electrochemical oxidation conditions they are readily transformed into cation-radicals and dications. With respect to their conductivity, dications have one obvious flaw: their high positive charge may deform the regularity of the stacking owing to the increased internal repulsion between neighboring stacks. Hence, the electron transfers depicted in Figure 10.22b would seem to be the preferred mechanism: here the stacks include cation-radicals together with neutral donor molecules also residing in the crystalline lattice of the ion-radical salt. In this case conductivity is almost devoid of dication generation. Such a mechanism is supported in that electron transfer from the donor to the acceptor is incomplete in all of the 'organic metals'. Even in the salt $TTF^+ TCQDM^-$, electrons are only about 60% transferred.

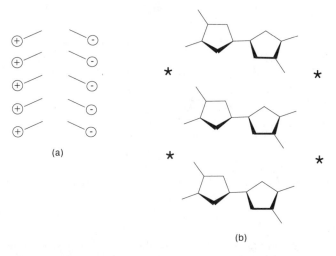

(a)

(b)

Figure 10.21 Orientation of ion-radical complexes in 'organic metals': (a) general view and (b) arrangement of $TMTSF^+ PF_6^-$ from X-ray data.

Since metallic conductivity along the chains of organic ion-radicals is provided by π-electrons, the molecular π-orbitals of adjacent particles should partially overlap and thus preferably be parallel to each other. Indeed, both these conditions are met: (i) all effective organic donors and acceptors have planar structures and hence a favorable orientation; and (ii) the ionic particles in the stacks are separated by distances which are much smaller than the sum of their van der Waals radii.

Although existing 'organic metals' cannot presently compete with nonorganic conductors, in particular the recently discovered high temperature superconductors, we can expect further development in the science and technology of 'organic metals'.

Fully conjugated, linear heterocyclic polymers have recently attracted attention. Of these, polypyrroles and polyquinolines are of special interest (Figure 10.23). These compounds are prepared by electrochemical oxidative polymerization of the parent heterocycles. Films formed from the

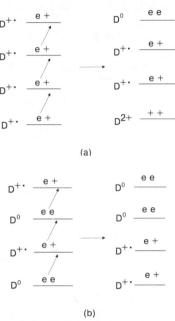

(a)

(b)

Figure 10.22 Mechanisms of conductivity in 'organic metals': (a) stacks containing cation-radicals only and (b) stacks containing cation-radicals together with neutral donor molecules.

heterocyclic polymers are good conductors and have found extensive applications in the manufacture of chemically modified electrodes and sensors, solid batteries and diverse composite materials in particular. Synthetic electroconductive nylon or polyester fibers covered with a thin layer of polypyrrole have been used as a coating for airplane fuselages to prevent the reflection of radar signals. The same textile has also been used to shield electronic equipment including computers and control systems.

The discovery of high conductivity in sulfur–nitrogen $(SN)_n$ polymers resulted in much effort being directed toward the synthesis of heterocyclic structures containing conjugated nitrogen–sulfur bonds in π-electron systems (Figure 10.23, structures **48–50**).

The idea of creating so-called molecular wires, the conductors necessary for molecular electronics, is of great interest. Bispyridinium salt (**51**) (Figure 10.24), a vinyl analogue of methylviologen (Figure 8.3), can be regarded as a prototype. The role of the wire in this salt is played by the conjugated polyene chain. Heterocyclic cations are the specialized molecular contacts. When a potential difference appears between the two ends of the chain, one of the pyridine cycles is reduced and the electrons entering this molecular system will migrate to the opposite end. Thus, a current arises in the molecular 'wire'. When placed across a cellular membrane, such wires can function as electron channels.

This idea has been extended to the development of molecular light transformers. One such compound (52) is also a polyene connected to an anthracene molecule at one end and a tetraphenylporphyrin system at the

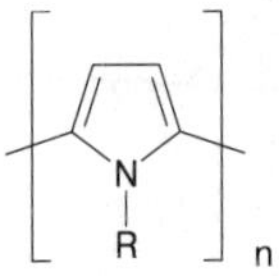

Polypyrroles

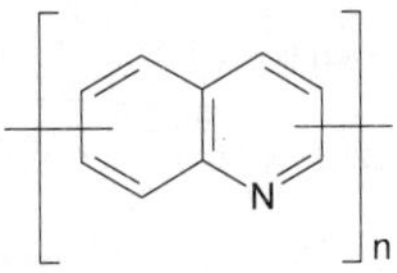

Polyquinolines

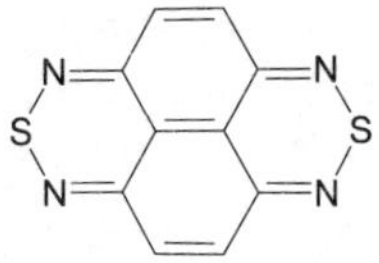

48

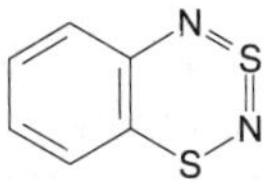

49

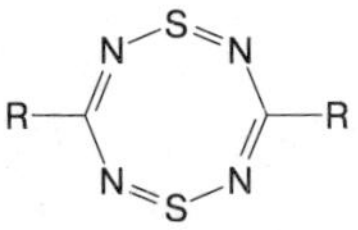

50

Figure 10.23 Electroconductive heterocyclic polymers and conjugated heterocycles containing nitrogen–sulfur bonds (**48–50**).

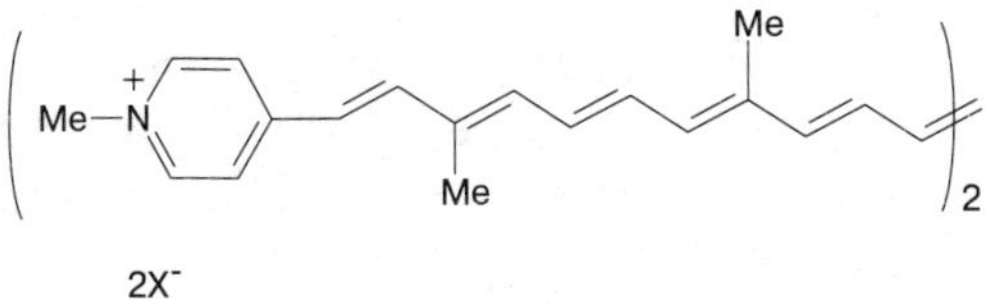

51

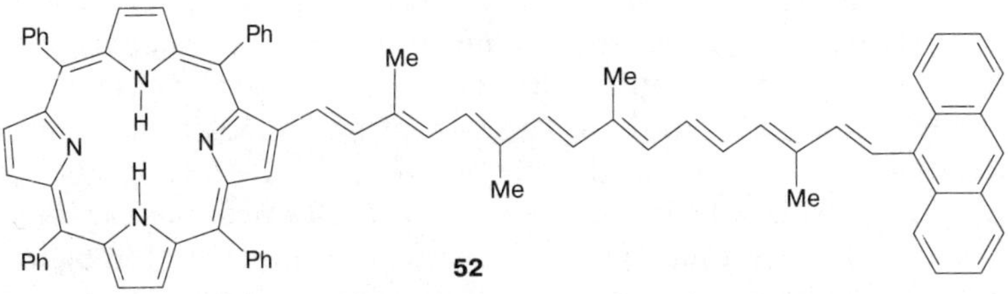

52

Figure 10.24 Molecular wires to convey electric current (**51**) and light energy (**52**).

other. Light of wavelength 256 nm first excites the anthracene unit. The energy absorbed is carried along the conjugated chain to the tetraphenylporphyrin ring, resulting in dramatically intense carmine red (656 nm) light emission. Similar systems can be used for transmitted of various types of signal, separating charges, locating biological objects by luminescence and so on.

10.6   Design of New Heterocycles

The construction of new chemical compounds by the introduction of aesthetic elements into their structures is now referred to as molecular design. This technique was applied in the synthesis of many compounds mentioned in the preceding sections of this chapter. The beauty of the molecular structures of cryptands, cavitands and artificial photosynthetic systems cannot be denied. To create such structures requires both artistry and skill on the part of the chemist. However, their original synthesis was motivated chiefly by practical needs. Notwithstanding, the history of science does provide examples where aesthetic enhancement was the sole objective of the investigation, an end in itself. Chemists often engage in complex syntheses simply to satisfy their curiosity—to achieve a goal which may be the determination of structure and properties, or the discovery of something unexpected. In effect, the heterocyclic chemist performs on a molecular level tasks similar to those of a jeweler on the macroscale. Jewelers enhance their works of art (rings, bracelets, necklaces, etc.) by modifying the shape, size and decoration, and by incorporating various precious stones. In the same manner, chemists synthesize heterocycles differing in the type and number of heterorings, in the nature of the ring fusion, in the pattern of substitution and so on.

An impressive example of such work was performed by the French chemist Sauvage, who synthesized a series of catenanes (Latin: *catena*, chain). These compounds contain two or more interlocked rings similar to the links in a chain. Such molecules were first obtained in the 1950 and 1960s. Over the next 20 years, however, developments in catenane chemistry were slow, apparently because of laborious synthetic methods and disappointing yields. The idea of statistical synthesis reawakened interest in the catenanes as this approach (represented in Figure 10.25) was simple and therefore more attractive. With this method, the cyclization of linear molecules containing reactive groups X and Y at the termini (one nucleophilic, the other electrophilic) is carried out in the presence of preformed macrocycles of a definite size. At the exact moment the new ring is formed by intramolecular reaction of the linear molecules, a statistically predetermined number will have 'threaded the needle' to form a catenane assembly (or dicatenane in the case of two initial rings being joined by the newly formed ring). The probability of such an event is very low. Moreover, molecular models indicate that the would-be catenane ring has to be constructed of no less than 21 atoms for it to be sterically possible for the rings to become linked.

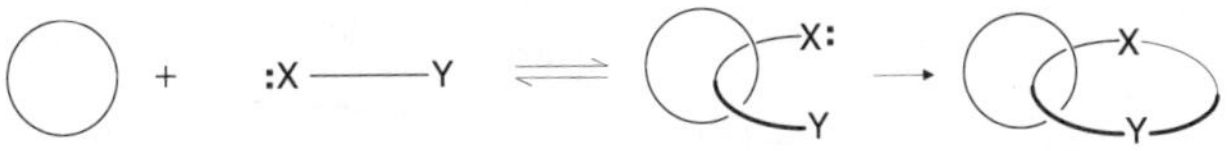

Figure 10.25 Statistical catenane synthesis.

In the early 1980s Sauvage *et al.* revived the idea of statistical synthesis on the basis of a new principle. A many-fold increase in the probability of closure of the interlocking rings was achieved during the alkylation of the phenanthroline bisphenol (**53**) with the polyether dichloride (**54**) in the presence of azacrown ether (**55**) and a copper(I) salt (Figure 10.26). Owing to the stable 2:1 complex formed by *ortho*-phenanthroline with the Cu⁺ ion (see Section 9.6), both reagents containing phenanthroline residues coordinate to form a complex of type (**56**) prior to cyclization. This accomplishes the formation of catenane (**57**) with yields of up to 60%.[†] Di- and tricatenanes were synthesized in a similar manner.

The synthesis of versatile macrocyclic systems analogous to porphyrins (Figure 10.27) is an intriguing task. Porphycene, an aromatic isomer of porphin which produces blue solutions with a red-violet luminescence in organic solvents, has recently been prepared. Rapid tautomeric interconversion between (**58a**) and (**58b**) is characteristic of porphycene. The aromatic extended analogue of porphyrin (**59**) also forms dark blue solutions but has surprised researchers with its magnetic properties. Analogues of porphin and porphycene in which some or all of the nitrogens have been substituted by oxygen or sulfur atoms have also been synthesized.

The discovery in the early 1950s of ferrocene (Figure 10.28, structure (**60**)), the first aromatic metallic π-complex with a 'sandwich' structure, promoted chemists to investigate heterocyclic analogues. For a long time, no success was forthcoming because the metal ions tended to coordinate more readily with the heteroatom than with the π-system. More recently, ferrocene heteroanalogues such as the tetramethylpyrrole anion complex of iron(II) (**61**), dipyridinechromium (**62**), pyridine–benzenechromium (**63**) and a number of other sandwich-like compounds have been prepared. The two nitrogen atoms in structures (**61**) and (**62**) are held rigidly in opposite orientations, whereas in complex (**63**) the rings slowly rotate.

---

†Preparative procedures in which reagents are localized spatially by metal ions are referred to as template syntheses.

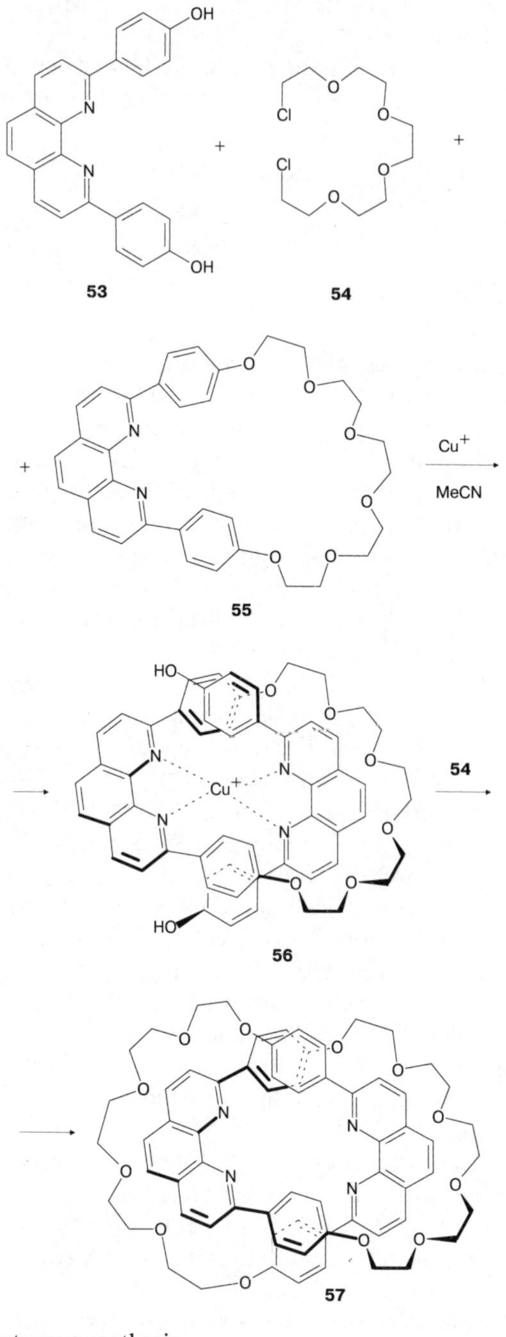

Figure 10.26 Dicatenane synthesis.

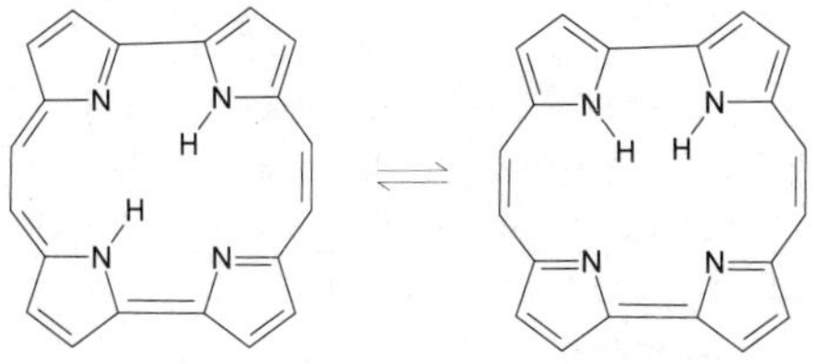

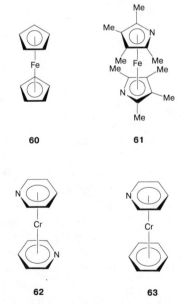

58a

58b

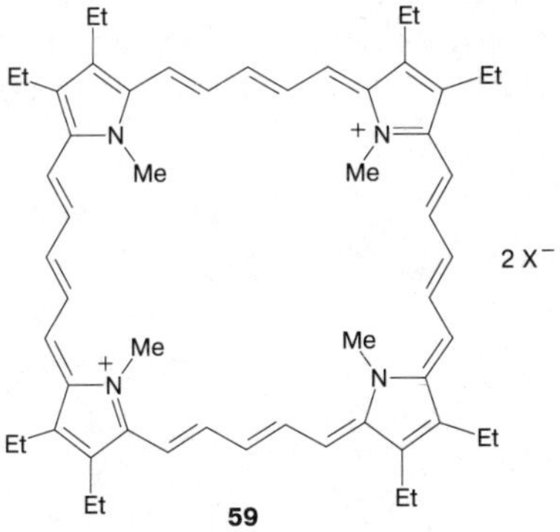

59

$2 X^-$

Figure 10.27   Porphyrin analogues.

60

61

62

63

Figure 10.28   Ferrocene (**60**) and some heterocyclic analogues (**61**)-(**63**).

The attention of heterocyclic chemists has always been attracted to heterocycles which contain unusual heteroatoms and to completely inorganic analogues of benzene (e.g. borazine, shown in Figure 1.7). In 1988 the synthesis of sodium pentaphosphacyclopentadienide (Figure 10.29, structure **64**) was reported. Compound **64** forms golden orange solutions which are rapidly decomposed by oxygen. The high sensitivity of its solutions is eliminated by the addition of 16-crown-6, which draws the Na^+ ions away from the anionic cycle. Such charge separation results in stabilization of the inorganic heterocycle. The complete symmetry of the five-membered ring has been proven, thus providing evidence for the identity of all of the phosphorus atoms, for the delocalization of the negative charge and for the aromaticity of this anion (**64**).

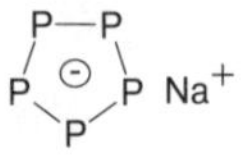

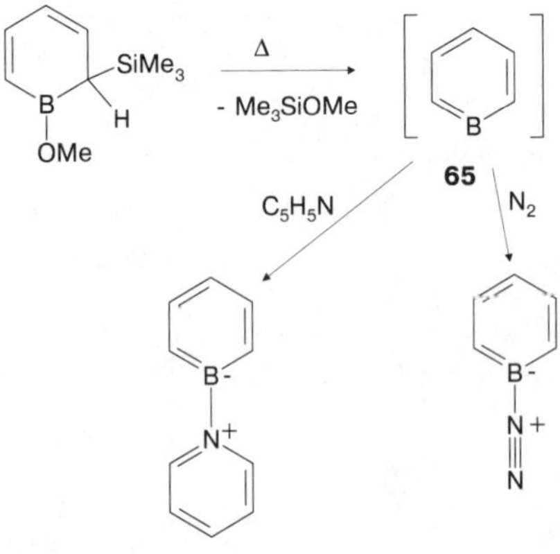

Figure 10.29   Heterocycles containing phosphorus or boron as the heteroatom.

For long, borabenzene (Figure 10.29, structure **65**) remained unknown despite the reports of a number of derivatives. However, chemists finally succeeded in generating (**65**) by thermolysis of 1-methoxy-2-trimethylsilyl-1,2-dihydroborabenzene (Figure 10.29). The presence of a vacant orbital in the boron atom makes borabenzene so reactive that it combines instantly with many basic compounds such as pyridine and even molecular nitrogen to form ylides (**66**) and (**67**) which are stable only at low temperatures.

The rapid development of science and technology does not allow us to cover all of the current trends concerning heterocyclic chemistry in one chapter. The authors are aware of this and hope that our readers will go on to follow new important achievements in this field.

10.7 Problems

1. Pyrido[3,2-g]indoles are good receptors for various derivatives of urea. Suggest a structure for the 1:1 molar ratio complex between compound A and urea. Account for the significant differences between the values of the association constant K_a ($1\,mol^{-1}$, $18\,°C$) for the complexes of A with (a) N,N'-dimethylurea ($K_a = 118$), (b) 2-imidazolidone ($K_a = 13\,000$) and (c) barbituric acid ($K_a = 74\,200$).

A

2. The so-called 'expanded porphyrin' B can form complexes with one or two molecules of methanol ($K_1 = 120$ and $K_2 = 301\,mol^{-1}$). Suggest structures for these complexes.

B

3. Macrocyclic receptor C forms a strong 1:1 complex with 9-methyladenine. What is the structure of the complex (hint: hydrogen bonding and stacking interactions stabilize the complex)?

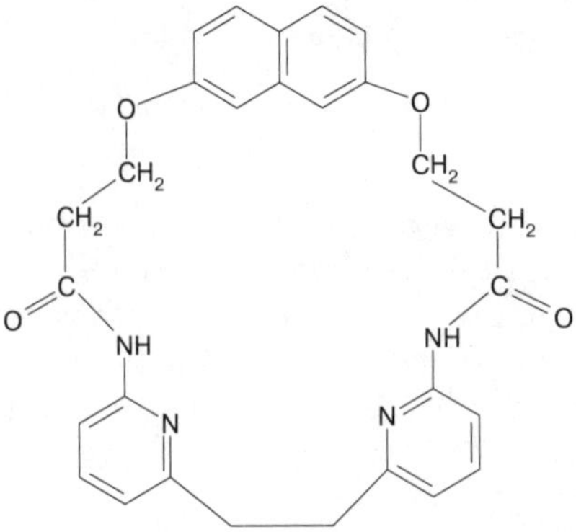

C

4. Crown ethers of suitable sizes (e.g. 18-crown-6) solubilize arenediazonium salts in nonpolar media (e.g. chloroform). At the same time they decrease the rate of diazonium salt decomposition. Explain.

5. Complexes of the bidentate ligands D–F with the copper(II) ion are decomposed by acid in 0.006 s, 0.02 s and 295 min, respectively. Account for the differences in stability.

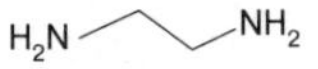

D

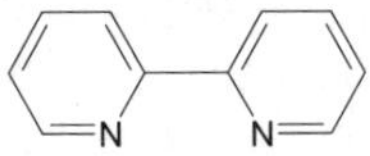

E

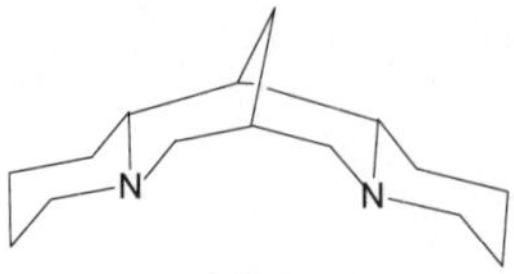

F

6. The CH_2Cl_2 solution of compound G cannot extract sodium or potassium picrates from their aqueous solutions. By contrast, the CH_2Cl_2 solution of

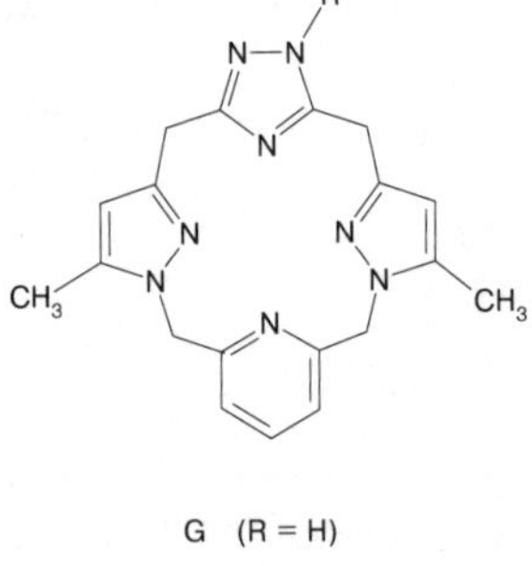

G   (R = H)

H   (R = CH₃)

compound H is able to remove these species. Explain.

7. N,N'-Dicyano-1,4-naphthoquinodiimine (J) forms a 1:1 charge transfer complex with tetrathiafulvalene (TTF) which possesses high electrical conductivity. However, tetracyano-1,4-naphthoquinodimethane (I) does

not form a complex with TTF. Explain.

10.8 Suggested Reading

1. Lehn, J.-M. (ed), *Comprehensive Supramolecular Chemistry*, Elsevier, New York, 1996.
2. Cram, D. J. and Cram, J. M., *Container Molecules and Their Guests*, Royal Society of Chemistry, Cambridge, 1994.
3. Williams, J. M., Ferraro, J. R., Thorn, R. J., Carlson, K. D., Geiser, U., Wang, H. H., Kini, A. M. and Whanglo, M.-H., *Organic Superconductors (Including Fullerenes). Synthesis, Structure, Properties, and Theory*, Prentice-Hall, Englewood Cliffs, NJ, 1992.
4. Bryce, M. R., *Chem. Soc. Rev.*, 1991, **20**, 355.
5. Curran, D., Grimshaw, J. and Perera, S. D., *Chem. Soc. Rev.*, 1991, **20**, 391.
6. Gokel, G. W., *Crown Ethers and Cryptands*, Royal Society of Chemistry, Cambridge, 1991.
7. Schneider, H.-J. and Durr, H. (eds), *Frontiers in Supramolecular Organic Chemistry and Photochemistry*, VCH, Weinheim, 1991.
8. Vögtle, F., *Supramolecular Chemistry: an Introduction*, Wiley, New York, 1991.
9. Cram, D. J., *Angew. Chem., Int. Ed. Engl.*, 1988, **27**, 1009.
10. Lehn, J.-M., *Angew. Chem., Int. Ed. Engl.*, 1988, **27**, 90.
11. Pedersen, C. J., *Angew. Chem., Int. Ed. Engl.*, 1988, **27**, 1021.
12. Vögtle, F. and Weber, E. (eds), *Host Guest Complex Chemistry: Macrocycles: Synthesis, Structures, Applications*, Springer, Berlin, 1985.
13. Hiraoka, M., *Crown Compounds: Their Characteristics and Applications'*, Elsevier, Amsterdam, 1982.

11 THE ORIGIN OF HETEROCYCLES

> The whisper might be born before the lips.
> In the woodlessness the leaves were falling down,
> And those to whom we dedicate our wisdom,
> Before this wisdom had their features crowned.
>
> O. Mandelstam

In this final chapter we try to answer the question posed in Pushkin's poem *Water-nymph*

> Where art thou from, sweet child of beauty?'

How did the heterocycles essential for life such as the purine and pyrimidine bases, porphyrins, 1,4-dihydronicotinamide, indoles and amino acids first appear on Earth? The philosopher Heraclites of Ephesus wrote 2500 years ago:

> Only those who know the origin and development of matter understand its nature.

The same idea was offered more simply by the popular Russian writer Koz'ma Proutkov:

> Find the roots of everything and thou willst understand much.

We know today that many of the above-mentioned heterocyclic compounds can be synthesized by living organisms, and the mechanisms of their biosynthesis are well established. However, to date, the nonbiological origin of heterocycles on Earth remains controversial. It seems obvious that prior to the appearance of the first primitive life forms on Earth, a stockpile of versatile raw materials must have developed. Thus, the question of the origin of heterocycles is a vital part of the overall mystery concerning the origins of life. While we are far from knowing the whole story, scientific comprehension of the issue has advanced significantly toward the original source of living matter. The origin of life on primitive Earth (protoearth) is presumed to be a result of chemical evolution. We now trace events back to the very beginning, relying on the data gained from cosmochemistry, astrophysics, geology and biology.

251

11.1   Interstellar Molecules

Heterocycles, like all other organic compounds, are formed predominantly from six elements (C, H, N, O, P and S), which are sometimes called 'organogens' as all living species are constructed from them. Of these, hydrogen, carbon, nitrogen and oxygen are widespread in the galaxy. It is of interest to consider how and why these elements originated and what their purpose was. Table 11.1 summarizes the main phases in the evolution of the universe.

Table 11.1 Astrochronological scale of chemical evolution (adapted from Voitkevitch, G. V., *Origin and Chemical Evolution of Earth* (in Russian), Nauka, Moscow, 1973, p. 95, Scheme 2; and Silk, J., *The Big Bang*, Freeman, New York, 1989, Chap. 4, p. 72, Table 4-1 with permission).

Evolutionary stage	Time after 'Big Bang' (10^9 of years)	Characteristic processes
Big Bang		Theoretical calculations indicate matter composed of neutrons, electrons and positrons with infinitely high density. Temperature between 10^9 and 10^{28} K.
Formation of the universe and of the first stars	0–1	Nuclear synthesis of hydrogen and helium
Explosions of supernovas; formation of gas and dust nebulas.	5–10	Nuclear synthesis of heavy elements; beginnings of chemical evolution in the cosmos.
Evolution of gas and dust nebulas; formation of small stars and planets. Appearance of the solar system.	10–12	Nuclear synthesis (radioactive 'clock' commences). Synthesis of interstellar particles and molecules.
Formation and development of the Earth	12–15	Molecular evolution on Earth. Synthesis of heterocycles and other biologically important substances. Biogenesis and Biological evolution

Modern astrophysics suggests that the universe appeared about 15–20×10^9 years ago as the result of a gigantic, even on a cosmic scale, event known as the 'Big Bang'. This involved the explosion of a high density, high temperature nucleus of matter. The fireball which first formed was composed of elemental particles and quanta and had a temperature of the order of 10^{28} K. One minute after the explosion, the main components of the universe were neutrinos, electrons and photons. After 10^6 years the universe had cooled to about 3000 K as a result of its expansion. Electrons and protons began to combine together to form hydrogen, and this new substance became separated from the radiation (one of the most important contemporary difficulties of astrophysics is understanding primordial radiation). The resulting matter, under the action of gravitational forces, began to form into stars and galaxies. The first stars were composed of hydrogen (75%) and helium (25%) and did not include heavy elements. According to our present theories, further development involved condensation (collapsing) of the center of the stars into superdense nuclei. This chapter ended with the explosion of these stars, the destruction of their atmospheric envelopes and the consequent formation of gaseous nebulas. Exploding stars are astronomically referred to as 'supernovas'. Their explosion is accompanied by extraordinarily bright flashes of light. A supernova flash is an extremely rare event on the scale of the human lifespan. The first supernova to be recorded was observed in the middle of the eleventh century. This star could be seen, even during the day, for an entire year. A second supernova was observed in 1987, but the radiation received on Earth was of much lower intensity.

The temperatures and pressures which develop during the explosions of supernovas are sufficiently high to trigger thermonuclear synthesis, during which hydrogen is consumed and helium is produced. These events are believed to have marked the start of chemical evolution. Hydrogen and helium were transformed by the capture of protons, neutrons and α-particles into carbon, oxygen and other heavy elements, the total weight of which now approximates 1% of the universe. The following is a scheme depicting the synthesis by nuclear fusion of some of the organogenic elements

$$^1H + {}^1H \longrightarrow {}^2D \qquad\qquad {}^{12}C + {}^1n_0 \longrightarrow {}^{13}C$$

$$^2D + {}^1H \longrightarrow {}^3He \qquad\qquad {}^{13}N + {}^1H \longrightarrow {}^{14}N$$

$$2\,{}^3He \longrightarrow {}^4He + 2\,{}^1H \qquad {}^{12}C + {}^1H \longrightarrow {}^{13}N$$

$$3\,{}^4He \longrightarrow {}^{12}C \qquad\qquad {}^{12}C + {}^4He \longrightarrow {}^{16}O$$

Nebulas in the skies represent the characteristic remains of extinct supernovas. These gas and dust residues, which gradually diffused from the site of the catastrophe, are evidence of the ancient explosions. However, gravitational forces then gathered many of these residues together and agglomerated them into dense star-like formations (suns) of smaller sizes. The evolution of gas and dust nebulas enriched with heavy elements led to the appearance of planets. Our galaxy is thought to have about 10^9 stars with planetary systems similar to our solar system.

Diatomic molecules such as C_2, CN, CH, CO, NH and OH are produced in the atmospheres of suns and other relatively cold stars. They are then ejected into interstellar space in pulses which derive from pressure variations between the surface of the star and space. Low temperature synthesis of more complicated molecules takes place in cosmic gas and dust nebulas. For example, hydrogenation and hydroxylation of the above-mentioned diatomic species provide multiatomic molecules such as ammonia, hydrogen cyanide, water, methane, methanol and others.

When the sky was observed only by optical spectroscopic methods, we could not detect organic molecules in space. 'We waited, staring at the heavens, to catch reflections of their being.' However, in the late 1960s radiotelescopes were introduced into practical astronomy. Immediately, in place of the previous silent 'darkness of the night', a multivoiced chorus of interstellar molecules was heard. Formaldehyde was the first organic molecule to be registered in far-off space by radioastronomers. The first signal was detected in 1969 at a wavelength of 6.2 cm. Formaldehyde was later shown to be one of the most abundant compounds in space, and was thus dubbed the 'universal molecule'. More than 100 cosmic substances have since been discovered, and the most important (as likely precursors of heterocyclic compounds) are listed in Table 11.2. The majority of these molecules were found in the center of our galaxy and in the Orion nebula.

Table 11.2 Cosmic molecules found in space between 1968 and 1976 (adapted from Lazcano-Araujo, A. and Oro, J., in *Comets and the Origin of Life*, (ed. C. Ponnamperuma), Reidel, Dordrecht, 1981 and Fox, S. W. and Dose, K., *Molecular Evolution and the Origin of Life*, Dekker, New York, 1977, Chap. 11, p. 330, Table 11-1, with permission).

Formula	Name	Formula	Name
NH_3	Ammonia	$MeC{\equiv}N$	Acetonitrile
H_2O	Water	$MeCHO$	Acetaldehyde
H_2CO	Formaldehyde	$CH_2{=}NH$	Methyleneimine
HCO^+	Formyl cation	$MeNH_2$	Methylamine
$\cdot C{\equiv}N$	Cyano radical	$CH_2{=}CHC{\equiv}N$	Acrylonitrile
$HC{\equiv}N$	Hydrogen cyanide	$MeOMe$	Dimethyl ether
HCO_2H	Formic acid	HCO_2Me	Methyl formate
$HC{\equiv}CC{\equiv}N$	Cyanoacetylene	$EtOH$	Ethanol
$MeOH$	Methanol	$HC{\equiv}C\cdot$	Ethinyl radical
$HCONH_2$	Formamide	$H_2NC{\equiv}N$	Cyanamide
$HC{\equiv}CMe$	Methylacetylene	$HC{\equiv}CC{\equiv}CCN$	Cyanodiacetylene
$H_2C{=}S$	Thioformaldehyde	H_2NCONH_2	Urea

Interstellar molecules are thought to result from homogeneous synthesis during the collision of ions and simple molecules in the vapor phase, as well as from heterogeneous reactions which take place on the surface of dust particles and granules that make up the interstellar dark clouds. Graphite or silicates covered with dirty ice formed from frozen methane, ammonia, formaldehyde and so on are the constituents of these granules. In laboratory experiments designed to reproduce space conditions (high vacuum, 10 K) many of the known interstellar molecules including formaldehyde, formamide, formic acid and others have been synthesized by the UV irradiation of mixtures of NH_3, H_2O, CO and CH_4. It has also been established that during solid phase polymerization induced by radiation, formaldehyde produces polyoxymethylene and polysaccharides at temperatures from 4 to 140 K. Radioastronomers have also discovered formaldehyde polymers in interstellar dust clouds. This observation became the foundation of the 'cold prehistory of life' theories. Heterocycles have not yet been detected in the cosmos, although there is indirect evidence for the existence of porphyrin-like compounds in interstellar space.

11.2 Organic Compounds in Comets and Meteorites

The abundance of organic molecules in interstellar space suggests that they could have been transported to the Earth's surface during the passage of the solar system through the harsh environment of gas and dust nebulas. Another potential source of simple precursors for more complicated molecules is the carbon-based material from which comets and meteorites are composed. We underline in this context the similarity between the isotopic compositions of comet and terrestrial carbon. The Earth is thought to have collided with comets more than one hundred times in its history. The data in Table 11.3 illustrate the role of cosmic organic carriers in 'seeding' our planet with carbon-based matter. We draw attention to the footnote of Table 11.3 which contains some 'food for thought', including a comparison of the carbon content in Earth sedimentary rocks with the quantity of carbon derived from the cosmos.

Table 11.3 Cosmic contributions in 'seeding' the Earth with carbonaceous matter during the first 2×10^9 years (adapted from Lazcano-Araujo, A. and Oro, J., in *Comets and the Origin of Life* (ed. C. Ponnamperuma) Reidel, Dordrecht, 1981, with permission).[a]

Delivery system	Weight of carbon (t)
Comets	10^{16}
Interplanetary dust (particle size less than 1 mm)	2×10^{13}
Meteorites	10^{12}–10^{18}
Interstellar clouds	10^9
Solar wind	1.5×10^8

[a]In the sedimentary layer of modern Earth the carbon content is assessed to be between 1.2×10^{16} and 1.9×10^{16} t; all living matter on Earth contains about 10^{12} t of carbon.

To date, heterocyclic compounds have been found neither in the nuclei of comets nor in their diffuse gaseous atmospheres. However, comets do contain carbon-containing neutral molecules such as C_2, C_3, CO, CO_2, CS and HCN; radicals such as $\cdot$CN, $\cdot\cdot$CH$_2$ and $\cdot\cdot\cdot$CH; and ions like C^+, CO^+ and CN^+, together with a number of other low molecular weight compounds. Investigations conducted in 1986 by means of the 'Vega' and 'Giotto' space stations determined that the nucleus of Halley's comet contains frozen methane under a black asphalt-like layer. This shell may itself be the product of radiation-induced methane polymerization, as has been demonstrated experimentally.

Exciting results have been obtained in the search for abiogenic organic compounds in meteorites. In the early morning of September 28, 1969, a large meteorite fell to Earth near the Australian town of Merchison. This stone meteorite, thus named 'Merchison', was of the carbonaceous chondrite type,[†] and its extractable matter was found to contain an array of organic substances. Among the heterocyclic compounds observed were three amino acids (pipecolinic acid, histidine and proline), porphyrins and pyrimidines, triazines and purines. The pyrimidines detected, shown in Figure 11.1, were found to be quite different from the known 'terrestrial' pyrimidines of biological origin. Although the purine bases found in meteorites have

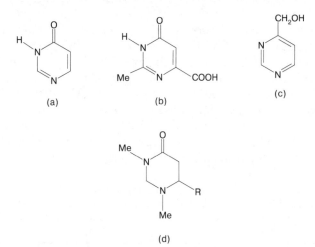

Figure 11.1   Pyrimidine bases found in the 'Merchison' meteorite: (a) pyrimidin-4-one, (b) 2-methylpyrimidin-4-one-6-carboxylic acid, (c) 4-hydroxymethylpyrimidine and (d) 1,3-dimethyl-6-alkylhexahydropyrimidin-4-ones (R = Et, Pr^n).

---

†Such meteorites are composed of spherical silicate formations called chondrules. Carbonaceous chondrites contain much greater quantities of carbon-based particles (up to 7%) than other types of meteorites (e.g. iron analogues). Chondrites are readily broken up. About 30% of the carbonaceous matter can be extracted by solvents; the remaining unextractable matter is polymeric in nature.

familiar structures (for example, adenine and guanine were separated from the carbonaceous matter of the 'Orguey' meteorite), scientists have no doubt that all of this carbonaceous matter has been formed abiogenically. These compounds appear to have been produced in the cosmos by the condensation of carbon dioxide and ammonia with hydrogen under the catalysis of iron and its alloys.

11.3   Do Heterocycles Exist on the Moon and Mars?

Specimens of lunar rocks and dust retrieved between 1969 and 1972 by the American spaceship 'Apollo' and the Russian robotic station 'Luna', contained only tiny quantities of carbon (2–200 ppm). The rocks contained small amounts of CO_2, methane, ethane, propane, acetylene, benzene and toluene, and trace levels of amino acids. Heterocyclic compounds were not detected, except for the extraction of an insignificant quantity of porphyrin-like substances ($0.005–0.1\,mg\,g^{-1}$) in one instance. There has been no satisfactory explanation to date for the low content of organic matter in the lunar materials.

The chemical probing of Martian soil by the American craft 'Viking' in 1976 demonstrated the complete absence of organic matter (the sensitivity of the analysis was $10^{-10}\,g\,g^{-1}$). Chromatographic–mass spectrometric determinations detected the presence of water and CO_2 only. While the first lunar and Martian expeditions revealed 'lifeless' deserts, hope exists that organic oases will one day be found.

11.4   The Atmosphere of Earth and Other Planets

To understand the modes of origin of biologically important organic compounds, we need first to be aware of the significant changes over geological time in the chemical compositions of the atmospheres of the Earth and the other planets within the solar system. Since all of the planets were formed by the agglomeration of gas and dust clouds consisting mainly of hydrogen and helium, there is no doubt that these gases provided the first atmospheres. Each of the heavy planets (Jupiter, Saturn, Uranus and Neptune) retains such a primeval atmosphere. Their gravitational fields are sufficient to hold these light gases within their atmospheres. The smaller planets (Mercury, Venus, Earth and Mars), on the contrary, lost their primeval atmospheres to the cosmic vacuum. As a result of volcanic eruptions and other gaseous discharges, their atmospheres have gradually become heavier because of the presence of carbon dioxide, water vapor, ammonia, nitrogen and the heavier inert gases such as argon. Such atmospheres are referred to as 'secondary'.

The evolution of the gas envelope on Earth progressed further. Scientists believe that the secondary atmosphere on Earth consisted mainly of water vapor, the condensation of which initiated the formation of oceans. Water vapor in the upper layers of the atmosphere, under the influence of lightning discharges and ultraviolet radiation, dissociated to give oxygen and a thin layer of ozone. Up to this time, although the chemical precursors essential for the origin of life already existed on the Earth's surface (see Section 11.5), intense ultraviolet radiation had prevented further progress. It is well known that ozone absorbs ultraviolet rays, and therefore formation of the protective ozone blanket allowed the synthesis of complex biomolecules and the development of primitive living organisms such as photosynthetic bacteria, simple algae, mosses, lichens and so on.[†] The development of these early life forms and the evolution of more complex plants led to a 'refueling' of the Earth's atmosphere with oxygen through photosynthesis, which gradually resulted in the atmospheric composition we know today.

The atmospheric compositions of the planets in our solar system are listed in Table 11.4.

Table 11.4 Compositions of the atmospheres of solar system planets (adapted from Marov, M. Ya., *Planets of the Solar System* (in Russian), 2nd Ed, Nauka, Moscow, 1986, Chap 5, p. 240, Table 4, with permission).

Planet	Surface temperature (°C)	Main components of atmosphere (vol%)
Mercury	−185 to 510	H_2 (<18), He (<20), Ne (40–60), Ar (<2), CO_2 (<2) Atmosphere is highly rarefied
Venus	500	CO_2 (95), N_2 (<3), H_2O (0.1–1.0), O_2 (<0.01), NH_3 (<0.01)
Earth	−80 to 50	N_2 (78), O_2 (21), Ar (0.9), H_2O (0.1–1.0), CO_2 (0.03)
Mars	−80 to 16	CO_2 (95), N_2 (2.7), Ar (1.6), O_2 (0.13), CO (0.077), H_2O (0.03)
Jupiter	−140	H_2 (87), He (12.8), NH_3 (0.01), CH_4 (0.01), PH_3
Saturn	−180	H_2, traces of CH_4, NH_3, N_2, He
Uranus	−170	H_2, CH_4
Neptune	−220 to −160	H_2, CH_4

[†]Some scientists believe that the first living organisms on Earth appeared prior to the introduction of oxygen to the atmosphere. Anaerobic bacteria may have been able to develop in a layer of water thick enough to protect them from the ultraviolet radiation prior to the formation of the ozone shield.

11.5   Heterocycles and the Origin of the Biosphere

Until the middle of the twentieth century the question of chemical evolution on Earth attracted the attention chiefly of biologists, geologists, paleontologists, poets and philosophers. The French writer de Saint-Exupery once wrote

> from a boiling lava-flow, from a stellar substance our life is born . . . a noble cosmic flower.

Poetic intuition is deep and a philosopher's mind is inquisitive and perceptive. However, a researcher in the ´ natural sciences must probe ideas by experimentation. Chemists have succeeded in applying this concept by successfully reproducing in the laboratory the conditions which existed on primitive Earth. Under these circumstances, practically all of the biochemically important complex organic molecules (including heterocycles) that provided the basis for life on Earth could be prepared from simple precursors.

11.5.1   SIMPLE PRECURSORS OF HETEROCYCLES

The data in Table 11.3 indicate that the Earth contains an enormous quantity of organic matter, but early in the history of the universe the variety of its components was not sufficient for the formation of all of the biological compounds. For example, the carbon sources did not include any pyrimidine bases or amino acids. Scientists are now certain that almost all of the compounds necessary for the generation of life could have arisen on the protoearth both in the surrounding atmosphere and on its surface, especially in the ocean. Abiotic syntheses could have taken place on primitive Earth utilizing one or more of the following energy sources: (i) solar ultraviolet radiation; (ii) electrical lightning discharges; (iii) ionizing radiation emanating from the upper layers of the terrestrial crust (at depths of up to 1 km; calculations show that the greatest source of energy was fission of the potassium-40 isotope); (iv) volcanic heat; and (v) shock waves, particularly those resulting from meteorite showers.

It is now generally accepted that the primeval Earth possessed a reducing atmosphere, the chief components of which were methane, ammonia, water and hydrogen. In the 1950s the American scientist Miller demonstrated at the University of Chicago that under the action of electrical discharges a mixture of these gases contained in a sealed sterile apparatus could be converted into an array of organic products. Formaldehyde, acetaldehyde, formic acid, acetic acid, succinic acid, lactic acid, fumaric acid, urea, a number of amino acids and HCN were among those found. Variations of the primary mixture composition (e.g. increasing the oxidizing character by the addition of CO_2 or CO, substituting nitrogen for ammonia, etc.) or the use of a different energy source (ultraviolet radiation, bombardment with high speed α-particles or electrons) did not significantly affect the experimental results.

Some of these gas phase reactions must be radical in nature. The initial molecules are broken down into highly reactive fragments: atomic hydrogen and Me·, NH_2· and OH· radicals. These species recombine, dimerize or dissociate further into still smaller particles. Thus, a methyl radical can successively lose all of its hydrogen atoms and ultimately become atomic carbon, as has been proven by the formation of graphite grains in model experiments. Combination of Me· with OH·, or of Me· with NH_2·, would give methanol or methylamine, respectively. Methanol can be transformed by a similar radical mechanism into formaldehyde, which in turn can be converted to acetaldehyde, formic acid, formamide, glyoxal and other products. In these transformations, the formyl radical ·CHO is an intermediate (Figure 11.2).

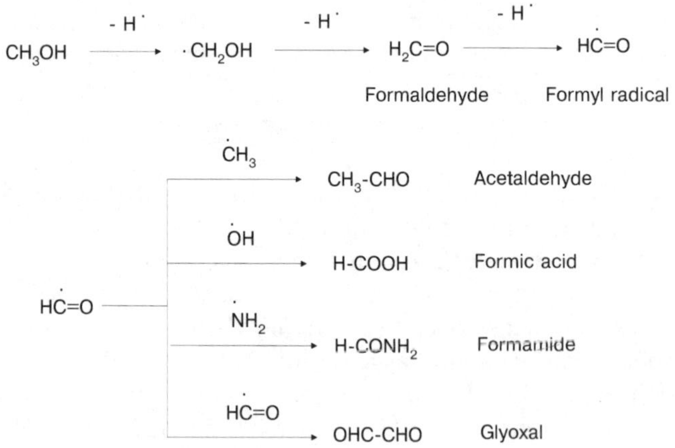

$$CH_3OH \xrightarrow{-H\cdot} \cdot CH_2OH \xrightarrow{-H\cdot} H_2C=O \xrightarrow{-H\cdot} HC=O$$

Formaldehyde   Formyl radical

HC=O

- $\dot{C}H_3 \longrightarrow CH_3\text{-}CHO$ Acetaldehyde
- $\dot{O}H \longrightarrow H\text{-}COOH$ Formic acid
- $\dot{N}H_2 \longrightarrow H\text{-}CONH_2$ Formamide
- $H\dot{C}=O \longrightarrow OHC\text{-}CHO$ Glyoxal

Figure 11.2 Probable origin of formaldehyde and some of its transformations in the model experiments carried out by Miller.

Hydrogen cyanide is a potent animal poison. However, there is striking evidence that this compound played a key role in the formation of a number of important biomolecules including amino acids, purines, pyrimidines and imidazoles on Earth. Significant quantities of HCN are produced in almost all of the model experiments irrespective of the source of the nitrogen (ammonia or molecular nitrogen) and carbon (methane or CO_2). The general reactions can be represented by

Methylamine is believed to be an intermediate which is subsequently

$$CH_4 + NH_3 \longrightarrow HCN + 3 H_2$$

$$3 H_2 + N_2 + 2 CO \longrightarrow 2 HCN + 2 H_2O$$

dehydrogenated to HCN

$$CH_3NH_2 \longrightarrow CH_2{=}NH \longrightarrow HCN$$

Hydrogen cyanide is a unique species in that the high polarization of the neutral molecule results in a strong electrophilic character while the CN$^-$ anion possesses high nucleophilicity. This doubly reactive compound thus readily polymerizes to produce dimers, trimers, tetramers and so on. Carbon–carbon chains bearing highly reactive functional groups such as C≡N, NH$_2$ and =NH are thus constructed (Figure 11.3).

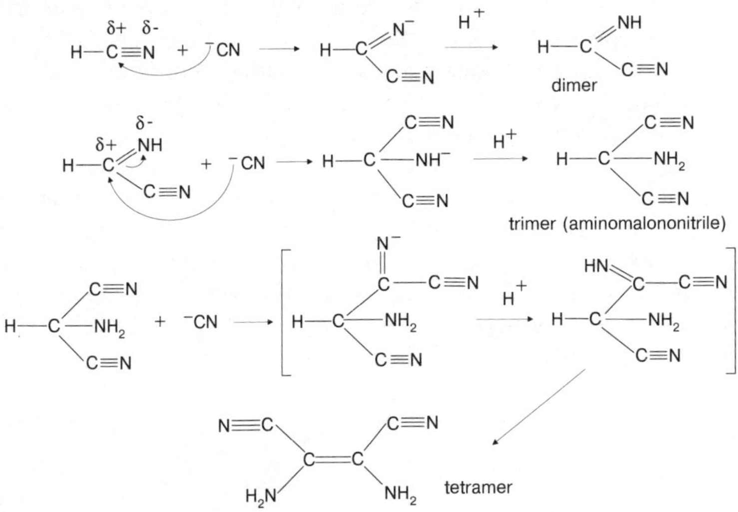

Figure 11.3   Formation of hydrogen cyanide oligomers.

Cyanoacetylene and acrylonitrile (cyanoethylene) are two further compounds of importance in chemical evolution. These compounds have also been found as products in many model experiments. Their formation can be envisioned as the result of the combination of a CN· radical with an ethinyl or vinyl radical, respectively

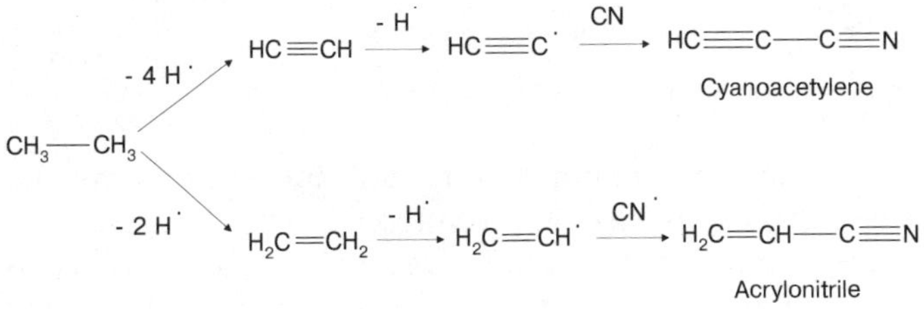

All of the above-mentioned radicals, in addition to the simple compounds formed in model experiments, have been detected in interstellar space, in the heads of comets and in the atmospheres of several planets. This provides support for the belief that the model experiments accurately depict chemical evolution on Earth, and reflects the universality of the chemical reactions which take place in various regions of the universe.

11.5.2 HETEROCYCLIC AMINO ACIDS

We have previously mentioned that α-amino acids of nonbiological origin are constantly transported to the Earth via cosmic carriers such as meteorites. However, all of the conditions necessary for their spontaneous appearance also appear to have existed on the primitive Earth. In fact, many α-amino acids, including those essential for living organisms, have been prepared in laboratory experiments in which the gaseous mixtures imitate the primeval Earth's atmosphere. The major pathway for their formation seems to be the Strecker synthesis; as shown in Figure 11.4, an aldehyde reacts with a mixture of hydrogen cyanide, ammonia and water. The aldehyde and ammonia first produce an aldimine which subsequently undergoes addition of hydrogen cyanide to its C=N bond. This is followed by hydrolysis of the aminonitrile thus formed to give the corresponding α-amino acid.

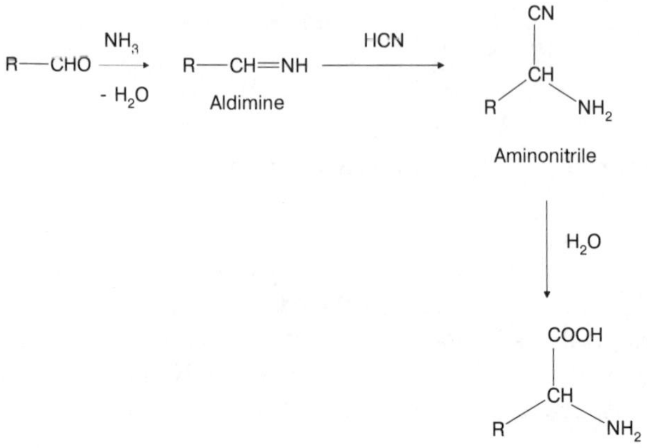

Figure 11.4   Strecker synthesis of α-amino acids.

Unique chemical mechanisms for the abiogenic formation of the heterocyclic amino acids histidine, tryptophan, proline and hydroxyproline (Figure 3.11) are not as yet generally accepted. Histidine has been found in mixtures formed by heating an aqueous solution of formaldehyde with ammonia at 185 °C. The formation of methyleneimine, hydrogen cyanide, acetaldehyde, glyoxal and formic acid is highly probable under these conditions. The subsequent reaction of glyoxal with ammonia and formaldehyde, well known to preparative chemists, can produce imidazole (Figure 11.5a)[†]. Imidazole contains two carbon atoms (at positions 4 and 5) with partial negative charges (Figure 2.1) which can be readily attacked by various electrophilic agents. Thus, the addition of glyoxal to imidazole can produce the hydroxy aldehyde (1), which can be converted (for example, by the formate anion) to imidazolylacetaldehyde (2). The Strecker synthetic pathway then converts (2) to histidine (Figure 11.5b).

A fundamentally similar scheme can be proposed for the synthesis of

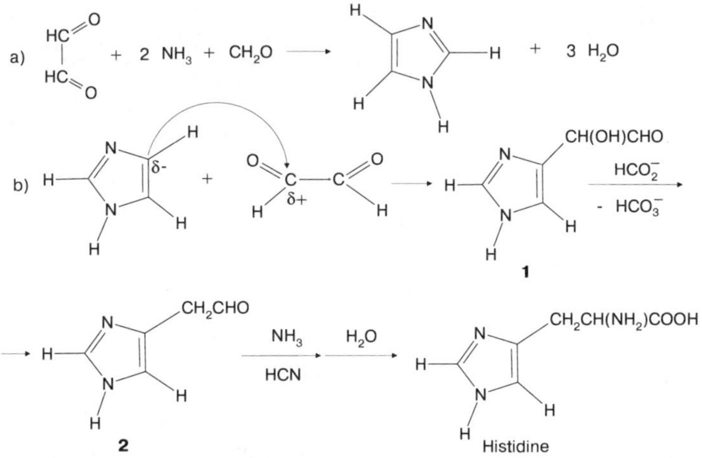

Figure 11.5 Possible scheme for the abiotic synthesis of histidine.

tryptophan from indole. The reaction with glyoxal occurs at the C-3 position of indole, the position of highest electron density. Evidence for the abiotic synthesis of indole was obtained by the irradiation of aqueous solutions of formaldehyde and ammonium nitrate. Moreover, appreciable quantities of indole are produced during the pyrolysis of methane (or other low alkane) and ammonia mixtures. The latter pathway seems to involve the production of acetylene which reacts further with ammonia in a number of steps to form various substances including indole

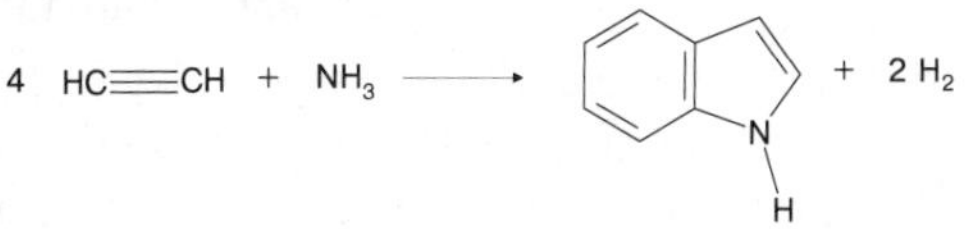

†The formation of imidazole (often along with indole, tryptamine, urea and other nitrogen bases) was also demonstrated in model syntheses. Imidazole formed when mixtures of CH_4, NH_3, H_2 and H_2O were bombarded with a stream of electrons. Irradiation of an aqueous solution of ammonium nitrate and formaldehyde similarly produced imidazole.

Tryptophan can also be produced from serine. Serine can be readily prepared from glycine and formaldehyde and is a frequent by-product of abiotic synthetic procedures aimed at providing amino acids. In the formation of tryptophan, serine is believed to be first converted into the unstable dehydroalanine which then rapidly undergoes a Michael reaction (addition to an activated double bond) with indole (Figure 11.6).

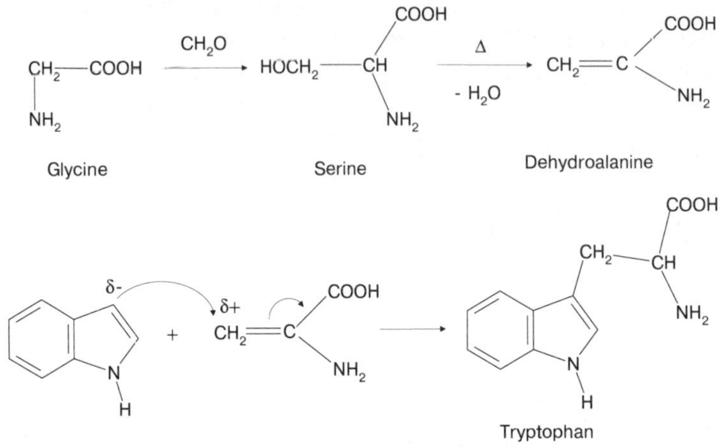

Figure 11.6 Formation of tryptophan from indole and serine.

Proline (2-pyrrolidinecarboxylic acid) was also found amongst the abiotic reaction products. Proline is formed with other amino acids by the application of electrical discharges to mixtures of CH_4, N_2 and H_2O or by heating mixtures of CH_4, NH_3 and H_2O at 900 °C. Interestingly, proline was separated from the aqueous extracts of lava samples taken during the eruption of Maunu Ulu in the Hawaiian islands.

11.5.3 PYRROLES AND PORPHYRINS

Definitive evidence for the theory of the abiogenic origin of porphyrins, pyrimidine bases and a multitude of other heterocyclic compounds was provided when they were detected in ashes and rocks brought up from below the Earth's crust during volcanic eruptions. Traces of porphyrins were also identified amongst the products in model CH_4–NH_3–H_2O mixtures treated with electrical discharges. The greatest variety of laboratory-synthesized porphyrins was achieved using pyrrole as the precursor. Mixtures containing formaldehyde (or other aldehydes) and pyrrole were treated with ultraviolet light, γ-rays, electrical discharges and heat (100–180 °C). The formation of porphyrins was observed in all cases, but the yields were increased by the presence of oxygen. Oxygen was demonstrated to promote oxidation of the originally produced, partially hydrogenated porphyrins (porphyrinogens) to the final products. Various polypyrrylmethanes, particularly, dipyrrylmethanes (Figure 11.7), were also found to be precursors of the porphyrinogens.

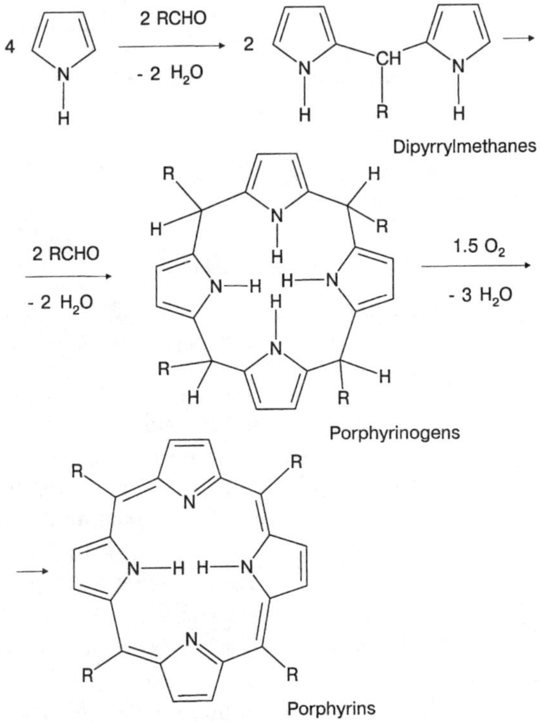

Dipyrrylmethanes

Porphyrinogens

Porphyrins

Figure 11.7 Formation of porphyrins from pyrrole.

Pyrrole itself is produced in many different abiogenic conversions such as the interaction of acetylene with ammonia (Chichibabin reaction in the presence of natural clays), of acetylene with hydrogen cyanide, of glucose with ammonia, and in the pyrolysis of the diammonium salt of mucic acid. However, porphobilinogen (Figure 11.8) is the most important pyrrole system from the point of view of the beginning of life on Earth. This compound is the biosynthetic precursor of practically all the naturally occurring porphyrins such as chlorophyll, hemoglobin, and vitamin B_{12}. The abiotic synthesis of porphobilinogen occurs readily by reaction of succinic acid with glycine. -Aminolevulinic acid is an intermediate product in this reaction, as in the biochemical synthesis carried out by living organisms.

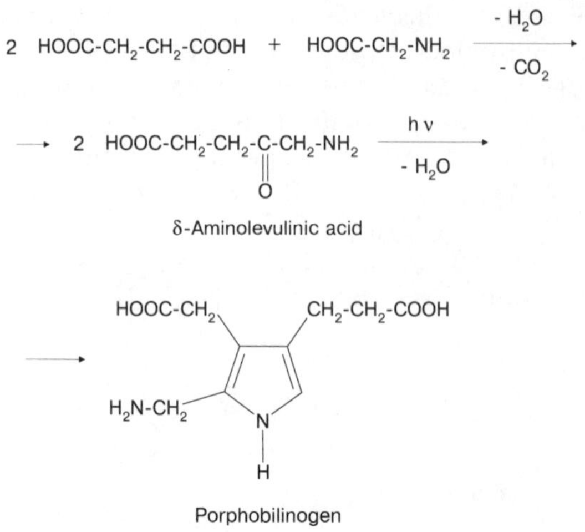

$$2 \quad HOOC\text{-}CH_2\text{-}CH_2\text{-}COOH \quad + \quad HOOC\text{-}CH_2\text{-}NH_2 \quad \xrightarrow[\;-\;CO_2\;]{\;-\;H_2O\;}$$

$$\xrightarrow{} \quad 2 \quad HOOC\text{-}CH_2\text{-}CH_2\text{-}\underset{\underset{O}{\|}}{C}\text{-}CH_2\text{-}NH_2 \quad \xrightarrow[\;-\;H_2O\;]{h\,\nu}$$

δ-Aminolevulinic acid

Porphobilinogen

Figure 11.8   Abiotic synthesis of porphobilinogen.

11.5.4   FURANOSE SUGARS

Monosaccharides are essential biological building blocks just as are the amino acids and the purine and pyrimidine bases. Since saccharide biomolecules (furanoses and pyranoses) exist preferentially in the cyclic or, more precisely, the heterocyclic form, it is pertinent to investigate the possibility of their arising abiotically. Almost all of the research in this field has focused on an observation by the Russian chemist Butlerov who noted in 1861 that heating aqueous solutions of formaldehyde with barium or calcium hydroxides resulted in the formation of mixtures of carbohydrates. Almost 30 compounds including trioses, tetroses, pentoses and hexoses were later identified among the products of this condensation, including such important molecules as ribose, fructose and glucose. The base-catalyzed transformation shown in Figure 11.9 involves a sequence of consecutive aldol autocondensations of formaldehyde and intermediate hydroxyaldehydes.

What is the probability of such a pathway 'seeding' the Earth with carbohydrates during the prebiological period? For some time there was a significant discrepancy in the theory of abiotic synthesis in that under laboratory conditions a highly basic medium was required even though such conditions were generally believed never to have existed in the Earth's oceans. However, this early stumbling block was eliminated when aqueous solutions of formaldehyde were heated in the presence of apatite and kaolinite and the same transformations were observed. The formation of formaldehyde was always detectable following the irradiation of methane–water-ammonia mixtures. These solutions, upon treatment with γ-rays, electron beams or ultraviolet light, evolve further to form sugars

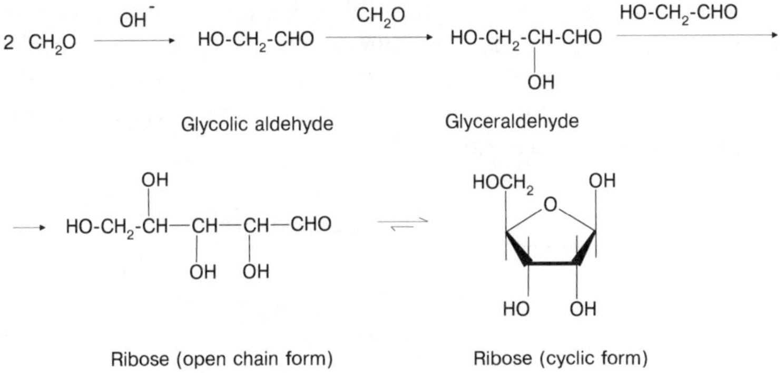

Ribose (open chain form)   Ribose (cyclic form)

Figure 11.9   Formation of monosaccharides by the base-catalyzed condensation of formaldehyde.

including ribose and deoxyribose. Addition of formaldehyde to the original mixtures augments the yields of these pentoses. Thus, the scheme shown in Figure 11.9 is the most probable explanation for monosaccharide formation on the primitive Earth.

Recently, it was proposed that the additional (potentially prebiotic) component, glycolaldehyde phosphate, could be involved in the aldol condensation of formaldehyde. Such coaldomerization leads to the direct formation of phosphorylated sugars including aldopentose 2,4-diphosphates. The reaction is diastereoselective, the main product being ribose 2,4-diphosphate with the same configuration of substituents as in natural RNA ribose

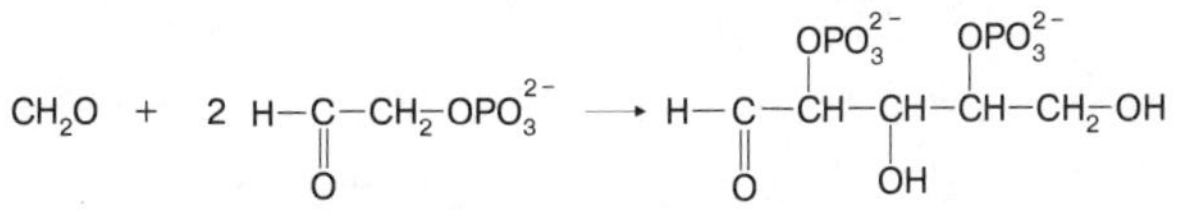

11.5.5 NICOTINAMIDE

Life on Earth in the primary 'broth' is unlikely to have commenced in the absence of vitamins (coenzymes) and their precursors. We first consider nicotinamide, the simplest and most important coenzyme. Nicotinonitrile, whose formation was noted in a number of model experiments, was undoubtedly the direct precursor. The action of electrical discharges on CH_4–N_2–H_2 mixtures yields 1% of nicotinonitrile. It is believed that acrylonitrile may be an intermediate in the synthesis (see Section 11.5.1): two molecules combine by a Diels–Alder interaction to form a pyridine ring. Hydrolysis of the nitrile group leads successively to nicotinamide and nicotinic acid (Figure 11.10). Nicotinonitrile can also be prepared by condensation of cyanoacetylene, propenal and ammonia, which themselves could have readily been formed in the Earth's protoatmosphere. The existence of nicotinamide in the primitive ocean suggests that the coenzyme NAD^+ and its reduced form NAD-H may have arisen during the early stages of chemical evolution and controlled some biochemically important redox reactions.

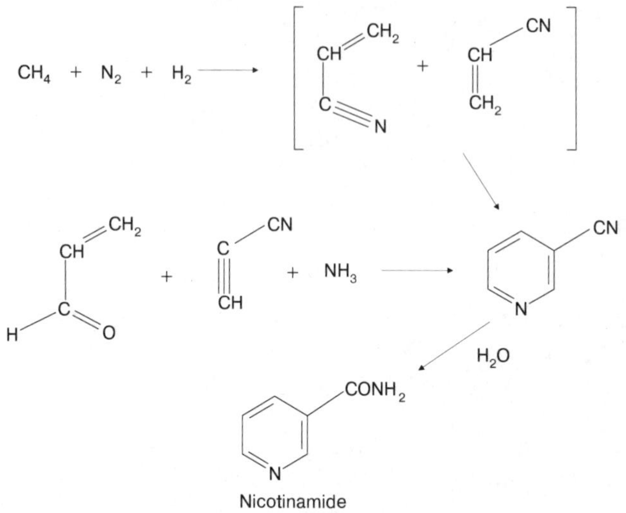

Figure 11.10   Abiotic pathways for nicotinamide formation.

11.5.6 PURINES AND PYRIMIDINES

On the protoearth, purine bases such as adenine and guanine are assumed to have been formed from hydrogen cyanide. This assumption is consistent with the empirical formula of adenine, $C_5H_5N_5$, which formally represents five molecules of HCN. Moreover, the sequence of carbon and nitrogen atoms in the molecule allows adenine to be viewed as a pentamer of hydrogen cyanide. The Spanish researcher Oro obtained experimental evidence for this hypothesis in 1963. He observed the formation of adenine while heating concentrated aqueous solutions of ammonium cyanide. Interestingly, this theoretical investigation later provided an industrially important method for the production of adenine.

We have already provided evidence for the formation of hydrogen cyanide in considerable quantities under the conditions which existed on the primitive Earth. Hydrogen cyanide is now considered to be the source of a number of important compounds including aminomalononitrile (Figure 11.3), a key precursor of the purines. Reaction with one molecule of HCN readily converts aminomalononitrile into 4-cyano-5-aminoimidazole, which formally represents four molecules of HCN. Upon condensation with one further molecule of HCN this substituted imidazole produces the desired adenine structure. The sequence of these transformations is given in Figure 11.11. Alternatively, hydrolysis of the cyano group converts 4-cyano-5-aminoimidazole into 5-aminoimidazole-4-carboxamide; subsequent addition to dicyanogen and further cyclocondensation provides guanine.

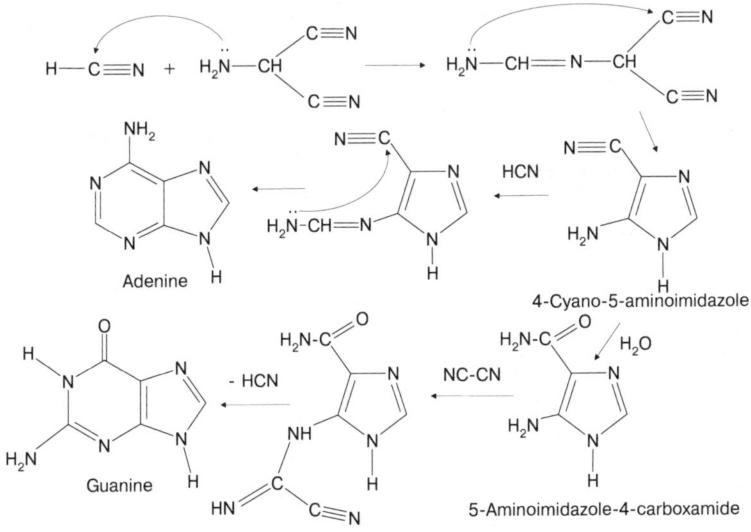

Figure 11.11 Putative scheme for the abiotic synthesis of purine bases.

Adenine and guanine are also formed in 1.0 and 0.5% yield, respectively, by ultraviolet irradiation of solutions containing HCN or NaCN. It was further established that they can result from the application of either high voltage electrical discharges or cold plasma to gaseous mixtures of CH_4, NH_3 and H_2 in the presence of apatite; in these conversions hydrogen cyanide appears to be an important intermediate. Glycine (a common product of abiogenic reactions) is transformed into purines upon heating.

Many of the details regarding the abiotic origin of pyrimidines are still unknown. So far, the only established synthetic precursor is urea. Heating urea with cyanoacetylene (100 °C, 5 h) leads to cytosine in a yield of 5% (Figure 11.12a). A record yield (14%) for abiotic transformations was obtained in the preparation of uracil by fusion of urea with malic acid in the presence of polyphosphoric acid (PPA) (Figure 11.12b). 5,6-Dihydrouracil is the putative intermediate in this reaction. It is not yet clear which oxidizing agent causes the dehydrogenation to uracil. However, neither polyphosphoric acid nor polyphosphates are formed under natural conditions; moreover, polyphosphates are rapidly hydrolyzed in water. Therefore, further evidence is required to support this proposed abiotic synthesis of uracil.

Thymine was probably produced via methylation of uracil with formaldehyde and formic acid (Figure 11.12c). A common scheme for the formation of all three biologically important pyrimidines under realistic conditions has been determined. The approach is based on β-aminoacrylonitrile or β-aminomethacrylonitrile (Figure 11.12d). Hydrolyses

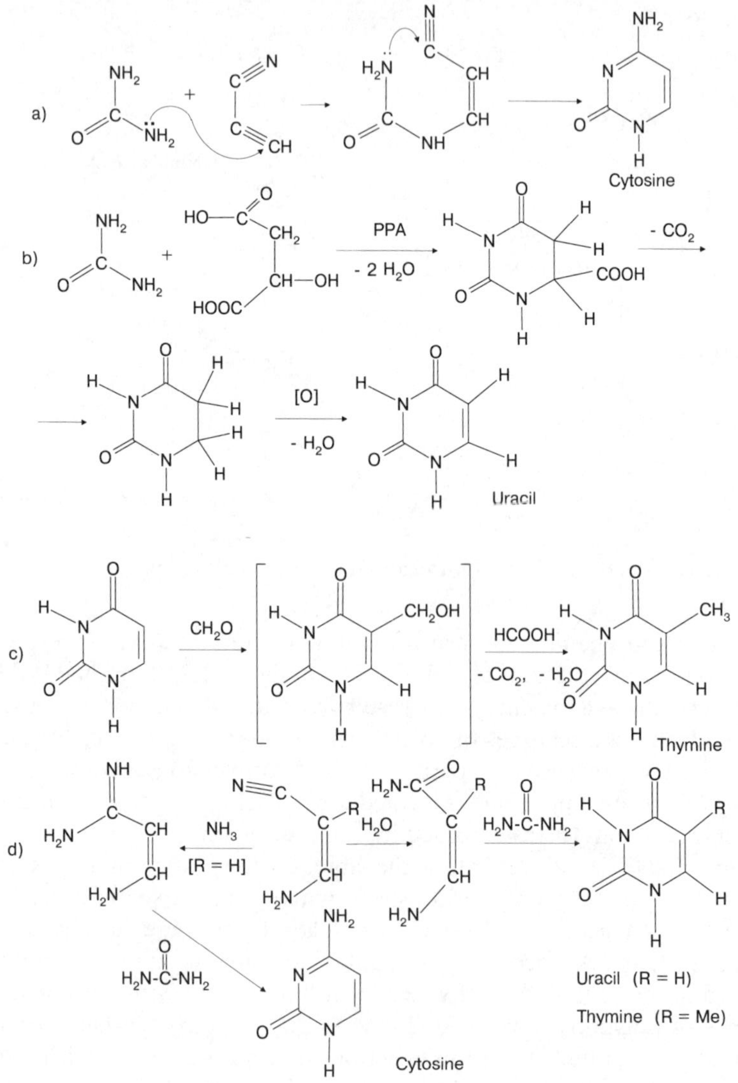

Figure 11.12   Probable routes for the abiotic synthesis of pyrimidine bases.

of the cyano groups give the amides of the respective acids which react further with urea to produce uracil or thymine. When ammonia (instead of water) is added to β-aminoacrylonitrile the result is an intermediate amidine which cyclizes with urea to give cytosine.

11.5.7 NUCLEOSIDES, NUCLEOTIDES AND POLYNUCLEOTIDES

We now have a general concept of how the heterocyclic compounds necessary for life appeared on Earth. Of course, these compounds represent only the initial phases of chemical evolution. The origin of the first biological substances, i.e. molecules capable of autoreproduction, is associated with the subsequent stages of chemical development, and especially with the appearance of polymeric compounds such as polynucleotides, polypeptides and polysaccharides. Our description of this epoch in molecular evolution is, to a large extent, speculative. However, a number of its stages, especially the initial phases, have been successfully imitated experimentally. We commence with an examination of the syntheses of nucleosides, nucleotides and polynucleotides from purine and pyrimidine bases via the elimination of water (the same principle also applies to the polypeptides and polysaccharides)

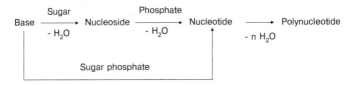

There are several possible modes by which such dehydration could have been promoted. For example, organic compounds, when washed ashore, are adsorbed on alumina minerals. Together with other dehydrating agents, the alumina could function as a catalyst for the polycondensation. Liquid phase polymerization is another inviting possibility which takes into account the high humidity of our planet and the fact that the intracellular biosynthesis of nucleic acids, proteins and polysaccharides is carried out in aqueous media. Of course, biosyntheses are now performed by complicated enzymes which diminish the activation energy, and it is unlikely that such enzymes existed in the primeval oceans. Such reactions would have been catalyzed in those times by other much more simple catalysts. The polycondensations would have been facilitated by solar heating of the organic solutions located in shallow reservoirs. As these bodies of water gradually evaporated, the equilibrium could have shifted toward the formation of polymeric products. The feasibility of both the liquid phase and solid phase hypotheses has been supported by contemporary experimental evidence.

A model experiment showed that adenosine could be synthesized in a dilute aqueous solution containing adenine, ribose and phosphoric acid. If deoxyribose was substituted for ribose, deoxyadenosine was obtained. It is of interest that the simplest nucleotide, adenosine monophosphate, is not formed under these conditions. However, in the absence of phosphoric acid, the nucleoside itself cannot be prepared; evidently, phosphoric acid somehow catalyzes the formation of nucleosides. It may transform ribose into ribose 1-phosphate which further bonds to adenine because the phosphate anion is a better leaving group than hydroxide (Figure 11.13).

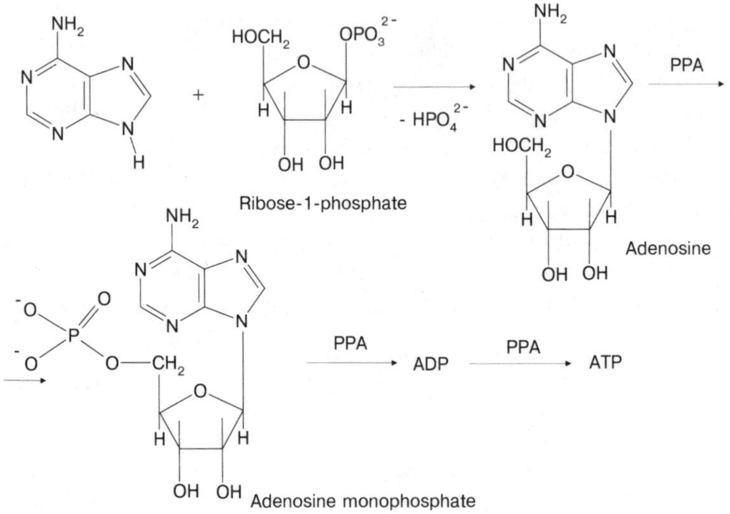

Figure 11.13 Abiotic synthesis of adenosine and its nucleotides (PPA = polyphosphoric acid).

Ultraviolet radiation of wavelength 250–260 nm is thought to activate purine and pyrimidine bases. Such harsh radiation is nowadays absorbed by the ozone layer, but the reductive atmosphere of the protoearth did not contain ozone and in those times ultraviolet radiation reached the Earth's surface. Among the heterocyclic bases, adenine is the most stable as far as the destructive effect of ultraviolet light is concerned. This may account for the increased content of adenine in nucleic acids and its presence in many other biologically important compounds such as ATP, NAD, NADP, FAD and CoA. An abiotic synthesis of adenine was achieved when ultraviolet light was substituted for low temperature plasma in the case of a model mixture containing CH_4-NH_3-H_2O-apatite.

Imitating the abiotic syntheses of nucleotides proved to be a more complex task. We have already noted that phosphoric acid does not phosphorylate nucleosides in aqueous media. Polyphosphoric acid and ethyl metaphosphate (polymeric $EtOPO_2$) proved to be successful phosphorylating agents. Thus, if a mixture of adenosine and ethyl metaphosphate is irradiated with ultraviolet light, adenosine monophosphate, adenosine diphosphate and adenosine triphosphate are formed. The preparation of nucleotides was also carried out successfully under heterogeneous conditions designed to imitate a dried-up ocean floor. While heating different nucleosides with inorganic and organic phosphates at temperatures ranging from 50 to 160 °C, formation of the corresponding nucleotides was observed in yields which increased on increasing the temperature. Similar transformations would seem to have ensured the emergence of ATP and other nucleotides on Earth in concentrations sufficient for a transition to the final stages of chemical evolution.

The abiotic synthesis of polynucleotides and the origin of nucleic acids with a defined sequence of bases are the least understood areas of chemical evolution. The ability to synthesize such compounds defines the boundary between nonliving organic matter and living systems. This Rubicon was crossed when, by an act of the Creator, nonliving matter was animated and acquired the primitive capacity to reproduce; that is, to record and transfer chemical information, to possess enzyme-like catalytic properties, and to have the ability to store and transmit energy; in other words, to show the characteristics of primary metabolism. In the course of molecular evolution these qualities were further adapted and refined.

All of the essential nucleotides based on uridylic, cytidylic and adenylic acid (Figure 3.3) have been successfully polymerized in model experiments by heating at 65–150 °C with polyphosphoric acid or ethyl metaphosphate in nonaqueous media. The polynucleotides obtained ranged in molecular weight from 15000 to 50000, corresponding to 60 to 200 nucleotide units. However, these polynucleotides could not serve as the basis for a genetic code as they contained only one purine or pyrimidine base. Moreover, under such condensation conditions, the molecules were bound to each other not only by the naturally occurring phosphodiester 3'-5'-linkage but also by linkages involving esterification of the other hydroxy groups of the sugar residue. Under the conditions which existed on the protoearth, the inclusion of various bases into the polynucleotide chain was supposedly ensured by their high concentration in certain reservoirs, or on solid surfaces which could serve as catalysts of the polycondensation reactions. This has been demonstrated by experiments in which kaolinite and a number of other clays not only absorbed significant amounts of the purine and pyrimidine bases, but also facilitated nucleotide polycondensation. Various metal cations such as magnesium(II), calcium(II), zinc(II), iron(III) and so on are believed to function as adsorptive centers in the clays. As the cations differ in their affinities toward each of the bases, they became a means of regulating the selection of nucleotides to be attached to the growing chain. Thus, naturally occurring clays containing lead(II) ions catalyze the formation of oligonucleotides with natural phosphodiester 3'-5'-bonds. It is thought that the first polynucleotides were probably of relatively low molecular weights and consisted of a limited number of bases. Adenine–thymine oligonucleotides may have prevailed and played a role in the primitive transport of RNA. Their matrix sequence of bases may have catalyzed the reproduction of similar polynucleotides and even, perhaps, polypeptides. The efficiency of the syntheses would have been enhanced after the transition from open aqueous media systems (oceans, lakes and lagoons) to closed, phase-limited systems (such as bubbles, microspheres and droplets) had been made. The latter are considered to be protocytes—the first models of living cells. Owing to polymerization processes, concentration gradients of nucleotides and amino acids could appear in protocytes isolated from the environment by a pellicle. As a consequence of the osmotic pressure which developed as a result of the concentration gradient, these substances were pumped from the external aqueous solution (the so-called 'primary broth') to the inside of the microspheres. In this manner a primitive metabolism accompanied by promatrix synthesis of biologically significant polymers may have originated.

The question of which came first, the proteins themselves or the nucleic acids which control the synthesis of proteins in modern living systems, is a fascinating one. We have already seen that various experimental abiotic conditions yield complex arrays of amino acids. Such conditions also lead to the formation of peptides. It was recently established that certain proteins (e.g. the enzyme replicase) are themselves capable of catalyzing the synthesis of polynucleotides from ribonucleoside triphosphates. In this way, RNA capable of responding to various changes resulting from chemical evolution and natural selection has developed despite its random sequence of nucleotides. Such RNA can be used as a matrix for the syntheses of other long chain RNAs. The evidence gathered thus far suggests that during the early stages of molecular evolution, polypeptides and polynucleotides came into existence independently of each other; and in the case of their appearance in the same phase-isolated system, evolutionary selection factors came into play. Macromolecules with the inherent ability to store and transfer information were more stable and adapted to new environmental situations and were thus naturally selected. Polypeptides and polynucleotides seem to comply with the rule of cross-stereocomplementarity. This means that a mutual specific recognition should exist between them, based on hydrophobic and other nonbonding interactions and also on steric relationships. Initially, peptides appeared to play the dominant role but control gradually shifted to the nucleic acids which carry more precise information and are more stable.

Undoubtedly, the original nucleic acids were single-chain molecules which could include both ribose and deoxyribose with a variable number of heterocyclic bases. However, natural selection eventually retained those structures and fragments which ensured the greatest thermodynamic stability of the macromolecules, the highest stability of complementary base pairs, the best selectivity and the optimum rate of matrix replication. Eventually, separation of macromolecules into DNAs and RNAs also occurred. The former, being more stable, were useful for the storage of information; the latter, less stable owing to the interaction of the two hydroxyl groups at the C-2 and C-3 positions of the ribose residue, were nevertheless more efficient in the transport of amino acids and in the synthesis of proteins. Thus, nature separated the 'chaff from the wheat' and the 'sheep from the goats' at the molecular level. The principle of nitrogen base complementarity was achieved only when the appropriate asymmetry of the sugars in the nucleic acids had developed. Over time it became possible for the D-form of ribose to be separated from the L-enantiomer. The development of the double-helix structure endowed a higher stability towards random modifications and mutations than the single-chain macromolecule. The structural peculiarities of the polynucleotides were enhanced by the helical conformation, thus making these molecules hereditary. The evolutionary process led to a remarkable result—the genetic code of the modern bioworld.

In addition of their role as carriers of genetic information, it was recently found that RNAs also possess some of the catalytic properties of enzymes. Hence, they have been called 'ribozymes'; for example, they can hydrolyze phosphodiester bonds and effect esterification. Thus, one RNA can cause the enzymatic cleavage of another RNA. After the discovery of this enzymatic activity, it was proposed that RNAs alone could have preceded the familiar DNA–protein regime in self-reproduction. Thus, the idea of an early 'RNA world' emerged. However, neither the prebiotic synthesis of ribonucleotides nor RNA replication would have been easy in the conditions of the primitive earth. Indeed, the RNA world itself may not be the first coding system to have arisen on the protoearth. In 1995 Orgel and coworkers described a possible prebiotic information system which might have preceded RNA. They constructed information-carrying molecules in which the backbone was a decameric oligopeptide composed of N-(2-aminoethyl)glycine units with attached side chains consisting of nucleic acid bases

This achiral amide decamer was shown to behave as an analogue of peptide nucleic acid (PNA) with attached cytidine bases (PNA-C_{10}) in functioning as a template for the synthesis of a complementary RNA decamer by successive ligation of guanosine mononucleotide units, each activated by a 2-methylimidazolyl group (2MeImpG) (Figure 11.14).

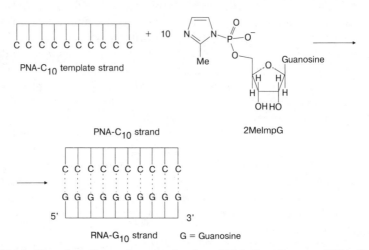

Figure 11.14   Template-directed synthesis of an RNA-G_{10} oligomer (PNA-C_{10} as a template and 2MeImpG as a substrate).

The PNA template accelerates the intermolecular condensation of the guanosine mononucleotide substrate, probably through a complex with Watson–Crick base pairing. In the course of this transformation all the possible n-mers are produced. The di- and trimers initially formed possess predominantly the nonnatural 2′-5′-phosphodiester linkage. However, the steps of further elongation to tetramers, pentamers and so on are highly specific and give the natural 3′-5′-linkages. This means that a generic transfer has been effected from PNA ('enzygenes') to RNA ('ribozymes') without loss of information. Significantly, a cytidine DNA decamer can direct the ligation of guanosine PNA dimers in this experiment, which suggests the possibility of reverse transcription and the probability of the partial evolution of more than one self-replicating protosystem. So might the gap between contemporary information-carrying polymers and the prebiotic chemistry of organic molecules be filled (Figure 11.15).

Natural selection at the nucleic acid level continues to function at the present time. Without this line of reasoning it is difficult to understand the origin of bacteria that live in the cooling water of atomic piles in which the radioactivity reaches 1000 roentgens. Without natural selection we could not account for mutant viruses, bacteria and lower fungi becoming resistant to once highly effective antibiotics.

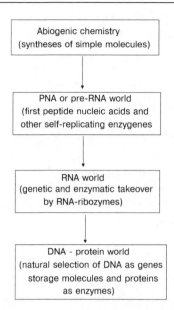

Figure 11.15 General scheme of evolution of information-carrying polymers from abiotic molecules.

The most enigmatic chapter of the history of molecular evolution involves the origin of the simplest single-cell structures and the subsequent formation of single-cell organisms. In order for life to commence, all of the necessary components, like proteins, carbohydrates, nucleic acids, fats and so on, had to be accumulated in a confined phase of limited area. Scientists have spent great effort in modeling a variety of procellular structures based on this requirement: bounded droplets (automatically formed from polypeptide and polynucleotide solutions), proteinoid microspheres, foam bubbles, armored microspheres (grains of alumoferrosilicate coated with a lipoid layer), and so on. It has been demonstrated that such systems possess a number of features characteristic of primary cells. In particular, they are distinguished by their high stability, specific catalytic properties and, most importantly, by their ability to grow (owing to their exchange of low molecular weight substances with the surrounding media), divide and degrade. Evidently, during the evolution of such structures, optimization of the growth trends and self-reproduction modes took place, giving rise to the first microorganisms with biological metabolism.

In this chapter, we have traced the course of chemical evolution from the primeval universe of atoms through simple molecules to highly complex polymers. We have observed how molecular structures spontaneously, but with natural regularity, became more and more complicated, and the manner in which the natural selection of structures occurred. We focused on the natural selection of heterocyclic compounds which were the most feasible to perform biologically important roles. This self-improvement is the essence of chemical evolution, the inescapable, preprogrammed drive for life. As Albert Einstein wrote

> Life is predetermined by the existence of atoms, and the mystery of all existence is contained in the very lowest step.

The general scheme of chemical evolution and the corresponding time scale are depicted in Figure 11.16 and Table 11.5. The data indicate that nature took $0.5-1 \times 10^9$ years to play the game of chemical evolution and to evolve to the simplest forms of life. This is not considered long term on the geological scale of time. Indeed, imprints of algal biofossils 3.4×10^9 years old have been found on rock samples extracted in Swaziland (southern Africa). Analogous natural antiquities in the schistose quartzites of Western Australia are dated at 3.5×10^9 years.

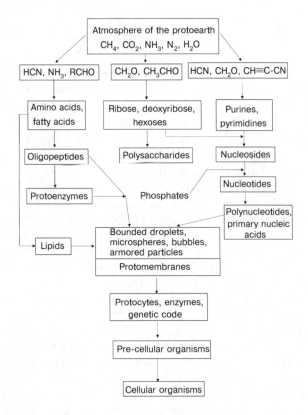

Figure 11.16   General scheme of chemical evolution (adapted from Lazcano-Araujo, A. and Oro, J., in *Comets and the Origin of Life* (ed. C. Ponnamperuma), Reidel, Dordrecht, 1981, with permission).

Table 11.5 Geochronological scale of chemical evolution (time scales in × 10⁹ of years) (adapted from Oparin, A. I., *Problems of Appearance and Essence of Life* (in Russian), Nauka, Moscow, 1973, p. 171, Figure 6, with permission).

Period of evolution	Age[a]	Geological era (duration)	Characteristics of stage (duration)
Origin of the solar system	5–6		
Formation of Earth	4.5–5.0		Beginning of molecular evolution on Earth
Formation of Earth's crust	4.5–4.0		Abiogenic synthesis of organic compounds (5.0–3.0)
Formation of secondary reductive atmosphere	4.0–3.5	Archaean (4.5–2.8)	Formation of ocean and primary 'broth' (4.0–3.0)
End of reductive and beginning of transitional atmosphere	3.5–2.0		Formation of protocytes and protoenzymes; development of primary nucleic code (3.0–2.0)
			Anaerobic epoch begins
Formation of oxygen atmosphere	2.0–1.0	Proterozoic (2.7–0.7)	Start of photosynthesis
			Appearance of aerobes
Beginning of Cambrian era	1.0 to present time	Paleozoic (0.7–0.23)	Biogenesis, biological evolution
		Mesozoic (0.23–0.067)	Appearance of mammals
		Cenozoic (0.067 to present)	Appearance of anthropoids

[a]Radio-dating by uranium lead (scale up to 5 × 10⁹ years).

11.6 Problems

1. An alternative hypothesis for the origin of life on Earth is based on the assumption that chemical evolution (up to the development of peptides and oligonucleotides) occurred not in the primordial ocean but in the atmospheric clouds of the protoearth. Provide ideas to support this cloud droplet chemistry theory. (Hint: consider the availability of initial reactants, energy sources and conditions required to carry out the

chemical transformations leading to the macromolecules, and see Figure 11.14.)

2. What inherent structural properties of hydrogen cyanide determined its key role in the abiotic origin of the biologically important heterocyclic systems?

3. In early experiments in which mixtures of H_2 (20%), CH_4 (40%), NH_3 (40%) and water vapor were subjected to electrical discharges, a number of organic products including HCN and amino acids were obtained. However, purines and pyrimidines were not identified in the final mixtures. Suggest an explanation.

4. Suggest a possible scheme for the abiotic formation of tryptophan from indole and xanthine from available precursors.

5. Recently, the dipeptide histidylhistidine (His-His) was prepared by a prebiotic synthetic method. Small His-containing peptides, and histidine itself, are assumed to promote some prebiotic reactions. Suggest reaction products and postulate a mechanism involving the catalytic action of the dipeptide His-His when an aqueous mixture (pH 8) of 0.1 M 2′-deoxyadenosine-5′-monophosphate (dAMP) and 0.02 M His-His is heated at 90 °C for 64 h in 10 h cycles to evaporation (conditions consistent with a primeval evaporating pond under tidal influences).

6. Suggest a scheme for the abiotic formation of porphobilinogen. What role may this substance have played in the emergence of life on Earth?

7. What factors may cause an elevated content of adenine in DNA and other important biomolecules?

8. What was the role of phosphoric acid in the prebiotic emergence of nucleosides?

9. Recently, the coenzymes adenosine diphosphate glucose, guanosine diphosphate glucose and cytidine diphosphoethanolamine were synthesized by a nonenzymatic prebiological method designed to model the primitive metabolism in the first Archaean cells. Nucleotides of this type may be regarded as metabolic RNA fossils. What role might these coenzymatic molecules have played in the first living cells? (For hints see Chapters 4 and 5).

11.7   Suggested Reading

1. Böhler, C., Nielsen, P. E. and Orgel, L. E., *Nature*, 1995, **376**, 578.
2. Gesteland, R. F. and Atkins, J. F.(eds), *The RNA World: The Nature of Modern RNA Suggests a Prebiotic RNA World*, Cold Spring Harbor Laboratory Press, Cold Spring Harbor, NY, 1993.
3. Eschenmoser, A. and Loewenthal, E., *Chem. Soc. Rev.*, 1992, **21**, 1.
4. Mason, S. F., *Chemical Evolution. Origin of the Elements, Molecules, and*

Living Systems, Clarendon Press, Oxford, 1991.
5. Altman, S., *Angew. Chem.*, *Int. Ed. Engl.*, 1990, **29**, 749.
6. Cech, T. R., *Angew. Chem.*, *Int. Ed. Engl.*, 1990, **29**, 759.
7. Fox, S. W., *The Emergence of Life: Darwinian Evolution from the Inside*, Basic Books, New York, 1988.
8. Horowitz, N. H., *To Utopia and Back: The Search for Life in the Solar System*, Freeman, New York, 1986.
9. Shapiro, R., *Origins: A Skeptic's Guide to the Creation of Life on Earth*, Summit Books, New York, 1986.
10. Dyson, F., *Origins of Life*, Cambridge University Press, Cambridge, 1985.
11. Dickerson, R. E. and Geis, I., *Chemistry, Matter and the Universe: An Integrated Approach to General Chemistry*, Benjamin/Cummings, Menlo Park, CA, 1981.
12. Fox, S. W. and Dose, K., *Molecular Evolution and the Origin of Life*, Dekker, New York, 1977.
13. Miller, S. L. and Orgel, L. E., *The Origins of Life on the Earth*, Prentice-Hall, Englewood Cliffs, NJ, 1974.
14. Calvin, M., *Chemical Evolution: Molecular Evolution Towards the Origin of Living Systems on the Earth and Elsewhere*, Oxford University Press, New York, 1969.
15. Bernal, J. D., *The Origin of Life*, World Publishing, Cleveland, OH, 1967.
16. Oparin, A. I., *The Origin of Life on the Earth*, 3rd Edn, Academic Press, New York, 1957.

CONCLUSION

Heterocycles arose on our planet long before the first living creatures. Their appearance was predetermined by the fundamental laws of chemical evolution. Together with other classes of organic compounds, heterocycles promoted the formation of life on Earth. We can now synthesize these life-based materials and others not occurring in nature. Scientists have learned to use heterocycles to improve the quality of life and to explore the secrets of nature.

It can be questioned why it is appropriate specifically to emphasize the role of heterocycles since analogies to the roles of other classes of organic compounds are easily found. In fact, dyes, luminophores, pesticides and drugs do not necessarily have to be heterocyclic in structure. In a similar fashion there are many common features in chemistry and physics between such related compounds as pyrrole and aniline, or between pyridine and nitrobenzene. Nevertheless, nature selected the heterocycles pyrrole and pyridine, and not the homocycles aniline and nitrobenzene, as the basis of the most essential biological systems. We now know the reason for this: the introduction of a heteroatom into a cyclic compound imparts new properties. Heterocycles are chemically more flexible and better able to respond to the many demands of biochemical systems. This book was written with the prime intention of drawing attention to this specific nature of heterocycles. Of course, some of the examples used may become superseded. New types of heterocyclic structures and materials will undoubtedly be discovered. However, our perceptions of the basic trends of development in heterocyclic chemistry are likely to remain largely intact.

There is no doubt that the chemistry of heterocycles will continue to progress. In particular, the fields of receptor chemistry and supramolecular chemistry will continue to expand, and the creation of more effective drugs and various electronic structures for conveying and storing information at the molecular level will continue well into the future.

At the conclusion of this book we wish to backtrack to the beginning and to unite two different themes: one mentioned in the preamble to the first chapter, and the other here, in an allegory of the Russian poet-symbolist Hippius

> For a long time, of her, we were singing praises,
> Of the pretender queen.
> The whisps of fragrant fumes are still being felt in hazes,
> The twinkling shrine still seen.

283

The authors emphasize that in the creation of this book they have learned much about the role of heterocyclic chemistry in life and society. They hope that the reader has been likewise informed. Undergraduate, graduate and postdoctoral students are invited to consolidate and expand their knowledge of organic chemistry and the applications of heterocycles in the problem sections at the end of each chapter.

ANSWERS AND REFERENCES TO SELECTED PROBLEMS

Chapter 1

1. In a 1,4-disubstituted piperidine, four equilibrating conformers, i.e. (a)–(d), are possible (hydrogen atoms are omitted)

The equilibrium proportions of such isomers depend on the nature of the substituents and their positions in relation to the nitrogen atom (Eliel, E. L., Kandasamy, D., Yen, C.-Y. and Hargrave, K. D., *J. Am. Chem. Soc.*, 1980, **102**, 3698).

2. 4-Hydroxy-1-methylpiperidine can exist in a boat conformation owing to intramolecular hydrogen bonding. The boat conformation is also fixed by the existence of a bridge (e.g. CH_2CH_2) between the 1-position and 4-position.

3. See Lambert, J. B., Oliver, W. L. and Jackson, G. F., *Tetrahedron Lett*, 1969, 2027.

4. A, C, E and F.

5. See Huber, H., *Angew. Chem., Int. Ed. Engl.*, 1982, **21**, 64; and Saxe, P. Schaefer, H. F., *J. Am. Chem. Soc.*, 1983, **105**, 1760.

6. Excluding different tautomeric forms, there are four isomers. Two contain nitrogen at the bridgehead.

285

7. Owing to the flexibility of their rings, azaannulenes can exist in various conformations with the nitrogen lone electron pair or N–H bond oriented either inside or outside the cycle depending on the 'internal' or 'external' disposition of the nitrogen atom.

Chapter 2

1. Piperidine is more basic than pyrrole owing to participation of the lone electron pair of the pyrrole nitrogen in the formation of the pyrrole aromatic π-sextet. The higher basicity of piperidine compared to pyridine is accounted for by the different hybridizations of their nitrogen atoms: sp^3-hybridization in piperidine and sp^2 in pyridine. Orbitals of greater s-character display higher electronegativity and therefore lower affinity toward a proton.

2. The pK_a of 12.9 corresponds to ionization of the N–H bond in neutral benzimidazole to give the benzimidazole anion. The pK_a of 5.3 is for ionization of the benzimidazolium cation to give neutral benzimidazole.

3. $pK_a = 2.39$ (ionization of the N–H bond in the purine cation), $pK_a' = 8.93$ (ionization of the N–H bond in neutral purine).

4. Basicities of anions: imidazole > benzimidazole > purine > tetrazole.

5. The nitration of imidazole in a strongly acidic medium occurs on the relatively inert imidazolium cation, whereas bromination in organic solvents occurs via the much more active neutral molecule.

6. (a) Nitration occurs at the free α-position of the pyrrole ring. (b) The N-ethylpyridinium salt is formed, which upon reduction gives the corresponding 1,4-dihydropyridine derivative. (c) The α-position of the furan cycle is exclusively nitrated. (d) A mixture of 1-methyl- and 2-methyl-1,2,3-triazole is formed. (e) 2-Methoxypyridine. (f) The chlorine atom in the pyrimidine nucleus is more active toward substitution and is replaced by a dimethylamino group.

7. The dipole moment is a vector value. The vector is considered to be oriented toward the negative pole. In pyridine, the vector sum is obviously directed from the center of the ring toward the nitrogen atom. By adding such vectors oriented at 60° (in pyridazine), 120° (in pyrimidine) and 180° (in pyrazine) one can assess rather accurately the relative dipole moments for these diazines: pyridazine > pyrimidine > pyrazine.

8. See Alcalde, E., Dinares, I., Fayet, J.-P., Vertut, M.-C. and Elguero, J., *J. Chem. Soc.*, *Chem. Commun.*, 1986, 734.

10. See Pfleiderer, W., in *Physical Methods in Heterocyclic Chemisty*, (ed. A. R. Katritzky), Academic Press, New York, 1963, Chap. 4.

11. See Eisch, J. J. and Jaselskis, B., *J. Org. Chem.*, 1963, **28**, 2865.

12. See Staab, H. A., *Angew. Chem.*, *Int. Ed. Engl.*, 1962, **1**, 351.

13. K, L, M and O.

Chapter 3

1. (a) Poly(ThrValLeuTyrCys). (b) The dipeptide LeuAsp.
2. (a) There are 561 codons and 561 residues. (b) In the first chain [A] = 0.35, [G] = 0.29 and [T + C] = 0.36. In the complementary chain [T] = [0.35], [C] = 0.29 and [A + G] = 0.36. See Regier, J. C. and Pacholski, P., *Proc. Natl. Acad. Sci. USA*, 1985, **82**, 6035.
3. Campbell, J. A., *J. Chem. Educ.*, 1976, **53**, 447, Problem Q248.
4. (a) 5'-TCGAGTAGCCGATGATCATCGTCGACGAT-3'; (b) 5'-UCGAGUAGCCGAUGAUCAUCGUCGACGAU-3'; and (c) Ser-Ser-Ser-Arg, Ser-Ser-Ser-Thr (adapted from Lehninger, A. L., *Biochemistry*, Worth, New York, 1970, Chap. 31, p. 727, Problem 6, with permission).
5. See Huang, H., Solomon, M. S. and Hopkins, P. B., *J. Am. Chem. Soc.*, 1992, **114**, 9240.
6. Two (adapted from Ternay Jr, A. L., *Contemporary Organic Chemistry*, 2nd Edn, Saunders, Philadelphia, PA, 1979, with permission).
7. Hydroxyproline and hydroxylysine are derivatives of the coded proline and lysine; it is known that these two coded amino acids are first incorporated into the polypeptide chain and are subsequently enzymatically hydroxylated.
8.

   5'...- A C G G A T C C T T T T T C T T T C T T C T T T T C T T C...- 3'

   3'...- T G C C T A G G A A A A A G A A A G A A G A A A A G A A G...- 5'

   3'... - T T T T T C T T T C T T C T T T T C T T .....- 5'

9. (a) Glycine and alanine; total 68%. (b) AGR, GGL, GGQ, GAG, AAAAAA and GGAGQGGYGGXQG. (c) Guanine and cytosine; 82%.

Chapter 4

1. Pandit, U. K. and Mas Cabre, F. R., *J. Chem. Soc., Chem. Commun.*, 1971, 552.
2. Fukuzumi, S., Kuroda, S., Goto, T., Ishikawa, K. and Tanaka, T., *J. Chem. Soc., Perkin Trans. 2*, 1989, 1047.
7. The adenosyl moiety is found in NAD$^+$, NADP$^+$, FAD, CoA, ATP, ADP and so on.
8. Itoh, S., Ogino, M., Fukui, Y., Murao, H., Komatsu, M., Ohshiro, Y., Inoue, T., Kai, Y. and Kasai, N., *J. Am. Chem. Soc.*, 1993, **115**, 9960.
9. The rate would increase approximately 10-fold (Lewis, C., Kramer, T., Robinson, S. and Hilvert, D., *Science*, 1991, **253**, 1019). In protic

(aqueous) solvents the substrate is highly stabilized through hydrogen bonding. Aprotic solvents force charge delocalization, stabilize the transition state by dispersion interactions, facilitate passage of the substrate from solution into the predominantly hydrophobic active site of the decarboxylase, and thus accelerate decarboxylation of the carboxylate anion. X-Ray analysis has shown that the carboxylate-binding site is highly hydrophobic in the case of histidine decarboxylase (Gallagher, T., Snell, E. E. and Hackert, M. L., *J. Biol. Chem.*, 1989, **264**, 12 737).

10. (a) Histidine. (b, c) Ovothiols act as antioxidants by enzymatic consumption of H_2O_2. During this process the ovothiols are oxidized to the disulfides. Enzymatic reduction of the disulfides by NADP-H regenerates the ovothiols (Shapiro, B. M., *Science*, 1991, **252**, 533).

Chapter 5

1. N–P bond in A, N–COMe bond in B, N–NO$_2$ bond in C, C–COMe bond in D and C-CHO bond in E.
7. 11.1 kcal mol^{-1} (adapted from Lehninger, A. L., *Biochemistry*, Worth, New York, 1970, Chap. 14, p. 311, Problem 4, with permission).
9. There are about 40 ml of ethanol in 100 ml of vodka. Oxidation of this quantity of ethanol in the Krebs cycle and in the respiratory chain provides 8.4 moles of ATP, corresponding to around 76 kcal mol^{-1}.

Chapter 6

7. (a) Nine (b) seven (c) five molecules of ATP, respectively (adapted from Lehninger, A. L., *Biochemistry*, Worth, New York, 1970, Chap. 21, p. 480, Problem 1, with permission).
8. 10.6 kcal mol^{-1}.

Chapter 7

1. Campbell, J. A., *J. Chem. Educ.*, 1977, **54**, 309, Problem Q302.
3.

Cysteine   Valine

4. Campbell, J. A., *J. Chem. Educ.*, 1977, **54**, 369, Problem Q306.
5. The structural similarity of allopurinol to the purines suggests that it functions as an antimetabolite. Allopurinol supposedly inhibits the enzyme xanthine oxidase which is involved in the metabolism of purines into uric acid.
6. Campbell, J. A., *J. Chem. Educ.*, 1977, **54**, 247, Problem Q294.
7. (a) γ-Aminobutyric acid (Figure 7.18). (b) Between 0.6 nm (in the extended conformation) and 0.45 nm (in the folded conformation) in the zwitterionic form of γ-aminobutyric acid; in C and E between 0.55 and 0.59 nm; in D 0.44 nm; in G 0.56 nm (Sytinsky, I. A., Soldatenkov, A. T. and Lajtha, A., *Prog. Neurobiol.*, 1987, **10**, 89).

Chapter 8

2.

7. (a) *Cis–trans* isomerism around the external C=C bond and optical isomerism due to the presence of an asymmetric carbon atom give rise to a total of four isomers. (b) Intramolecular hydrogen bonding between the OH group and the N-2 atom of the triazole ring is possible. (c) The N-4 atom owing to its sterically uncrowded position and highest nucleophilicity (Katagi, T., *J. Agric. Food Chem.*, 1988, **36**, 344).

Chapter 9

2. Conjugation in indigoid dyes decreases the order of the C=C bond existing between the two nuclei, thereby facilitating rotation about the bond and, consequently, interconversion between the *cis* and *trans* forms. However, in indigo itself the *trans* form is additionally stabilized by intramolecular hydrogen bonds, making conversion to the *cis* form significantly more difficult.

3. Indigoid dyes are rather weak bases and NH acids and are ionized only under the action of strong acids and alkalis. Thus, concentrated sulfuric acid causes protonation at the C=O group, enhancing its electron-accepting properties and making conjugation more effective. The degree of conjugation is also increased when the N-anion is formed under the action of sodium *tert*-butoxide.

4. The conjugated 18 π-electron system, including the four nitrogen atoms and internal C–C and C=N bonds, is believed to be the main chromophore in the porphyrin system which provides annular conjugation and coloration. The chromophore does not appear to be affected by the hydrogenation of the external C=C bonds.

9. Photochromism in compounds A and B is due to migration of a proton from the CH₂ group to the *ortho*-nitro group with formation of a colored photoexcited isomer, e.g. I, which is blue in contrast to the tan isomer A. This proton transfer probably occurs as an intramolecular process, which is impossible in the case of compound C. However, the presence of the *para*-nitro group is also important since it enhances the acidity of the CH₂ protons (Gilfillan, E. D. and Pelter, M. W., *J. Chem. Educ.*, 1994, **71**, A4; and Prostakov, N. S., Krapivko, A. P., Soldatenkov, A. T., Furnaris, K., Savina, A. A. and Zvolinskii, V. P., *Chem. Heterocycl. Compd.*, 1976, **16**, 312).

I

Chapter 10

1. Hedge, V., Hung, C.-Y., Madhukar, P. *et al.*, *J. Am. Chem. Soc.*, 1993, **115**, 872.
2. Sessler, J. L., Mody, T. D. and Lynch, V., *J. Am. Chem. Soc.*, 1993, **115**, 3346.
3. Hamilton, A. D., *J. Chem. Educ.*, 1990, **67**, 821.
4. Gokel, G. W. and Cram, D. J., *J. Chem. Soc, Chem. Commun.*, 1973, **481**.
5. Busch, D. H., *Chem. Rev.*, 1993, **93**, 847.
6. Tarrago, G., Marzin, C., Najimi, O. and Pellegrin, V., *J. Org. Chem.*, 1990, **55**, 420.
7. Aumüller, A., Hädicke, E., Hünig, S., Schätzle, A. and von Schülz, J. U., *Angew. Chem., Int. Ed. Engl.*, 1984, **23**, 449.

Chapter 11

1. Prebiotic reactants such as CH_4, H_2O, NH_3 and N_2 were present in the atmosphere and simple, open chain and heterocyclic organic molecules might arise from such precursors under the action of UV light, corona discharge, lightning or heating from asteroid impacts. These molecules may have been prevented from falling to the Earth's surface by atmospheric moisture adhesion. Being entrapped by cloud droplets which contain catalytic clay particles, these molecules could yield biologically important macromolecules by polymerization and multiple evaporation–condensation processes. Oligomerization and polymerization of amino acids, purine bases and pyrimidine bases could have occurred via dehydration which would have been more difficult in the ocean than in the atmospheric droplets. Complex molecules might also be supplied to the Earth by comets and meteorites.

3. The likely cause is the tremendous excess of hydrogen in the initial mixtures. Thermodynamic calculations show that the formation of heterocyclic molecules is unlikely in highly reductive mixtures mimicking the protoearth's atmosphere.

5. Here dAMP is hydrolytically dephosphorylated. The reaction proceeds via a proton transfer mechanism (concerted general acid–base catalysis by the dipeptide)

9. The first two compounds might be important in (i) the activation of such nutrients as carbohydrates in the primordial 'broth' by means of phosphorylation, (ii) the extraction of energy from fuel molecules and the simultaneous storage in the energy-rich bonds of these high energy phosphorylated compounds, and (iii) the synthesis of biochemical building blocks. The third cofactor might be useful in the transfer of ethanolamine to diacylglycerol in the biosynthesis of phospholipids necessary for simple membrane formation and in providing the primeval cells with individual identities (Mar, A. and Oro, J., *J. Mol. Evol.*, 1991, **32**, 201).

INDEX

Index compiled by G. Jones